AF520887

# CONTENTS

## Section I: Principles of Biofumigation

*Chapter 1*

# INTRODUCTION

Soil-borne pests and pathogens, including weed propagules, nematodes, insects, fungi, bacteria and certain other agents, can be limiting factors in the production of horticultural and other crops. One of the principal strategies used by the growers of high-value horticultural crops to combat these organisms is pre-plant soil disinfestation, using pesticides or other physical or biological methods. Soil fumigants are the most effective soil disinfestation chemicls, and methyl bromide (MB) is the most important soil fumigant chemical used by growers around the world. It is a broad spectrum pesticide with excellent activity against most potential soil pests.

Apart from controlling major soil-borne pathogens and pests, soil fumigation with MB and other fumigants frequently provides increased crop growth and yield responses, even when pest control does not appear to be a factor. Although the mechanisms of these increases have not been fully explained, control of minor pathogens and liberation of soluble mineral nutrients in soil are involved (Chen *et al.*, 1991).

This stimulation of crop growth resulting from soil fumigation is important to producers of many high value agricultural commodities to assure maximum returns for their investments. Consumption of MB for pre-plant fumigation in California is estimated at 16 million lbs annually – nearly 50% of the total amount used in the United States. MB fumigant is used primarily for production of strawberries, fruit and nut crop orchards and vineyards, vegetable crops and nurseries. However, MB was identified as a risk to the stratospheric ozone layer in 1992 and targeted for worldwide phase out in 1997 by means of the Montreal Protocol, an international treaty. Under the current terms of the agreement and of the federal Clean *Air* Act, preplant consumption of MB in the United States is scheduled to be gradually phased out by 2005 (USDA 2000). The impending loss of MB as a soil fumigant has stimulated intensive efforts to develop and implement suitable replacement strategies. One of the difficulties in replacing MB is that the compound is very useful in a wide range of operations that are important in agriculture. MB is not the first major soil fumigant pesticide removed from use over the past 25 years due to regulatory actions by various governmental agencies; DBCP (1,2-dibromochloropropane) and EDB (ethylene dibromide) have both lost their registration, and 1,3- dichloropropene (Telone II) was suspended for several years.

Scientists have been continuously working to develop usable alternatives for soil disinfestation. Many of the replacement treatments for methyl bromide will be other chemical products that remain available, including 1, 3-dichloropropene, metham sodium and chloropicrin, applied alone or in combination. In addition, registration is being sought for chemical fumigants not currently allowed for use in agriculture, such as methyl iodide and propagyl bromide. Apart from synthetic chemical alternatives, numerous nonchemical strategies have also been researched, field validated and in some cases implemented commercially. For certain situations, soil solarization and biofumigation are among the most useful of the nonchemical disinfestation methods.

## 1.1. METHYL BROMIDE ALTERNATIVES

It is becoming clear that few, if any, alternatives are presently available that can directly replace MB for soil fumigation under a wide variety of environmental conditions and for a broad spectrum of pests. A few soil fumigants are being used commercially in place of MB in California. When applied under optimal conditions, these materials - including 1, 3-dichloropropene, chloropicrin, metham sodium and others - alone and in combination, may provide pest control efficacy similar to that of MB. However, these materials tend to provide a narrower spectrum of control than MB, have less predictable efficacy, and may have their own problems with environmental pollution and safety. In addition to soil fumigants, many nonfumigant chemicals (fungicides, herbicides, bactericides and nematicides) can be used for soil disinfestation, as well as an array of physical, cultural and biological methods including soil steaming (nursery production), host-plant resistance, crop rotation, antagonistic microorganisms and others. Where practical, soil solarization and biofumigation can be of benefit to users, whether or not they choose to employ soil pesticides.

In recent years the phasing-out of methyl bromide has had the effect of bringing into sharp focus the need for alternative strategies for the management of soil-borne pests and diseases, not just for users of methyl bromide but in a general sense. Less generally realized is the prospect that problems related to maintaining efficacy following the repeated application of different pesticides to soil are likely to arise where previously they did not occur. Although methyl bromide replacement is not the subject of this book, it provides a topical and relevant foundation for setting the context and theme.

Among the alternative control methods being touted to replace methyl bromide are the use of other fumigant-like pesticides that have a long history of use, such as the methyl isothiocyanate generator metham sodium, 1.3-dichloropropene (1,3-D), chloropicrin, or mixtures of these, and various biologically-based options. But can these agents and methods achieve the efficacy and all-important consistency of performance of the robust and forgiving methyl bromide in the highly intensive production systems where it has been used for so long? Additionally, are there issues with using such alternatives where history can provide warnings of potential problems that need to be recognized and managed?

Among the potential alternative control methods being touted to replace methyl bromide are soil solarization and biofumigation that are among the most useful of the nonchemical disinfestation methods.

## 1.2. SOLARIZATION

The use of clear polyethylene film to cover moistened soil and trap lethal amounts of heat from solar radiation was first reported by Katan and colleagues in Israel in the mid-1970s (Katan, 1987). DeVay and associates at University of California, Davis began an intensive research programme on the promising technique shortly thereafter, and the term "soil solarization" was soon coined to describe the process by cooperators in the San Joaquin Valley. Researchers found that solarization could be a useful soil disinfestation method, especially in areas with hot and arid conditions during the summer months, such as the Central Valley and southern deserts. In certain cases, the treatment has also been effective, primarily for weed management, in cooler coastal areas (Elmore *ct al.*, 1993). The pesticidal activity of solarization was found to stem from a combination of physical, chemical and biological effects, as described in several comprehensive reviews (Katan 1987; Chen *et al.*, 1991; DeVay *et al.*, 1991; Stapleton, 1998, 2000).

Although solarization can provide excellent soil disinfestation under suitable conditions, it has significant limitations and should not be considered a cure-all or universal replacement for MB. For example, solarization is most effective close to the surface of the soil under climatic and weather conditions of high air temperature and long days for soil heating. It will not control all pest organisms, may require that land be taken out of production for 3-to-6- week treatments during the summer months, and requires disposal of used plastic film. Therefore its practical value to the user must be assessed by several factors, including extent and predictability of pesticidal efficacy, effect on crop growth and yield, economic cost/benefit and personal pest management philosophy (Stapleton, 1997).

## 1.3. BIOFUMIGATION

Another useful tool for soil disinfestation is use of various organic amendments that have pesticidal activity when incorporated into soil. Various mechanisms of pesticidal activity have been reported for these amendments. Some may facilitate a shift in soil to a community of microflora that is antagonistic to certain soil-borne pests. Others release biotoxic compounds during their degradation in soil. This latter process, especially when volatile compounds are released into the soil atmosphere by the decomposing amendments, is popularly known as "biofumigation" (Stapleton, 1998).

The biofumigation phenomenon has been exemplified by studies done by various researchers, showing that combining solarization with residues of cabbage or other *Brassica* spp., as well as composted animal manures, provided improved disease management via enhanced release of biotoxic volatile compounds from the amendments in heated soil (Ramirez-Villapudua and Munnecke, 1987; Gamliel and Stapleton, 1993a, 1993b). Other studies have reported fungicidal activity of *Brassica* residues in coastal soils without solar heating (Subbarao *et al.*, 1999).

Biofumigation, the use of isothiocyanate-generating brassicas as biologically-active green manures to emulate the use of the synthetic pesticide metham sodium, has a genesis deep in history, but a level of adoption that belies a heavily documented potential. Similarly, the broadly related issue of enhanced biodegradation of soil-applied pesticides, to which metham sodium is susceptible, has been well documented. However, the warnings from history seem to have been rarely heeded. Why are these responses so?

Biofumigation is the agronomic practice of using volatile chemicals (allelochemicals), released from decomposing *Brassica* tissues, to suppress soil-borne pests and pathogens. The use of Brassicas such as canola (*Brassica napus*) as break crops to help control take-all (*Gaeumannomyces graminis*) in cereal rotations in Australia, is an example of biofumigation. The most common volatiles produced during the breakdown of Brassicas are the isothiocyanates (ITCs). ITCs are related to the active ingredient in the commercial fumigants metham sodium and dazomet and are highly toxic to pests and pathogens. They are released following tissue damage, when myrosinase enzymes, at neutral pH, hydrolyze glucosinolates (GCs). GCs are sulphur containing chemicals (thioglucosides) that are produced as secondary metabolites by Brassicas and most researchers believe their role is to provide resistance against pests and pathogens. The aim of this book is to describe current strategies to include biofumigation as part of an integrated approach to MB replacement in horticulture.

**Muscodor species for mycofumigation:** The endophytic fungus, *Muscodor albus,* produces a mixture of volatile organic compounds (VOC's) that are lethal to a wide variety of plant pathogenic fungi and bacteria. It is also effective against nematodes and certain insects (Strobel, 2008).

## 1.4. NATURALLY OCCURRING ISOTHIOCYANATES

Fortuitously, a characteristic of isothiocyanates is that they occur conveniently in nature, notably deriving from the glucosinolates common as secondary plant compounds in the diverse Brassicaceae (Rosa *et al.*, 1997; Fahey *et al.*, 2001). The pest suppression potential of isothiocyanates released from Brassica residues has long been generally recognized. However, the use of brassicas in an attempt to suppress noxious soil-borne organisms has had sporadic attention with generally mixed results, ranging from pest, disease, or weed suppression (Mojtaheedi *et al.*, 1993; McLeod and Steele, 1999) to no suppression (Johnson *et al.*, 1992) to in some instances pathogen stimulation.

The term biofumigation was coined by Kirkegaard. As originally conceptualized, it was a crystallizing term for that part of the suppressive effects of Brassica species on noxious soil-borne organisms that arose quite specifically through liberation of isothiocyanates from hydrolysis of the glucosinolates that characterize the Brassicaceae (Kirkegaard and Matthiessen, 2004). Since isothiocyanates are biologically active, and the methyl isothiocyanate generator metham sodium is widely used as a broad-spectrum fumigant like pesticide, there existed a logical link and an impetus to the notion of harnessing, and further developing, a biological source of isothiocyanates for suppression of soil-borne pests and diseases (Brown and Morra, 1997). The term biofumigation attempted to ascribe, in a simplified way, a mechanistic name to this particular part of a general phenomenon of allelopathic effects that have been observed in the Brassicaceae for centuries and given them a reputation as poor companion plants .

## 1.5. SOURCE OF NUTRIENTS

*Brassica* defatted meal materials also supply the soil with nutrients, which may interact with crop nutrition. Potentially mineralizable N (PMN) is a measure of soil ability to release inorganic N useful for crop N uptake. The aim of this work was to compare the PMN of soil amended with *Brassica* meals, with that of soil after incorporation of other plant materials or inorganic N.

## 1.6. IMPROVED SOIL QUALITY

The important benefits of biofumigation include reduced soil crusting, improved infiltration, increased organic matter and the subsequent increase in beneficial microflora and microfauna are often overlooked. Soil improvement following a *Brassica* may produce longer lasting effects on disease suppression and crop health, as well as soil 'health'. Field studies were conducted in an attempt to quantify these beneficial properties. Soil penetration resistance is a measure of soil compaction, which increases over time with cultivation. Generally, at field capacity, growth of plant roots is not affected at <1000 kPa; while seedling emergence will be retarded and root growth restricted at 1000-2000 kPa; and root growth is impeded other than through cracks, channels and points of weakness at >2000 kPa.

*Chapter 2*

# BIOFUMIGATION

## 2.1. WHAT IS BIOFUMIGATION?

Biofumigation is an agronomic technique that makes use of some plants' defensive systems. The main plant species in which this is found are the Brassicaceae (cabbage, cauliflower, kale, and mustard), Capparidaceae (cleome) and Moringaceae (horse-radish). In suitable conditions the biofumigation technique is able to efficiently produce a number of important substances. In the above plant families, one of the most important enzymatic defensive systems is the myrosinase-glucosinolate system. With this system, tissues of these plants can be used as a soft, eco compatible alternative to chemical fumigants and sterilants. In a number of countries over the past few years, several experiments have been carried out to evaluate the effectiveness of the myrosinase-glucosinolate system, in particular using the glucosinolate -containing plants as a biologically-active rotation and green manure crop for controlling several soil-borne pathogens and diseases. The use of this technique is growing, and it is studied in several countries at a full-field scale (USA, Australia, Italy, The Netherlands and South Africa), thus triggering the interest of some seed companies, with a positive effect on the "biofumigation" seed market, which is significantly growing year after year. New potential has also been found for the dehydrated plant tissues and/or for defatted meal pellets production and use. An intense discussion amongst researchers of this topic in the various countries seems to be of fundamental importance particularly to define and develop future common strategies.

Biofumigation can be defined as the incorporation of biomass into soil, resulting in the release of toxic volatiles that reduce soil pests. Other benefits of biofumigation include: improved soil texture, increased water holding capacity, and improved soil microbial community structure. The Brassicaceae family is a source of potential biofumigation material. Family members contain secondary plant metabolites called glucosinolates, which are believed to be involved in plant defense. When tissues are damaged, glucosinolates are enzymatically broken down by myrosinase to produce nitriles, thiocyanates, isothiocyanates and other products. Isothiocyanates, the predominant breakdown product, have biocidal activity on fungi (Harvey and Sams, 2000; Charron and Sams, 1999), bacteria (Delaquis and Mazza, 1995) and other pests.

Bio-fumigation is based on incorporating soil amendment (fresh plant mass, manure) into the soil, which will release several substances, known as isothiocyanates (ITC's), able to suppress soil-borne pests, plus a soil heater to enhance biological activities. Plants from Cruciferae family (Cabbage, radish, cauliflower etc.) release large amount of these toxic substances in the soil and are considered the best material for biofumigation.

## 2.2. WHAT BENEFIT CAN IT ADD TO EXISTING FARMING SYSTEMS?

Brassica cover crops are known to reduce runoff and preserve nitrogen, and now researchers want to know how these relatives of broccoli and cauliflower might also suppress weeds, nematodes and pathogens in tomato fields.

Brassicas, including numerous mustard species, provide not only biomass to the soil but when incorporated they release chemicals known as glucosinolates, which further break down into isothiocyanates having fumigant properties similar to Vapam.

The biofumigation achieved by disking brassica residues into the soil can significantly suppress weeds, nematodes, and soil-borne pathogens.

The most conspicuous example of commercial success of this approach has been in the suppression of lettuce drop, a disease caused by *Sclerotinia minor*, after the incorporation of broccoli residue prior to planting lettuce.

In similar fashion, potato growers in the Pacific Northwest report considerable success in reducing Verticillium wilt with winter cover crops of brassicas.

Cucumber, pepper and tomato crop residues are effective to control root-knot nematode, *Meloidogyne incognita* populations and weeds, but increasing efficacy with organic residues is recommended. Biofumigation also improves physical and biological soil characteristics. The recommended dose is 100 t/ha for the first year, when the problem is serious, decreasing in subsequent years to 45 t/ha for the fourth year. Higher doses could be reduced by using green manure. No phytotoxic effects are observed, or any other restriction on production.

Soil characteristics as well as crop, region and type of organic matter, should be taken into consideration for field application.

## 2.3. BIOFUMIGATION AND SOLARIZATION

Based on the field observations, most organic growers who use solarization combine it with biofumigation, primarily by adding various composts, manures, teas and other organic amendments to the soil before solarizing. Initial results indicate that bio-fumigation (adding chopped broccoli to nematode infested soil) increases the efficacy of solarization. Thus, higher levels of control are achieved at lower temperatures or over shorter periods. In addition to commercial applications, the widespread use of solarization in home and community gardening should be noted. Although most home gardeners do not have access to chemical soil disinfestants, solarization has become a main stream cultural practice and contributes to improved plant health and production in this setting.

## 2.4. COMPOSTS AND BIOFUMIGATION

Composts improve plant health and productivity due to enhanced soil nutrient levels and improved physico-chemical properties (increased bulk density, porosity, and aeration) and increased water-holding capacity (Corti *et al.*, 1998). Disease suppression due to the advantageous changes to the soil microbial composition, including increases in the populations of known biological control agents (*Trichoderma, Bacillus,* and *Pseudomonas* spp.), has been noted by many researchers (Hoitink and Boehm, 1999; Lyons *et al.*, 2001).

Composting is an ecologically sound way to recycle organic farm wastes, such as animal manures and plant debris. It avoids the environmental hazards associated with burning, burying, or spreading these agricultural by-products on fields. These methods are not sustainable and can lead to problems such as increased particulate matter in the air, leaching and ground water pollution, and toxic excesses of phosphorous and nitrogenous compounds in the soil.

Potential problems with compost as a pre-plant treatment include using a product that has not been composted properly and, hence, is not mature. Compost testing is to be encouraged. With the assistance of extension agents and university resources maturity levels can be gauged by measuring C: N ratios and pathogen levels assessed via assays. Feedback from this type of testing can direct more effective composting methods.

Many researchers throughout the world have investigated the relationship between the addition of composts to soil and plant health. The pre-plant application of composts has many advantages. They have the ability to provide biological control of plant diseases while simultaneously improving physical soil properties and micronutritional levels (Corti *et al.*, 1998). The composted material used is often derived from agricultural waste products and, hence, serves to "recycle" these voluminous compounds. Composted soil amendments act to create a suppressive environment in the soil that can biologically control soil-borne pathogens through microbial competition, antagonism, antibiosis, hyperparasitism, and possibly induce systemic resistance in plants (Hoitink and Grebus, 1994). In particular two types of composts will be studied, poultry waste compost (PWC) and spent mushroom substrate (SMS).

The three phase composting process ultimately leads to the recolonization of the substrate with mesophilic microflora and microfauna (Hoitink, 1986). Microorganisms commonly found in composts include: *Bacillus* spp., *Enterobacter* spp., *Pseudomonas* spp., *Streptomyces* spp., *Trichoderma* spp., and *Gliocladium* spp. (Hoitink, 1994). Some of these species are recognized to exhibit properties of biological control.

Biological control of plant pathogens has been observed and studied for many years. As defined by Baker (1987), biocontrol is "the action of parasites, predators, or pathogens in maintaining another organism's population at a lower average than would occur in their absence."

## 2.5. GREEN MANURING AND BIOFUMIGATION

Soil-borne diseases are often suppressed by the use of methyl bromide to sterilize the soil, especially for strawberry cultivation in UK. This chemical will be banned from use after 2008, so there is great pressure to develop an alternative. In USA, brassica biofumigant crops are being used for potatoes, and in Italy improved cultivars have been developed for green manuring. There are a number of different Caliente Mustards available for this use, and a comprehensive description is available.

It is necessary to chop the plant into small pieces before ploughing into the ground. The plant cells emit a natural gas which helps to destroy various diseases, especially *Verticillium dahliae*, a fungus that produces 'wilt' in strawberries. The incorporation of 'green manure' into the soil also has the well known advantages like improving soil fertility and structure, the addition of nutrients, aeration and organic matter and organic culture.

Green manuring is a traditional technology for providing general benefits in terms of nutrients, soil properties, organic matter and break-crop effects, but it has been largely supplanted in modern agriculture by the advent of inorganic fertilizers. Fig. 2.1 is a diagrammatic summary of the many ways in which an incorporated Brassica green manure can influence the yield and quality of a subsequent commercial crop. While the primary focus will be on the pathway shown for biofumigation as an isothiocyanate-based concept for suppression of soil-borne pests and diseases, indicated by the large arrow, it is appropriate to briefly mention some of the related aspects.

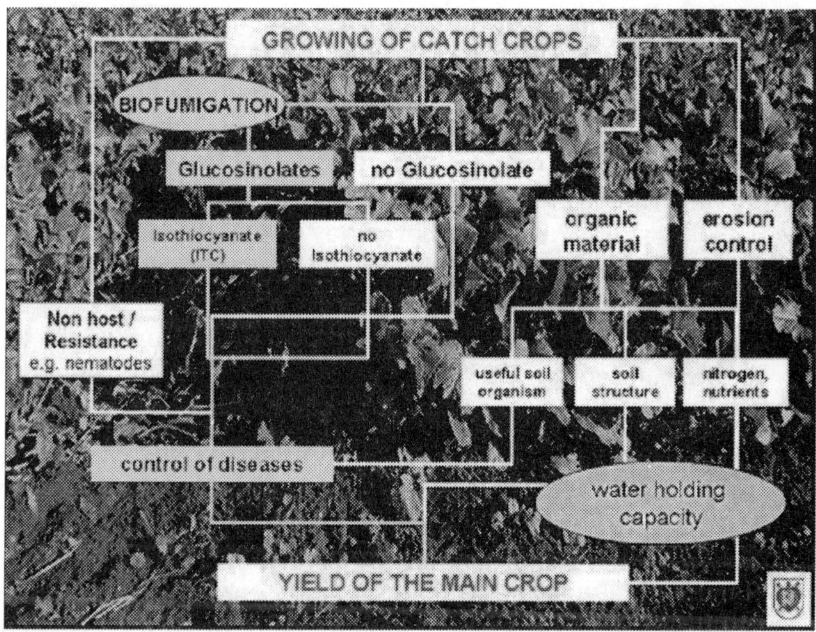


**Fig. 2.1.** Pathways for control of soil-borne pests and diseases.

The pathways shown in Fig. 2.1 include several that are unrelated to soil-borne pest and disease control and which are common to most green manures, including beneficial impacts on organic matter, nutrition, soil structure and erosion control. Those related to pest and disease suppression include non-host or trap crop effects, and indirect effects on the pest or pathogen associated with changes in the populations of antagonistic organisms, as well as the more direct suppressive effects of compounds released from the tissues that may not be related to glucosinolates or isothiocyanates. It is clear from Fig. 2.1 that the isothiocyanate-related biofumigation pathway is only one of many ways in which Brassica green manures can influence the growth of a crop. It is worth highlighting some of the notable, previously reported non-biofumigation impacts of green manures before a more detailed consideration of the ways in which the biofumigation pathway has, particularly in the last 5-10 years, been refined for pest suppression.

### 2.5.1. Green Manures

Green manures, tillage practices, and organic amendments can provide benefits to following crops and to farming systems in general associated with maintenance of soil cover and soil integrity, soil sanitation effects, reduced erosion, greater soil organic matter and soil structural improvements that improve water penetration. Brassica green manures are no exception, and the combination of a large tap root with a dense network of fine surface roots appears to make some Brassica species particularly useful in this regard. For example, on the sandy irrigated soils in the western United States the inclusion of mustard green manure crops has reduced the level of wind erosion and improved the water infiltration of the soils, with a steady improvement in soil structure where they are included in the system for some years. These impacts of brassicas on soil structure and water penetration have also been observed in hard-setting soils in the Australian wheat belt following canola rotation crops, where they appear to be related to the influence of the canola roots on soil aggregation.

Brassica green manure crops are effective at capturing soil mineral nitrogen that may otherwise leach from the soil, and when incorporated into the soil can provide a source of organic nitrogen that can become available to subsequent crops. The recent review provides a very comprehensive account of the use of green manures and catch crops in this regard, and the significant body of literature reviewed includes many studies involving Brassicaceous species such as radish (*Raphanus sativus*), white mustard (*Sinapis alba*), and winter rape (*Brassica napus*) in comparison with leguminous and grass species.

### 2.5.2. Trap Crops and Non-Hosting

Brassicaceous green manures can also act as trap crops. Perhaps the most well documented case involving their use for this purpose is for the control of sugar beet nematodes (*Heterodera schactii*) in northern Europe (Muller, 1999). Fodder radish (*Raphanus sativus*) and white mustard (*Sinapis alba*) cover crops have been selected and developed to be grown as green manures preceding sugar beet crops.

The brassicas are invaded by the nematodes, which develop within the roots, but have their sexual differentiation disrupted. This results in very low numbers of females in the subsequent generation, causing significant decline in the population and reducing the infection of subsequent sugar beet crops. It is a unique example related to specific nematode resistant brassicas rather than a general non-hosting effect.

For a green manure to be effective in disease control it is generally desirable that it not host the pathogen in question so that a decline in population or inoculum occurs during its growth. Indeed the fact that Brassica species are generally moderate hosts to some other important plant parasitic nematodes (e.g., *Meloidogyne* spp.), has reduced their applicability as biofumigant green manures as very careful management is required to avoid population increases on the biofumigant crop, particularly in warmer climates (Stirling and Stirling, 2003). In such cases, it is possible that the suppressive effects of the incorporated tissues could still reduce the population to manageable levels, but it is generally desirable to select biofumigants that are poor hosts, or to grow them at times of the year when pathogens do not build up.

## 2.6. SOIL HEALTH AND FERTILITY

Soils from the on-farm research plots should be analyzed in order to document changes in soil pathogens and soil fertility. These samples should be used to detect changes in microbial composition over time with the proposed treatments. This will enable us to provide scientific evidence pertaining to increases in populations of beneficial soil microorganisms and decreases in pathogens.

This data will be used to develop methods of detecting changes in soil pathogen populations to allow better decisions on when biofumigation is needed. As part of on-going research, the information gathered from the farm plots should be used to develop diagnostic kits that measure microbial populations in the soil as an indicator of the general health of the soil system.

## 2.7. OPPORTUNITIES TO ENHANCE BIOFUMIGATION

Having established that suppression by incorporated *Brassica* tissues is associated with GSL hydrolysis products, and indications of which hydrolysis products are most toxic, there are significant opportunities to enhance the biofumigation potential of Brassicaceous green manures. Firstly, by selecting brassicas which produce the greatest amount of the GSL-precursors most toxic to the target organisms, and secondly by managing the incorporation process to maximize the exposure of the organisms to the toxic compounds at the most vulnerable stage. Strategies to increase the production of GSL hydrolysis products by brassicas have been summarized elsewhere (Kirkegaard and Sarwar 1998), and rely upon; (1) significant variation in the type and concentration in individual GSLs among different species, cultivars and plant parts, (2) independent variation in biomass production and (3) differential toxicity of different hydrolysis products to particular organisms. Together these represent opportunities to select or develop biofumigant types which may provide up to $10^5$-fold increase in biofumigation potential over varieties selected at random.

### 2.7.1. Improving Efficacy in the Field

The potential efficacy of ITCs present in *Brassica* green manures can be considered by comparison with amounts of the synthetic soil fumigant methyl ITC (MITC) which are applied (517–1294 nmol/g). Assuming a maximum biomass of 15 t/ha, a tissue GSL concentrations of 100 µmole/g and an incorporation depth of 10 cm (soil bulk density 1.4), the potential ITC production (assuming 100% conversion) would be equivalent to 1070 nmol/g, which is in the range of commercial MITC application. Although the efficiency of conversion from incorporated tissues can be as low as 15% (Borek *et al.*, 1997), several ITCs have been shown to be up to 10 times more toxic than MITC. It is difficult however, to predict the impact of the lower concentrations of ITC released over an extended time period relative to commercial fumigant applications. More information is required on the fate, persistence and efficacy of the biocidal compounds released from incorporated tissues.

The impact of the biocidal compounds released from the tissue can also be increased by matching the timing of incorporation and release of biocides to the most vulnerable stages of the pest organism's life-cycle. While sufficient time must be provided for the organic material to decompose to avoid both physical and potential allelopathic interferences in the following crop, delaying for too long may allow some pathogens to recover.

Further studies are in progress to improve our understanding of the accumulation of GSLs in *Brassica* plants, the fate and activity of the biocidal hydrolysis compounds released from incorporated tissue in the soil, and the most effective ways of incorporating biofumigant crops into integrated pest management strategies.

## 2.8. BIOFUMIGATION CROPS

Table 2.1. Brassica plant species used for biofumigation

| Species | Selection | Sowing time | Biomass yield (t/ha) | Contents in GLs (µmol/ g d.m.) | GLs yield (Moles/ ha) | Main GLs | % N content d.m. | Biocidal activity |
|---|---|---|---|---|---|---|---|---|
| *Brassica juncea* | ISCI 20 | Autumn | 93.8±1.1 | 12.9±0.4 | 201±10 | Singrin | 1.5 | 3 |
| | | Spring | 46.5±10.6 | 25.8±0.4 | 246±32.5 | | 2.0 | |
| *Brassica juncea* | ISCI 61 | Autumn | 133.5±23 | 11±1.1 | 183.6±26 | Singrin | 1.4 | 2 |
| | | Spring | 60±14.1 | 17.7±2.3 | 213±35 | | 2.0 | |
| *Brassica juncea* | ISCI 99 | Autumn | 109±16 | 16.9±0.5 | 300±49.9 | Singrin | 1.2 | 4 |
| | | Spring | 54.5±9.5 | 33.6±1.6 | 333.8±41 | | 1.6 | |
| *Rapistrum rugosum* | ISCI 4 | Autumn | 84±33.2 | 27.6±3.2 | 562±158 | Cheirolin | 1.9 | 4 |
| | | Spring | 65.3±21.4 | 24.7±1.0 | 160.7±45 | | 2.8 | |
| *Eruca sativa* | Nemat | Autumn | 87±19.2 | 9.4±0.6 | 186±15 | Erucin | 1.5 | 1 |
| | | Spring | 45.3±4.2 | 12.9±2.5 | 107.2±8.5 | | 2.0 | |
| *Brassica nigra* | ISCI 27 | Autumn | 102.8±25 | 17.1±0.3 | 255.5±8.1 | Singrin | 0.9 | 4 |
| | | Spring | 46±15.5 | 20.7±0.2 | 204.5±61 | | 1.5 | |

Fig. 2.2. Brassicas help in controlling soil-borne pathogens.

### 2.8.1. Rapeseed (Canola)

Two Brassica species are commonly grown as rapeseed, *Brassica napus* and *Brassica rapa*. Rapeseed that has been bred to have low concentrations of both erucic acid and glucosinolates in the seed is called canola, which is a word derived from Canadian Oil.

Annual or spring-type rapeseed belongs to the species *B. napus*, whereas winter-type or biennial rapeseed cultivars belong to the species *B. rapa*. Rapeseed is used as industrial oil while canola is used for a wider range of products including cooking oils and biodiesel. Besides their use as an oil crop, these species are also used for forage. If pest suppression is an objective, rapeseed should be used rather than canola since the breakdown products of glucosinolates are thought to be a principal mechanism for pest control with these cover crops.

Rapeseed has been shown to have biological activity against plant parasitic nematodes as well as weeds. Due to its rapid fall growth, rapeseed captured as much as 120 lb of residual nitrogen per acre in Maryland. In Oregon, aboveground biomass accumulation reached 6,000 lb/acre and N accumulation was 80 lb/acre.

Some winter-type cultivars are able to withstand quite low temperatures (10°F). This makes rapeseed one of the most versatile cruciferous cover crops, because it can be used either as a spring-or summer-seeded cover crop or a fall-seeded winter cover crop. Rapeseed grows 3 to 5 feet tall.

### 2.8.2. Mustard

Mustard is a name that is applied to many different botanical species, including white or yellow mustard (*Sinapis alba*, sometimes referred to as *Brassica hirta*), brown or Indian mustard (*Brassica juncea*, sometimes erroneously referred to as canola), and black mustard (*B. nigra*) (Fig. 2.3) (Koch, 1995).

Fig. 2.3. *Sinapis alba* and *Brassica juncea* used for biofumigation.

The glucosinolate content of most mustards is very high compared to the true *Brassicas*. In the Salinas Valley, California, mustard biomass reached 8,500 lb/acre. Nitrogen content on high residual N vegetable ground reached 328 lb N/acre. Because mustards are sensitive to freezing, winter-killing at about 25°F, they are used either as a spring/summer crop or they winter-kill except in areas with little freeze danger. Brown and field mustard both can grow to 6 feet tall.

In Washington, a wheat/mustard-potato system shows promise for reducing or eliminating the soil fumigant metham sodium. White mustard and oriental mustard both suppressed potato early dying (*Verticillium dahliae*) and resulted in tuber yields equivalent to fumigated soils, while also improving infiltration, all at a cost savings of about $66/acre. Mustard green manures can be used to replace fumigants to improve infiltration in potato cropping systems. Mustards have also been shown to suppress growth of weeds.

### 2.8.3. Radish

The true radish or forage radish (*Raphanus sativus*) does not exist in the wild and has only been known as a cultivated species since ancient times. Cultivars developed for high forage biomass or high oilseed yield are also useful for cover crop purposes (Fig. 2.4). Common types include oilseed and forage radish.

**Fig. 2.4.** Daikon radish at full canopy closure, planted in August in Blacksburg, VA (Appalachian region), photographed approximately 60 days after planting.

Their rapid fall growth has the potential to capture nitrogen in large amounts (170 lb/acre in Maryland) and from deep in the soil profile. Above ground dry biomass accumulation reached 8,000 lb/acre and N accumulation reached 140 lb/acre in Michigan. Below ground biomass of radishes can be as high as 3,700 lb/acre.

Oilseed radish is less affected by frost than forage radish, but may be killed by heavy frost below 25°F. Radish grows about 2 to 3 feet tall. Radishes have been shown to alleviate soil compaction and suppress weeds.

In an Alabama study of 50 cultivars belonging to the genera *Brassica*, *Raphanus*, and *Sinapis*, forage and oilseed radish cultivars produced the largest amount of biomass in central and south Alabama, whereas winter-type rapeseed cultivars had the highest production in North Alabama.

### 2.8.4. Turnips

Turnips (*Brassica rapa* var. *rapa*) are used for human and animal food because of their edible root. Turnip has been shown to alleviate soil compaction. While they usually do not produce as much biomass as other brassicas, they provide many macrochannels that facilitate water infiltration. Similar to radish, turnip is unaffected by early frost but will likely be killed by temperatures below 25°F.

### 2.8.5. Processed Brassica Amendments

Another mode for achieving biofumigation effects is to use Brassica-derived isothiocyanate-rich materials such as seed meals (Fig. 2.5) or oils as soil amendments to achieve pesticidal effects (Lazzeri *et al.*, 2004). There may be a niche for such products, but they are likely to be regarded as pesticides by regulatory authorities and will therefore potentially face significant hurdles in implementation that the growing of a Brassica crop rich in appropriate isothiocyanates and manipulating it to maximize their release would not face.

Fig. 2.5. Brassica seed meal used for biofumigation.

## 2.8.6. Non-Brassica Plant Species

***(i) Grasses:*** Members of the grass family (Gramineae) also produce a rich diversity of bioactive chemical compounds, including phenolics, glycosides, benzoxazinones and amino acids. Although many are primarily known for their allelopathic activity against other plants (Putnam and DeFrank, 1983), they may also possess properties deleterious to a broad range of fungi, bacteria, nematodes and insects in soil. For example, residues of several Graminaceous crops of agronomic importance, including cultivars of barley *(Hordeum vulgare)*, wheat *(Triticum aestivum)*, triticale *(X. triticosecale)* and oats *(Avena sativa)*, all demonstrated significant, deleterious effects on soil-borne nematodes during their decomposition in soil. Phytotoxicity was evident in many test plants when they were subsequently established in the soil shortly after amendment with the residues (Stapleton, 2006).

A similar phytotoxic effect was even more pronounced when field and greenhouse studies were conducted using sudex, a hybrid of sorghum and Sudan grass *(Sorghum bicolor × S. sudanense)*, as a cover crop. Severe allelopathic effects occurred on subsequently planted tomato, broccoli and lettuce transplants, unless a waiting period of at least 6 weeks was observed between the incorporation of sudex residues in soil and planting of the following crop (Summers *et al.*, 2009).

Allelochemicals are plant-produced compounds (other than food compounds) that affect the behaviour of other organisms in the plant's environment. For example, Sudan grass (and sorghum) contains a chemical, dhurrin that degrades into hydrogen cyanide, which is a powerful nematicide (Luna, 1993; Forge *et al.*, 1995; Wider and Abawi, 2000). Some cover crops have exhibited nematode-suppressive characteristics equivalent to aldicarb, a synthetic chemical pesticide (Grossman, 1990).

***(ii) Garlic and Onions:*** The biocidal properties of garlic, onion and leek are attributed to sulfur volatiles produced during degradation of *Allium* tissues. The primary emitted compounds are thiosulfinates and zwiebelanes mainly converted in soil or in *Allium* products (extracts) to disulfides. The activities of these compounds were studied *in vitro* on soil pathogenic fungi and insects in order to measure their disinfection potential. These studies show a good potential for three disulfides: dimethyl disulfide (DMDS), dipropyl disulfide (DPDS) and diallyl disulfide (DADS) to inhibit several fungal species: *Aphanomyces euteiches, Colletotrichum coccodes, Fusarium moniliforme, Fusarium oxysporum radicis cucumerinum, Phytopht-*

*hora cinnamomi, Pythium aphanidermatum, Rhizoctonia solani, Sclerotium rolfsii* and *Sclerotinia slerotiorum.* The insecticidal activity of two disulfides was also evaluated on termites. The sensitivity varied with termite phenotypes and DMDS was the most toxic disulfide. This study shows that disulfides exhibit greater toxicity for insects than for fungi.

Yet another group of bioactive plants are those in the onion family (Alliaceae). Garlic and onion, especially, have been known for their bioactive properties since ancient times. Feasibility studies examined the decomposition of garlic and onion residues in moist soil, as related to the seed inactivation of four important agricultural weeds: black nightshade *(Solanum nigrum)*, common purslane *(Portulaca oleracea)*, London rocket *(Sisymbrium irio)* and barnyard grass *(Echinochloa crusgalli)*. The inhibitory and herbicidal effects of the Alliaceous residues were generally mild or inconsistent when tested at soil temperatures of 23°C. However, at 39°C, which by itself was mildly inhibitory to weed seed germination, the activity of the decomposing residues was far more potent (Mallek *et al.*, 2007).

The term 'biofumigation' has been extended to the use of *Allium* species producing compounds derived from sulfur amino-acids. The primary active compounds produced in crushed *Allium* spp are thiosulfinates but they are rapidly transformed into active disulfides, particularly dimethyl disulfide (DMDS) which is also produced by Brassicaceae.

A lot of other plants have shown interesting activities against nematodes and fungi.

*Chapter 3*

# SOLARIZATION

Soil-borne pests can be controlled in crop plants by pre-plant application of pesticides, including the fumigants methyl bromide, chloropicrin, and metham sodium. The use of these materials, however, is often undesirable due to their toxicity to animals and people, their residual toxicity in plants and soils, the complexity of soil treatment, and their high cost. Furthermore, restrictions on the use of soil-applied pesticides seem imminent as existing environmental legislation is implemented. As a result, there has been an increased emphasis on reduced-pesticide or non-pesticidal control methods.

Soil solarization is a non-pesticidal method of controlling soil-borne pests by placing plastic sheets on moist soil during periods of high ambient temperature. The plastic sheets allow the sun's radiant energy to be trapped in the soil, heating the upper levels. Solarization during the hot summer months can increase soil temperature to levels that kill many disease-causing organisms (pathogens), nematodes, and weed seed and seedlings. It leaves no toxic residues and can be easily used on a small or large scale. Soil solarization also improves soil structure and increases the availability of nitrogen (N) and other essential plant nutrients.

Solarization is a simple, safe, and effective method that has been used with field, vegetable, and flower crops and in orchards, vineyards, greenhouses, gardens, and landscapes in for over 12 years. It can be combined with organic soil amendments or reduced rates of pesticide application for greater effectiveness. Large increases in plant growth, harvestable yield, and crop quality often occur in solarized soil and may continue for more than one growing season. The potential for using soil solarization to control diseases and pests in the warmer areas is excellent.

Solarization is a technique that elevates soil temperatures beneath a clear plastic layer to reduce soil-borne pests. The capacity of soil solarization to suppress propagule numbers of soil-borne pathogens relies on many factors. The temperatures obtained in the moistened soil covered by the transparent sheeting and the exposure time of the organisms to these elevated temperatures are both important characteristics of this pre-plant soil treatment. Solarization has been effective in disease control in many geographical locations around the world. It is most successful in regions with the appropriate meteorological parameters such as high air temperatures and extended periods of high radiation.

In the United States, successful studies have been conducted in several states. A study in California showed that solarization alone increased strawberry yield by 12% over non-treated plots (Hartz *et al.*, 1993). In northern Florida, research showed that soil solarization decreased densities of *Phytophthora nicotianae* and *Ralstonia solancearum* to depths of 25cm and 15cm, respectively (Chllemi and Olson, 1994). In an Alabama study, temperatures attained via solar heating were 48°C at the soil surface and 34°C 20cm below the surface (Himelrick and Dozier, 1991). In studies previously conducted at The University of Tennessee, temperatures of 49.5°C at 5cm depth and 41.5°C at 25cm depth have been recorded under solarization treatments.

## 3.1. RESULTS OF SOLARIZATION

### 3.1.1. Increased Soil Temperature

The heating effect of soil solarization is greatest at the surface of the soil and decreases with depth. The maximum temperature of soil solarized in the field is usually from 42° to 55°C at a depth of 5 cm and from 32° to 37°C at 45 cm. Control of soil pests is usually best in the upper 10-30 cm. Higher soil temperatures and deeper soil heating may be achieved inside greenhouses or by using a double layer of plastic sheeting. Soil solarized in greenhouses may reach 60°C at a depth of 10 cm and 53°C at 20 cm. Soil solarized in black plastic nursery sleeves under a single or double layer of clear plastic can exceed 70°C.

### 3.1.2. Improved Soil Physical and Chemical Features

Solarization initiates changes in the physical and chemical features of soil that improve the growth and development of plants. It speeds up the breakdown of organic material in the soil, resulting in the release of soluble nutrients such as nitrogen ($N0_3$, $NH_4+$), calcium (Ca++), magnesium (Mg++), potassium (K+), and fulvic acid, making them more available to plants. Improvements in soil tilth through soil aggregation are also observed.

### 3.1.3. Control of Soil-borne Pathogens

Repeated daily heating during solarization kills many plant pathogens, nematodes, and weed seed and seedlings. The heat also weakens many organisms that can withstand solarization, making them more vulnerable to heat-resistant fungi and bacteria that act as natural enemies. Changes in the soil chemistry during solarization may also kill or weaken some soil organisms.

Although many soil pests are killed at temperatures above 30° to 33°C, plant pathogens, weeds, and other soil-borne organisms differ in their sensitivity to soil heating. Some pests that are difficult to control with soil fumigants are easily controlled by soil solarization. Other pests are also affected but cannot be consistently controlled by solarization. These may require additional control measures.

***(i) Fungi and Bacteria:*** Solarization controls populations of many important soil-borne fungal and bacterial plant pathogens, including *Verticillium dahliae*, which causes *Verticillium* wilt in many crops; certain *Fusarium* spp. that cause *Fusarium* wilt in some crops; *Phytophthora cinnamomi*, which causes *Phytophthora* root rot; *Agrobacterium tumefaciens*, which causes crown gall disease; *Clavibacter michiganensis*, which causes tomato canker; and *Streptomyces scabies*, which causes potato scab (Table 3.1). Other fungi and bacteria are more difficult to control with solarization, such as certain high temperature fungi in the genera *Macrophomina*, *Fusarium*, and *Pythium*, and the soil-borne bacterium *Ralstonia solanacearum*.

Table 3.1. Some diseases and nematodes controlled by soil solarization

| Disease | Crop |
|---|---|
| Verticillium wilt | Tomato, potato, egg-plant, cotton, strawberry |
| Fusarium wilt | Tomato, melon, onion, cotton |
| Pink root rot | Onion |
| Southern stem rot (white mold) | Peanut |
| Rhizoctonia seedling disease (sore shin or damping off) | Potato, onion, bean |
| Crown Gall | Walnut |
| Phytophthora root rot | Ornamentals |
| Nematodes (lesion, root knot, reniform, cyst, sting, ring, stubby root and dagger) | Various crops |

***(ii) Nematodes:*** Soil solarization can be used to control many species of nematodes (Table 3.1). However, soil solarization is not always as effective in controlling nematodes as it is in controlling fungal diseases and weeds because nematodes are relatively mobile and can recolonize soil rapidly. Nematode management may therefore, require yearly treatment. Control by solarization is greatest in the upper 30 cm of the soil. Nematodes deeper in the soil profile may survive solarization and damage plants with deep root systems.

Nematode control by solarization is usually adequate to improve the growth of shallow-rooted, short-season plants. It is particularly useful for organic gardeners and home gardeners. Solarization may also be a beneficial addition to an integrated nematode control system. For example, excellent control of root knot nematode (*Meloidogyne incognita*) was obtained in the San Joaquin Valley by combining solarization with the application of composted chicken manure (Gamliel and Stapleton, 1993b).

***(iii) Weeds:*** Soil solarization controls many annual and perennial weeds. While some weed species are very sensitive to soil solarization, others are moderately resistant and require optimum conditions (good soil moisture, tight-fitting plastic and high radiation) for control. Winter annual weeds seem to be especially sensitive to solarization, and control of winter annuals is often evident for more than one year following treatment. Soil solarization is especially effective in controlling weeds in fall-seeded crops such as onions, garlic, carrots, broccoli and other brassica crops, and lettuce. White sweet clover (*Melilotus alba*) is one of the few winter annuals that are poorly controlled. Although summer annual weeds are less temperature-sensitive than winter annuals, most summer annuals are relatively easily controlled by soil solarization. Control of purslane (*Portulaca oleracea*) and crabgrass (*Digitaria sanguinalis*) may be more difficult to achieve. If purslane is controlled, it is a good indicator that the soil has been adequately heated.

Solarization generally does not control perennial weeds as well as it controls annual weeds because perennials often have deeply buried underground vegetative structures such as roots and rhizomes that may resprout. Seeds of Bermuda grass (*Cynodon dactylon*), johnsongrass (*Sorghum halepense*) and field bindweed (*Convolvulus arvensis*) are controlled by solarization. Rhizomes of Bermuda grass and Johnson grass may be controlled by solarization if they are not deeply buried. Solarization alone is not effective for the control of the rhizomes of field bindweed. Yellow nutsedge (*Cyperus esculentus*) is only partially controlled by soil solarization. Purple nutsedge (*Cyperus rotundus*) is not significantly affected; marginal solarization has actually induced purple nutsedge to grow.

### 3.1.4. Encouragement of Beneficial Soil Organisms

Fortunately, although many soil pests are killed by soil solarization, many beneficial soil organisms are able to either survive solarization or recolonize the soil very quickly afterwards. Important among these beneficials are the mycorrhizal fungi and fungi and bacteria that parasitize plant pathogens and aid plant growth. The shift in the population in favour of these beneficials can make solarized soils more resistant to pathogens than non-solarized or fumigated soil.

***(i) Earthworms:*** The effect of soil solarization on earth worms has not received much attention, but it is thought that they retreat to lower depths and escape the effects of soil heating.

***(ii) Fungi:*** Beneficial fungi, especially *Trichoderma, Talaromyces*, and *Aspergillus* spp., survive or even increase in solarized soil. Mycorrhizal fungi are more resistant to heat than most plant pathogenic fungi. Their populations may be decreased in the upper soil profile but studies have shown that this is not enough to reduce their colonization of host roots in solarized soil.

***(iii) Bacteria:*** Populations of the beneficial bacteria *Bacillus* and *Pseudomonas* spp. are reduced during solarization but recolonize the soil rapidly afterward. Populations of *Rhizobium* spp., which fix nitrogen in root nodules of legumes, may be greatly reduced by solarization and should be reintroduced by inoculation of leguminous seed. Soil-borne populations of other nitrifying bacteria are also reduced during solarization. Population levels of actinomycetes are not greatly affected by soil solarization. Many members of this group are known to be antagonistic to plant pathogenic fungi.

During solarization of soil, populations of oxidize negative fluorescent pseudomonads and gram positive bacteria, including *Bacillus* species, may be reduced by 78 to 86% compared with non-solarized soil (Stapleton and DeVay, 1984); whereas, populations of Actinomycetes may be reduced from 45 to 58 percent in solarized soil (Stapleton and DeVay, 1984). Surprisingly, after solarization, *Pseudomonas* species quickly recolonize the soil and their populations reach high levels. Of great significance is the change in populations of *Bacillus* species during solarization; the percentage of colonies in solarized soil which exhibited antibiosis to *Geotrichum candidum* increased nearly 20-fold when compared with non-solarized soil (Stapleton and DeVay, 1984). These bacteria are among those which are rhizosphere competent and are believed to contribute to the increased growth response of plants grown in solarized soil (Katan, 1987). Although initial populations of *Bacillus* species are greatly reduced, they are spore formers and are a major component of the soil microflora.

In contrast to the studies in California and Israel, studies in Western Australia showed that solarization increased the total numbers of bacteria and actinomycetes in soil. However, as in the California study (Stapleton and DeVay, 1984) where there was an increase in the proportion of antagonistic gram positive bacteria in solarized soil. The Western Australia study showed that the proportion of bacteria (actinomycetes) antagonistic to *Fusarium oxysporum*, *F. solani*, and *Rhizoctonia solani* was increased compared with non-solarized soil. In other studies, *Actinomyces scabies* was controlled by soil solarization.

### 3.1.5. Increased Plant Growth

Plants often grow faster and produce both higher and better-quality yields when grown in solarized soil. This can be attributed, in part, to improved disease and weed control; but increases in plant growth are still seen when soil apparently free of pests is solarized. A number of factors may be involved. First, minor or unknown pests may also be controlled. Second, the increase in soluble nutrients improves plant growth. Third, relatively greater populations of helpful soil microorganisms have been documented following solarization, and some of these, such as certain fluorescent pseudomonad and *Bacillus* bacteria, are known to be biological control agents.

## 3.2. ADAPTABILITY OF SOLARIZATION

Solarization has considerable versatility, being adaptable to various agricultural production applications.

### 3.2.1. Protected Cultivation

Worldwide, probably the major commercial use of solarization is in conjunction with greenhouse/glasshouse/plastic-house culture, especially in regions where the protected crops are grown only in the winter. The empty structures can be closed in the heat of the summer and plastic film laid on the soil for solarization. This method of double insulation provides a still-air chamber above the solarized soil for added heating and greater efficacy. Greenhouse solarization is used primarily in Japan, the Near East and other Mediterranean countries (Stapleton, 1997).

### 3.2.2. Nursery Production

Solarization without combination with a chemical pesticide is not suitable for open field nursery production, due to zero tolerance for surviving pathogen propagules deeper in the soil. However, in containerized nursery production, solarization can provide complete eradication of pests. A "double-tent" solarization technique, which functions in much the same way as the greenhouse treatment just described, was approved by the California Department of Food and Agriculture in 1999 for nematode-free production of container-, flat- and frame-grown nursery stock. Among other mandated details, the approved treatment specifies solarization of soil at a minimum of 70°C for at least 30 contiguous minutes (Stapleton *et al.*, 1999).

### 3.2.3. Open Field Production (Annual Crops)

Open field cultivation of row crops is the setting in which solarization was first discovered and tested. Soil to be treated should be thoroughly moistened either by pre-irrigation or by drip irrigation beneath the clear plastic film. Soil can be solarized in fields by either complete coverage or by covering only the planting beds. This technique works best with shallow-rooted, late-season crops in warmer locations so that the solarization can be done between crops in the summer (Elmore *et al.*, 1997).

### 3.2.4. Open Field Production (Permanent Crops)

Solarization can be modified for use in managing certain weed, disease and nematode pests before or during establishment or replanting of orchard and vineyard crops, especially in warm areas such as the Central Valley. The method is most useful for managing shallowly distributed pests. It would not be expected to effectively control nematodes, fungi and other pests deep in the soil. When using solarization in conjunction with growing plants, treatment with black, rather than clear, plastic film may be preferable. For example, early studies indicated that solarization with clear plastic film beginning shortly after planting killed almond and apricot trees due to excessive heat. However, similar treatment with black film controlled the disease *Verticillium* wilt without damaging the trees. Other benefits included conservation of soil moisture and reducing humidity in the crop canopy (Elmore *et al.*, 1997). This type of solarization is mainly used on a commercial basis in the Central Valley for *Prunus* spp. and citrus.

Although solarization can be an effective soil disinfestant, the stand alone process, which largely depends on passive solar heating, has inherent limitations. Fortunately, it is compatible with other physical, chemical and biological methods of soil disinfestation to provide more efficacious and/or predictable treatment through integration. As MB is phased out, many current users will turn to other pesticides for soil disinfestation. Combining these pesticides (perhaps at lower dosages) with solarization (perhaps for a shorter treatment period) may be effective. When using solarization in conjunction prove to be the most popular option for users in warm climatic areas who want to continue using chemical soil disinfestants.

### 3.2.5. Non-conventional Users

Solarization is presently used on a relatively small scale in conventional agriculture, but its use will probably increase as MB becomes unavailable. On the other hand, solarization has become a widespread practice for organic growers, home gardeners and other users who cannot or will not use chemical soil disinfestants.

## 3.3. FACTORS THAT LIMIT EFFECTIVENESS OF SOLARIZATION

### 3.3.1. Location

Soil solarization is most effective in warm, sunny locations. It also has been used successfully, but less predictably, in the cooler coastal areas and in many cooler parts during periods of highest air temperatures and clear skies. Greenhouse, nursery, and seedbed (containerized) media solarization are more effective in cooler climates than field solarization.

### 3.3.2. Weather

Highest soil temperatures occur when days are long, air temperatures are high, skies are clear, and there is no wind. The soil heating effect may be limited on cloudy days. Wind or air movement across the plastic will rapidly dissipate the trapped heat. Also, strong winds may lift or tear sheets.

### 3.3.3. Timing

The best time for solarization of soil is from June to August, although good results may be obtained in May and September, depending on weather and location. The heat peak in many areas is around May 15. To maximize production, soil solarization should be done during a period in crop rotations when fields are idle. Soil can be solarized during summer (when summer temperatures are too hot for many crops) and planted during fall or winter.

### 3.3.4. Duration of Treatment

The longer the soil is heated, the better the control of pests will be. However, heating the soil longer than required for effective control (6 to 8 weeks) may be deleterious to the soil. Although some pest organisms are killed within 14 days, 4 to 6 weeks of treatment in full sun during the summer is recommended for field application. Solarization of containerized growth media and greenhouses may be done in a few days during the heat of summer. Some relatively heat-resistant organisms may require longer (up to 8 weeks) solarization for control. The combination of pesticides, fertilizers, and certain organic amendments with solarization may reduce the needed treatment time.

### 3.3.5. Soil Preparation

A smooth seedbed is ideal for solarization. Air pockets between the plastic and the soil greatly reduce soil heating. Solarization will be ineffective if the seedbed is not smooth and the plastic does not rest directly on the soil.

### 3.3.6. Soil Moisture Content

If the soil is too dry (less than 70 percent of field capacity), weed seed and pathogens may not imbibe enough water to make them vulnerable to the increased heat.

### 3.3.7. Soil Colour

Dark soils absorb more solar radiation than lighter coloured soils and reach higher temperatures during solarization. However, adding dark material, such as charcoal, to a light loam soil has only raised maximum temperatures 1° to 2°F. Organic material such as manure may give the same limited effect.

### 3.3.8. Orientation of Beds

The heating of soil in raised beds will be most uniform if the beds are oriented north to south rather than from east to west. More uniform heating gives better control of pests. Solarization is most effective when there is no slope or when the slope has a south or southwest exposure. Lower temperatures and poor control of pests will occur on north-facing slopes. Cultivation after solarization and cultivation deeper than 7.5 cm after soil solarization should be avoided because it may bring weed seed and pathogens to the upper soil layer, causing severe weed and disease problems.

Integrity of plastic sheeting holes or tears in the plastic will adversely affect solarization. Animals and people should be prevented or discouraged from walking on or otherwise disturbing the plastic.

## 3.4. COMBINING SOLARIZATION WITH OTHER CONTROL METHODS

Under conducive conditions and proper use, solarization can provide excellent control of soil-borne pathogens in the field, greenhouse, nursery, and home garden. However, under marginal environmental conditions, with thermotolerant pest organisms or those distributed deeply in soil, or to minimize treatment duration, it is often desirable to combine solarization with other appropriate pest management techniques in an integrated pest management approach to improve the overall efficacy of treatment (Stapleton, 1997). Solarization is compatible with numerous other methods of physical, chemical, and biological pest management. This is not to say that solarization is always improved by combining with other methods.

Combining soil solarization with pesticides, organic fertilizers, and biological control agents has led to improved control of pathogens, nematodes, and weeds and may be especially useful in cooler areas, against heat-tolerant organisms, or to increase the long-term benefits of solarization.

### 3.4.1. Solarization and Biofumigation

The primary focus of solarization research will be towards the indirect effects of solarization and the effectiveness of combining it with other pest management strategies such as biofumigation. Katan and Devay (1991) states that microbial processes, induced by solarization, may contribute to disease control beyond the physical effect of heating. Both the vulnerability of soil-borne pathogens to mycoparasitic attack and the activity of beneficial microorganisms could be enhanced by solarization. In particular, a study by Lifshitz *et al.* (1983) showed that sclerotia of *S. rolfsii* became heavily colonized by bacteria under the influence of the sub-lethal temperatures generated by solarization. Cracking of the rind and increased leakage of organic compounds from the sclerotia was reported. Hence, the germinability of these propagules was compromised.

Organic amendments may increase the effectiveness of solarization against pests (Gamliel and Stapleton 1993, Ozores-

Hampton *et al.*, 2004, Wang *et al.*, 2006). Wang *et al.* (2006) showed that following a cowpea cover crop with solarization was more effective than solarization alone. In fact, the effectiveness of this combined treatment was comparable to methyl bromide fumigation.

Cabbage residues may or may not improve solarization success. When cabbage residues were incorporated as large undecomposed pieces, solarization was not improved and the cabbage may have served as a reservoir for diseases (Chellemi *et al.*, 1997). However, grinding cabbage into powder and mixing it uniformly into the soil may improve solarization efforts (Gamliel and Stapleton, 1993).

### 3.4.2. Chemical Controls

Many field trials have shown that, under the prevailing conditions, pesticidal efficacy of solarization or another management strategy alone could not be improved upon by combining the treatments (Stapleton and DeVay, 1995). However, even in such cases, combination of solarization with a low dose of an appropriate pesticide may provide the benefit of a more predictable treatment which is sought by commercial users. For example, although combining solarization with a partial dose of 1, 3-dichloropropene did not statistically improve control of northern root knot nematode (*Meloidogyne hapla*) over either treatment alone; it did reduce recoverable numbers of the pest to near undetectable levels to a soil depth of 46 cm (Stapleton and DeVay, 1983).

Low application rates of fungicides, fumigants, or herbicides have been successfully combined with soil solarization to achieve better pest control. The elevated temperatures seem to increase the activity of fungicides such as metham sodium, so lower rates may be applied. Solarization speeds up the disappearance of EPTC (Eptam) and vernolate, either by increasing their volatility or their degradation. Other chemicals, such as terbutryn or carbendazim, have slower degradation rates after solarization, possibly because of changes in the populations of soil microorganisms after solarization. Although such pesticides may be effective for longer periods than normal, care must be taken that they do not harm the next crop. Chemical controls may be applied either before or after solarization. A possible disadvantage of combining soil solarization with chemical control is that the chemical control may reduce the long-term benefits of solarization.

### 3.4.3. Amendments and Fertilizers

Solarization can also be combined with a wide range of organic amendments, such as composts, crop residues, green manures, and animal manures and inorganic fertilizers to increase the pesticidal effect of the combined treatments (Ramirez-Villapudua and Munnecke, 1987; Gamliel and Stapleton, 1993a, b; Chellemi *et al.,* 1997). Incorporation of these organic materials by themselves may act to reduce numbers of soil-borne pests in soil by altering the composition of the resident microbiota, or of the soil physical environment (biofumigation). Combining these materials with solarization can sometimes greatly increase the biocidal activity of the amendments. However, this appears to be an inconsistent phenomenon, and such effects should not be generalized without first conducting confirmatory research. The concentrations of many volatile compounds emanating from decomposing organic materials into the soil atmosphere have been shown to be significantly higher when solarized (Gamliel and Stapleton, 1993b).

Many commercial users of solarization in California apply manures or other amendments to soil before laying the plastic. There is evidence that these materials release volatile compounds in the soil that kill pests and help stimulate the growth of beneficial soil organisms. For example, the southern root-knot nematode, which was incompletely controlled in lettuce by either solarization or application of composted chicken manure, was completely controlled by combining the two, resulting in a large yield increase (Gamliel and Stapleton 1993).

### 3.4.4. Biological Controls

The successful addition of biological control agents to soil before, during, or after the solarization process in order to obtain increased and persistent pesticidal efficacy has long been sought after by researchers. There have been great hopes of adding specific antagonistic and/or plant growth promoting microorganisms to solarized soil, either by inundative release or with transplants or other propagative material, to establish a long-term disease-suppressive effect to subsequently planted crops (Katan, 1987; Stapleton and DeVay, 1995).

It is considered that by combining solarization with biofumigation and composted amendments the re-introduction of biocontrol agents such as *Trichoderma* spp. and *Bacillus* spp. may be more effective than either treatment alone in controlling soil-borne diseases. Populations of these two microbial antagonists increase relative to other microorganisms in solarized soil (Stapleton and DeVay, 1985).

Soil solarization has also been successfully combined with the fungal biological control agents *Trichoderma harzianum* and *Talaromyces flavus*, which were added to the soil or planting material.

In another integrated pest management strategy, clear plastic applied to planting rows in summer for solarization may be left in place and painted silver to control aphid-borne viral diseases in fall vegetable crops.

## 3.5. ECONOMICS OF SOLARIZATION

The cost of soil solarization depends on the thickness of the plastic used, the area of soil covered, the method of irrigation, and the method of plastic application, connection, and removal. These costs should be balanced against alternative methods of pest control, and in some cases should be viewed over a period of more than one year or growing season. In general, the greatest economic return from soil solarization will be obtained from high-value crops grown in soils infested with pathogens, nematodes, or weeds.

### Advantages

- Non-pesticidal and simple.
- No health or safety problems associated with use.
- No registration is required.
- Crops produced are pesticide-free and may command a higher market price.
- Controls multiple soil-borne diseases and pests.
- Selects for beneficial microorganisms.
- Tends to increase soil fertility.
- Increases soluble $N0_3$, $NH_4$, Ca, Mg, K and soluble organic matter.
- May improve soil tilth.
- Can speed up in-field composting of green manure.

### Disadvantages

- Is restricted to areas with warm to hot summers.
- May be less effective in cooler coastal areas.
- Land must be taken out of production for 4 to 6 weeks during the summer.
- May not fit in with some cropping cycles.
- May be difficult for those using a small amount of land intensively.

- Limited number of retail outlets for UV-inhibiting plastics.
- Disposal of plastic may be a problem.
- Large amounts of plastic cannot currently be recycled.
- Some pests are not controlled or are difficult to control.
- No pest control in the furrows between strips (if applied in strip coverage).
- High winds and animals may tear the plastic.

*Chapter 4*

# MODE OF ACTION

## 4.1. BIOFUMIGATION

### 4.1.1. The Glucosinolate-Myrosinase System

Glucosinolates are a class of sulphur compounds occurring as secondary plant products almost exclusively in families of the Order Capparales: Brassicaceae, Capparaceae, Moringaceae, Tovariaceae and Resedaceae (Brown and Morra, 1997). They are particularly common in the Brassicaceae, plants that are widely cultivated as important vegetable, condiment, oilseed and forage crops (Fenwick *et al.*, 1983). Comprising some 350 genera and 2500 species (Rosa *et al.*, 1997), the Brassicaceae therefore present a rich potential pre-adapted source of material for investigation and application for beneficial uses in cropping systems.

Brassica species glucosinolates in particular have been the subject of many extensive reviews over the last half-century. Increasingly, from a plant sciences perspective, attention has been directed at brassicas in relation to their long-recognized allelopathic properties and hence their potential to be harnessed for use in the management of pests in agriculture (Brown and Morra, 1997; Rosa and Rodrigues, 1999). However, despite a large body of evidence indicating positive effects and considerable potential, there has until recently been a lack of significant adoption of brassicas for soil-borne pest and disease management in cropping systems.

**(i) Glucosinolate hydrolysis:** Glucosinolates are organic anions possessing a thioglucose moiety, a sulfonated oxime, and any one of a variety of aliphatic or aromatic R groups. At least 120 structurally different glucosinolates have been identified in 16 different families of angiosperms. Although glucosinolates themselves possess limited biological activity, enzymatic degradation by thioglucoside glucohydrolase or myrosinase, results in the formation of a number of compounds including isothiocyanates, nitriles, $SCN^-$, oxazolidinethione, epthionitriles, and organic thiocyanates. The products formed from any one glucosinolate are controlled by both the chemistry of the glucosinolate itself as well as reaction conditions such as pH, $Fe^{++}$ concentration, and the presence of specified proteins and coenzymes.

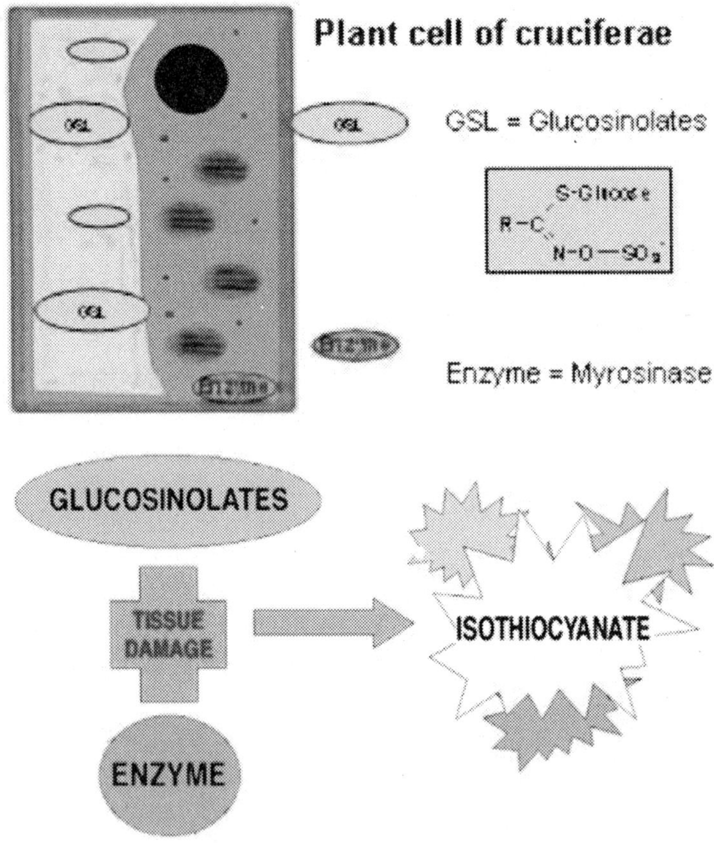


**Fig. 4.1.** Glucosinolate hydrolysis to release isothiocyanate.

Much of the research has focused on identifying and quantifying the biologically active molecules produced by mustard and rapeseed tissues once they are incorporated into soil. The most recent work showed that $SCN^-$ is produced from the primary glucosinolate in *Sinapis alba* meal by way of an unstable isothiocyanate intermediate.

The bioactive hydrolysis products of glucosinolates, particularly the isothiocyanates, can be used to control soil pests and weeds by incorporating glucosinolate-containing plant material in soil-a practice known as biofumigation. The fate of glucosinolates and their hydrolysis products in soil determines both the efficacy and environmental impact of biofumigation. Knowledge of the processes by which these compounds are sorbed, degraded or otherwise lost from the soil is fundamental to developing effective, but environmentally benign biofumigation strategies. Effective biofumigation relies on maximum hydrolysis of the glucosinolate in the plant tissue to generate high isothiocyanate concentrations in the soil after incorporation. This is favoured by maximum cell disruption, by addition of water, and a high soil temperature. Residual glucosinolates are very weakly sorbed, readily leached and are microbially degraded and mineralized in soil. In contrast, isothiocyanates are strongly sorbed by the organic matter in soil, react strongly with nucleophilic groups present in soil, and are prone to volatilization losses in addition to microbial degradation and mineralization. These loss processes are influenced by soil type, water content and temperature. Using appropriate incorporation strategies, sufficiently high isothiocyanate concentrations (>100 nmol $g^{-1}$) can be achieved in soil using biofumigation for effective suppression of susceptible pests. The relatively rapid sorption and degradation of the isothiocyanates in the period of days after incorporation minimizes the risks of persistence in the environment or leaching. Biofumigation is therefore a promising technique which can be further developed to form part of Integrated Pest Management (IPM) strategies to reduce reliance on synthetic pesticides with minimal unintended impacts on the environment.

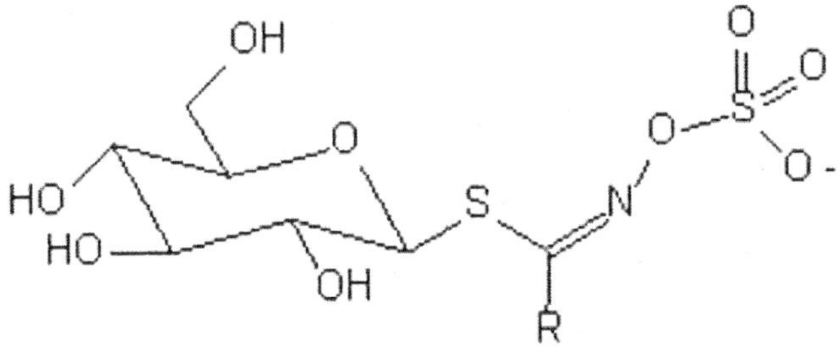

**Fig. 4.2.** General structure of glucosinolates.

Glucosinolates are characterized by a common core chemical entity, with a side-chain group of variable chemical structure that distinguishes individual types (Fig. 4.2), of which more than 120 have been identified in three groups: aliphatic, aromatic, and indolyl (Brown and Morra, 1997; Fahey *et al.*, 2001). Only the aliphatic and aromatic glucosinolates liberate isothiocyanates upon hydrolysis. Fahey *et al.* (2001) provide a comprehensive survey of the chemical structure of all glucosinolates known to that time, and an outline of their synthesis and biosynthesis. Glucosinolates consistently occur in plants in conjunction with a hydrolytic enzyme thioglucosidase hydrolase, known commonly as myrosinase. Of themselves, glucosinolates have no or very limited biological activity. But they are important because of the wide variety of active products that derive from them as a result of myrosinase action, a dynamic link that has led to the interaction commonly being dubbed the "glucosinolate-myrosinase system".

It is generally believed that the glucosinolate-myrosinase system is part of plant defense against insects and possibly pathogens and that it evolved from the more prevalent system of cyanogenic glucosides. Glucosinolates are very stable water-soluble compounds, while myrosinase is sequestered in vacuoles within myrosin cells, with both compounds separately coexisting distributed throughout the plant (Fahey *et al.*, 2001). It is only upon physical disruption of the plant tissue that the physical separation of the myrosinase from its glucosinolate substrate is overcome, triggering in the presence of water the hydrolysis reactions that release a variety of biologically-active products such as isothiocyanates, organic cyanides, oxazolidinethiones and ionic thiocyanates (Brown and Morra, 1997). Among the degradation products of glucosinolates, the isothiocyanates have been generally reported as the most biologically active, being recognized since early in the twentieth century as broad-spectrum biocides (Brown and Morra, 1997; Rosa and Rodrigues, 1999) (Fig. 4.3).

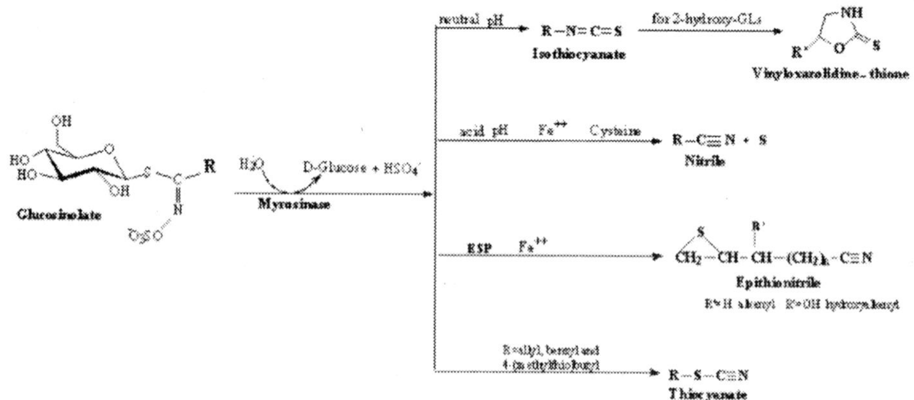


**Fig. 4.3.** General scheme of glucosinolates hydrolysis via myrosinase.

Reaction conditions that favour the production of isothiocyanates are pH around neutral, warm temperature and high dilution with water; at more acid pH, lower temperature and drier conditions the less biologically active nitriles predominate. The type of isothiocyanate generated corresponds to the type of glucosinolale substrate, as characterized by its side-chain. Interestingly, and perhaps a little ironic given that the biofumigation concept was partly founded on the basis that the synthetic methyl isothiocyanate generator metham sodium has found widespread commercial use as a pesticide, methyl glucosinolate does not occur in the Brassicaceac, although it is a characteristic component of the closely related Capparaceae (Fahey *et al.*, 2001).

**(ii) Glucosinolate Profiles:** Elements in the determination of biofumigation potential were to establish the glucosinolate profile of a diverse range of field-grown brassicas (Kirkegaard and Sarwar, 1998), to determine the effect of environment and ontogeny of the plants on glucosinolate production and the implications of that for screening, and the *in vitro* testing of the toxicity of some of the more common isothiocyanates occurring in brassicas to soil-borne fungal pathogens. The results of this systematic set of studies revealed a range of results relevant to the development of biofumigation as a pest management tool. In a slightly different, but conceptually similar approach, tests of the biocidal activity of various purified glucosinolates hydrolyzed with myrosinase were used to determine the most active hydrolysis products. The results were then used to find Brassicaceae with high levels of those glucosinolates and implement them in field use (Lazzeri *et al.*, 2004).

Key findings of Kirkegaard and Sarwar (1998) were that the variation in biomass, glucosinolate profiles and concentrations in both shoots and roots provide significant scope to select or develop brassicas with enhanced biofumigation potential. Total glucosinolate production on a ground area basis from almost eighty brassicas sampled at mid-flowering showed wide variation (0.8-45.3 mmol $m^2$) that derived equally from differences in biomass and glucosinolate concentration, and that were not correlated in either root or shoot tissues. Roots contributed an average of 23.6% (2 to 81%) of the total plant glucosinolates, with their contribution limited more by low biomass than glucosinolate concentration, which was usually similar or higher than that of shoots. The glucosinolate concentration in root and shoot tissue did not correlate with seed levels, ruling out seed glucosinolate levels as a selection parameter for higher biofumigation potential.

The types of glucosinolates present in the tissues varied considerably between species but were consistent within species. Shoots contained predominantly aliphatic glucosinolates while aromatic forms predominated in roots. Some species (e.g., *B. napus*) had predominately non-isothiocyanate liberating glucosinolates in the tissues while others (e.g., *B. juncea*) produced mostly isothiocyanate-liberating glucosinolates. Ultimately, what these results showed was that it is not possible to randomly use any Brassica (extending even to cultivars within a species) for biofumigation purposes. It is essential to be selective. Furthermore, the clear implication was that it is not helpful, and may even be counter-productive, to report studies imputing biofumigation effects without describing in some way the biofumigation potential of the species or cultivar used, as their capacity in that regard varies greatly. In this regard, only one of the seventeen field studies reported the glucosinolate concentration of the Brassica tissues.

Given that biofumigation effects by the definition used here are related to isothiocyanates, the use of glucosinolate concentration gives an indirect measure of the biofumigation potential of plants and is inherently predicated on a high level of conversion to isothiocyanates. To validate this assumption, and to develop a more direct and simpler analytical procedure, the glucosinolate concentration and isothiocyanate release by hydrolyzing tissue of matched samples of Brassica shoot and root tissues were analyzed. In around 80% of samples, the molar concentration of glucosinolates in tissues was greater than that of the isothiocyanates released upon hydrolysis. The reasons for the observed trends were obscure and remain speculative without further study, yet are nevertheless noteworthy in terms of realizing biofumigation potential. Not only was the direct measurement of isothiocyanate evolution by gas chromatography considered a desirable method because it was directed at the primary bioactive principles, but also it was a simpler analytical procedure that avoided several steps involving wet chemistry.

**(iii) Effects on Plant Growth and Environment:** In a study using a representative selection of brassicas from five species, the total glucosinolate concentration in root and shoot tissue varied considerably (3- to 10-fold) with environment. Interactions between species and environments meant that the ranking of the brassicas for total shoot and root glucosinolates changed with environment, while the added effect of environment on phenological development and biomass production further influenced overall glucosinolate production by a crop on a ground area basis. Total glucosinolate concentration generally declined with the onset of flowering and was lowest at maturity, whereas spring-sown plants generally had greater concentration than those sown in autumn.

These results suggested that there will be an optimum time when the balance of biomass and glucosinolate concentration yields the greatest glucosinolate production per unit of ground area. Usefully, the study was conducted simultaneously in the field and glasshouse, producing results suggesting that a glasshouse environment can be used to determine the types and proportions of glucosinolates present in a Brassica and to rank various types within, but not between, species for the total concentration in the tissues. Overall, the results indicated that both environment and ontogeny can influence all of the plant-related components of the biofumigation potential. Thus, the glucosinolate production and biofumigation potential has the capacity to vary significantly. An accurate estimate of biofumigation potential therefore requires measurement in the target environment. Again, few of the studies included more than superficial information on growing conditions or development stage of the plants, thus nullifying comparisons and assessments of mechanisms.

**(iv) Biological Activity of Various Isothiocyanates:** The *in vitro* toxicity of six pure isothiocyanates (four aliphatic and two aromatic) to the mycelial growth of five root pathogens was tested by adding them to the headspace above, or dissolving them in the agar growing medium. Aromatic isothiocyanates were less toxic in the headspace tests because their lower volatility lessened headspace concentration, but were more toxic than the aliphatic isothiocyanates when dissolved in the agar. In both the vapour exposure and direct contact experimental methods, the toxicity of the aliphatic isothiocyanates decreased with increasing length of the side chain, although there was little difference between methyl and propenyl isothiocyanate (the latter occurring commonly in mustard, *B. juncea* and *B. nigra*) (Kirkegaard and Sarwar, 1998) in the headspace experiment. Similarly, the shorter chain of the two aromatic isothiocyanates was more toxic in the agar. Thus, shorter-chained isothiocyanates within each molecular structure class are inherently more toxic, irrespective of volatility; this supported similar contact-test findings using an insect.

In other studies where only the volatiles released from hydrolysis of Brassica tissues contacted test fungi, tissues dominant in more volatile types of isothiocyanates such as propenyl isothiocyanate were more toxic (Angus *et al.*, 1994) while in experiments using isothiocyanates dissolved in agar, aromatic isothiocyanates were up to 20 times more toxic. In an *in vitro* study using an insect as the bioassay subject that compared vapour exposure of pure methyl (aliphatic) and pure 2-phenylethyl (aromatic) isothiocyanates, little difference was found in toxicity at warmer temperature, but toxicity of the less volatile aromatic isothiocyanate fell away particularly sharply as temperature was reduced across the lower part of the range 20, 15, 10 to 5°C. It was concluded that the equivalent toxicities at warmer temperature were a consequence of the more than 70-fold higher contact toxicity of 2-phenylethyl than methyl isothiocyanate overriding its much lower volatility. Similarly, when the insects were exposed to vapours from hydrolyzing plant tissues rich in either propenyl or 2-phenylethyl glucosinolates at 15°C, the absolute amounts of dry tissue required were similar for given mortality levels, indicating that the effect was consistent when plant tissue substituted for pure isothiocyanates.

A seven-fold variation in susceptibility was found of the individual fungi they tested. Smith and Kirkegaard (2002), seeking to explain whether the beneficial break-crop effects of canola in canola-wheat rotations in the Australian cereal belt (Angus *et al.*, 1991) may be associated with biofumigation effects, recorded considerable variation in the sensitivity of different fungi to 2-phenylethyl isothiocyanate, which occurs in the roots of canola (Kirkegaard and Sarwar, 1998). This is consistent with differences in sensitivity of fungal species to pure isothiocyanates that have been reviewed by Brown and Morra (1997) where fungicidal concentrations of isothiocyanates for given fungi may differ by an order of magnitude. The variation in sensitivity to isothiocyanates provides opportunities to select Brassica species that produce the largest quantities of the isothiocyanates most toxic to the target organism.

Tests carried out using hydrolysis of pure glucosinolates *in vitro* revealed that sulphur-substituted aliphatic types were the most toxic to *Fusarium*, and to *Pythium* and *Rhizoctonia*. Subsequent wide screening of several glucosinolate-containing plant species has led to the selection and recent commercialization of several biofumigants that produce high quantities of the most potent glucosinolates (Lazzeri *et al.*, 2004). Using this systematic approach to match the most potent brassicas with target organisms known to be sensitive to particular isothiocyanates provides the best possible chance of achieving successful suppression in the field. Not only do these results indicate that the glucosinolate profile of Brassica biofumigants will differentially affect pathogens, but they possibly help explain why the efficacy of metham sodium is widely regarded as variable. Again, the paucity of information in the studies make it difficult to determine the role of isothiocyanates in suppressive effects of brassicas or suggest methods to improve observed beneficial effects.

**(v) Implications for Adoption:** The effectiveness of brassicas for biofumigation will ultimately depend on many factors beside the biofumigation potential of the particular Brassica used. The timing of incorporation or exudation of the tissue needs to coincide with the susceptible stage of the pest organism and suppression must be of a sufficient level and persistence to provide protection for the crop of interest. Additionally, the effectiveness of biofumigation will be influenced by the effectiveness of incorporation, activity of the hydrolyzing myrosinase enzyme, efficiency of the hydrolysis and losses resulting from volatilization, sorption onto soil components, leaching and microbial degradation. Nevertheless, the concept of biofumigation potential has provided a foundation from which to consider strategies to manipulate its components to enhance suppression for particular management scenarios.

Some of the more speculative early conclusions of Kirkegaard and Sarwar (1998) (and others over the years), in particular those based on extrapolation of *in vitro* determinations of isothiocyanate toxicity to the field, have proven inapplicable once assessed *in vivo*. In general, however, subsequent findings by a range of researchers, particularly on the most appropriate Brassica species to use, the plant parts and types of isothiocyanates on which to focus and the conditions required to achieve maximum release of isothiocyanates from Brassica tissue in soil, have ultimately shown the prescience of this integrative approach. It has been valuable in taking research beyond the limitations and even misdirections stemming from what was arguably hitherto over extrapolation from *in vitro* laboratory and empirical field studies, to systematically go much closer to achieving field-applicable outcomes. Previously, few field studies reported this crucial information necessary to assess potential to improve field efficacy.

It is relevant to comment that optimistic extrapolation from *in vitro* results is commonplace in many areas of biological science that propose potential solutions to a problem. There are a variety of reasons for this. Unfortunately, extrapolation or over statement particularly from *in vitro* laboratory studies (which are much simpler to do than *in vivo* or field studies), that is not carefully qualified can slow progress or even be counter-productive to the development of a concept because its proponents can be seen by practitioners as unrealistic toward real-world situations and needs. It is believed that the concept of biofumigation has had its share of this and there is a need for care in presentation of results to avoid any proliferation of the issue. Similarly, poor or inconsistent results from empirical field studies that have not adequately represented crucial factors may stultify progress by being given inappropriate levels of credence.

**(vi) Identity and Production of ITCs between Brassicas:** GC profiles and the subsequent ITCs produced varies between *Brassica* species. Researchers have identified over 100 ITCs, 20 of which are commonly produced by Brassicas and known to have a biocidal effect. The difference in structure of individual GCs and ITCs depends on their organic side-chain (aliphatic, aromatic or indole), which also influences their biocidal activity.

A preliminary trial aimed to identify and quantify ITCs produced by the freeze-dried roots of Indian mustard (*Brassica juncea*) and a turnip/canola mix (*B. campestris* / *B. napus*) (Bc/Bn), before and after their incorporation into a light clay loam. Three main ITCs were identified from *B. juncea* roots (3-butenyl, 4-pentenyl, 2-phenyethyl) and five from the Bc/Bn mix (3- butenyl, 4-pentenyl, 2-phenyethyl, 5-methylthiopentyl, benzyl). In total, the Bc/Bn mix produced 8 times more ITCs ((7.5 mol ITC/g root tissue) than *B. juncea* (0.9 mol ITC/g root tissue). These results positively demonstrate that different *Brassica* species produce different types and concentrations of ITCs. Similar trends were evident when roots were incorporated into soil, although ITCs were recovered at lower concentrations. These levels were at least 15 times below those required to control most pests and pathogens (based on commercial fumigant levels). Clearly, further research needed to investigate methods of elevating the concentration of ITCs released into soils by Brassicas, particularly under field conditions.

### (vii) Increasing ITC Production

***(a) Using Plant Stress:*** Brassicas increase production of GCs in response to plant stress (e.g. nutrient stress, mechanical damage, disease, etc). For this reason, a field trial was conducted to determine if various stress treatments applied to *B. juncea* could increase its production and release of ITCs into soil. These stresses were applied prior to the incorporation of the crop into soil (12 weeks after sowing) and included low levels of contact herbicide (bipyridyl at 5mL/100 L water), mechanical damage (puncturing of roots and shoots) and a defense elicitor. Controls consisted of a non-stressed treatment and a fallow.

Although biofumigant treatments did not kill *Phytophthora cactorum*, they suppressed its growth by 10% compared to fallowed plots. Similarly, biofumigant treatments reduced weed biomass by an average of 30%. These results show a clear potential for biofumigation to suppress pathogen and weed populations in the field. There was no evidence that stress treatments increased the efficacy of biofumigation against *P. cactorum* or weeds, although ITC concentrations are yet to be fully analyzed.

Biofumigation is based on the use of glucosinolate-containing plants for the control of soil-borne pest and diseases. Upon tissue damage, glucosinolates are hydrolyzed by endogenous enzymes (myrosinase) and a range of biologically active compounds are formed. Isothiocyanates (ITCs) are the quantitatively dominating products formed at neutral pH. Most of these compounds are volatile and only sparingly soluble in aqueous systems, and depending on the R-group structure and the presence of nucleophiles, further transformation of ITCs occurs. At lower pH and in the presence of certain molecules able to deliver two redox equivalents, the proportion of nitriles increases at the expense of ITC.

The effect of ascorbic acid and glutathione on the production of nitriles at pH 5 was investigated by micellar electrokinetic capillary chromatography (MECC). The presence of 0.25 µmol ascorbic acid increased the production of nitriles although at higher concentrations the proportion of nitriles decreased. Increasing amounts of GSH favored the production of nitriles (40% of the total degradation products were nitriles in the presence of 2 µmol GSH). The oxidation of GSH gives the redox equivalents needed for the liberation of the sulphur from the unstable intermediate of the glucosinolate hydrolysis leading to the formation of the nitrile.

**(viii) Biofumigation Potential-Beyond Empiricism:** The type, concentration and distribution of glucosinolates in different plant parts vary between Brassica species and cultivars. Hence the capacity to generate isothiocyanates varies accordingly. Although the biocidal activity of various isothiocyanates released by Brassica tissues *in vitro* is well established (Brown and Morra, 1997), difficulty in interpreting the role of isothiocyanates in the varied suppressive effects reported has characterized most field studies of green manuring with brassicas. A significant number of previous studies have reported suppression of a range of pests and pathogens in the field using a range of different Brassicaceous green manures. While the levels of suppression reported in some experiments are encouraging, their variability, even for the same pathogen is problematic for the wider adoption of the approach. In many cases there is insufficient information on key aspects of the approaches used to evaluate the likely role of isothiocyanates in the responses observed, or to assess the likelihood that the level of suppression could be enhanced with better management.

The great difficulty with such empirical studies that characterized earlier work in the area of soil-borne pest and disease suppression by brassicas is that they provide limited scope to assess the basis of observed effects. More tellingly, it restricted the capacity to pursue opportunities to improve the positive effects that were implicit in the broad general knowledge that existed. In highlighting the need for a systematic approach to research in order to confirm and, if warranted, establish and further develop biofumigation as a harnessed, workable process, it has been pointed out that the risks of empiricism are two-fold: firstly, without information on glucosinolute types and concentration, biofumigation may be disregarded as an option when different species or cultivars may be suppressive; secondly, biofumigation may be pursued as an option when the observed impacts are unrelated to glucosinolate hydrolysis products (Potter *et al.*, 1998; Kirkegaard and Matthiessen, 2004).

It has been contended that results existing to that time made effective and consistent adoption of the concept too problematic. It was feared that a failure to adopt a systematic approach to biofumigation research may result in both lost opportunities and inappropriate applications. They went on to propose a systematic approach with two basic steps: establishing the potential for suppression and linking suppression to glucosinolate hydrolysis products, and to hold test efficacy in relation to other practices within the farming system. Several international research groups have adopted such a systematic and collaborative approach to assessing and developing the biofumigation potential of Brassicaceous amendments, most notably groups in Australia (Kirkegaard *et al.*, 1998), Italy (Lazzeri *et al.*, 2004) and the United States (Brown and Morra, 1997). In the light of recent findings, a third element, that of the actual biofumigation related to release efficiency of isothiocyanates and their fate in soil, which was initially implicitly part of the biofumigation potential component, has been teased out and placed separately into the systematic approach scheme.

### 4.1.2. Non-Glucosinolate or Isothiocyanate-Related Effects

There are several examples in the literature of significant suppression by incor-porated Brassica amendments that are not associated with the glucosinolate concentration of the tissue. The most consistent of these are those related to nematode suppression. A study by Potter *et al.* (1998) showed clearly that the significant suppressive effects (60 to 95%) of six diverse Brassica species were unrelated to either the total glucosinolate content of the tissues, or to the concentration of propenyl glucosinolate, the major glucosinolate found in mustard which is known to produce the biologically active propenyl isothiocyanate upon hydrolysis. Similar results have also been shown for *Meloidogyne javanica* by McLeod and Steele (1999), and in glasshouse studies of the microbial complex associated with apple replant disease. The most likely explanation for these observations are either that the incorporation of organic matter itself increased the populations of antagonistic organisms in the soil, or that non-glucosinolate compounds released by brassicas are toxic to nematodes. Indeed many potentially biologically active, non-glucosinolate sulphur-containing compounds are released from Brassica amendments, and other products of microbial decomposition of tissues including fatty acids can also be biologically active. Irrespective of the mechanism responsible, these levels of suppression that are unrelated to glucosinolates can significantly confound the interpretation of biofumigation studies if appropriate controls are not included in the experiments.

In addition to isothiocyanates, hydrolysis of glucosinolates in incorporated biofumigants can release a range of other potentially toxic products including nitriles, epinitriles, and ionic thiocyanates (Brown and Morra, 1997). Although isothiocyanates are generally considered the most toxic of these hydrolysis products, they may not be released in large concentrations in all circumstances following incorporation (Bending and Lincoln, 1999). Much of the weed suppression noted following incorporation of Brassicaceous seed meals is likely to arise from the ionic thiocyanate rather than the isothiocyanates. Often experimental protocols used to determine the toxicity of different amendments can favour different compounds depending on volatility (e.g., sealed containers), water solubility (irrigation/soil water content) and reactivity (soil organic matter content). To resolve the question of isothiocyanate related suppression it is generally desirable to correlate the level of pest suppression with measured levels of isothiocyanate released in soil, a task that is not trivial given the relatively rapid loss from soils as a result of many different processes (Brown and Morra, 1997).

During plant residue decomposition, concentrations of volatile chemical compounds in the soil tend to increase with increasing temperature (Fig. 4.4). This is important because toxic effects are a function of the toxicant concentration multiplied by the duration of exposure. The liberation of volatile compounds from decomposing crop residues generally occurs within a few days after their incorporation in moist soil (Fig. 4.5). Manipulation of the system via soil covers and/or heating to maximize biofumigant concentrations can be the difference between the effective and ineffective management of pests.

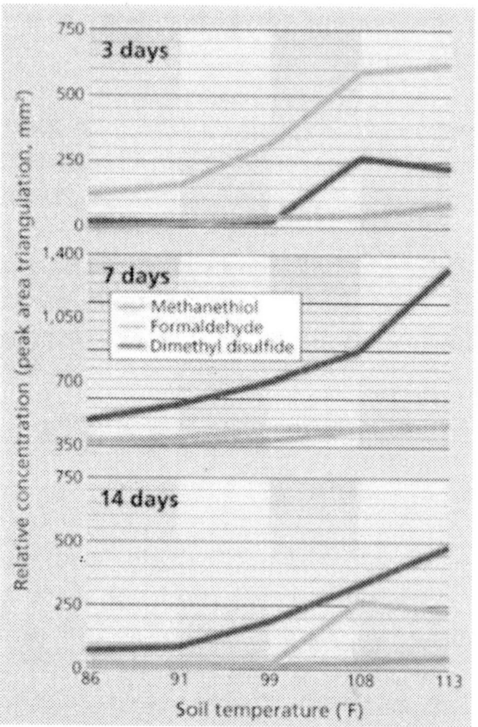


**Fig. 4.4.** Effects of soil temperature and time on relative concentration dynamics of three volatile chemicals in soil during a laboratory study. The chemicals are nonglucosinolate-derived decomposition products of cabbage plant residues, which were incorporated in soil microcosms 3, 7 and 14 days prior to headspace sampling and analysis by gas chromatography.

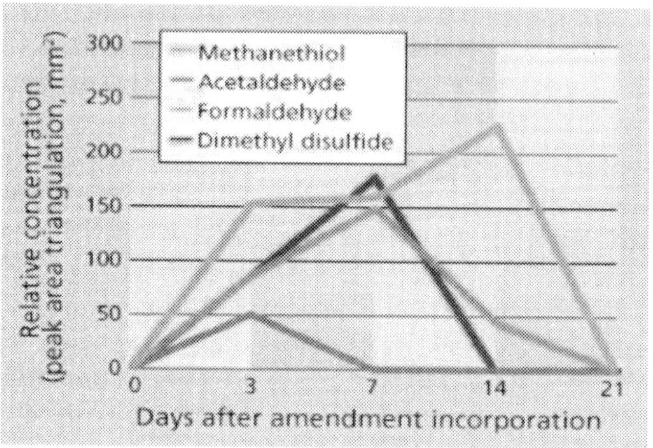


**Fig. 4.5.** Relative soil atmosphere concentration dynamics of four volatile chemical compounds produced during cabbage plant decomposition in a field study. The compounds, not derived from glucosinolate hydrolysis, were liberated from plant residues incorporated at day '0' in moist soil, which was then subjected to diurnal solar heating (solarization) prior to headspace sampling and analysis by gas chromatography.

### 4.1.3. An alternative pathway for glucosinolate degradation in soil

Soil disinfestation by incorporation of *Brassicaceae* plants materials is based on the exploitation of the glucosinolate-myrosinase system which releases inhibitory compounds such as isothiocyanate into the soil. However, the biocontrol effectiveness of this practice is often lower than expected, possibly due to microbial interaction with the bioactive compounds. In order to identify the possible function of microorganisms in this process, their interaction with glucosinolates and glucosinolate-derived products was investigated. We report the ability of a soil isolate of *Aspergillus flavus* to grow in liquid culture containing 2-propenyl and 2-phenylethyl glucosinolate or their desulfo-derivatives, and convert them to nitriles. This finding would suggest the existence in *A. flavus* of an arylsulfatase and a ß-thio-glucosidase, different from myrosinase, whose combined action could direct glucosinolate conversion in soil towards nitriles rather than isothiocyanates. Experiments are in progress to confirm in soil this pathway, which could at least partially explain the low concentrations of isothiocyanates recovered after biofumigation (Galletti *et al.*, 2008).

***(i) Onion and Garlic:*** *Allium* spp show pesticidal effects (Auger *et al.*, 2002) most often linked to volatile substances derived from sulphur amino acids.

When cell membranes are ruptured the amino acids, S-alk(en)yl-cysteine sulfoxides (RCSO), come into contact with alliinase enzymes present in vacuoles (Lancaster and Collin, 1981). RCSO are then cleaved to produce alk(en)yl sulphenic acids that rearrange to form unstable thiosulfinates (Ti) in all *Allium* spp. In the case of onion (*Allium cepa)* and leek (*Allium porrum)*, in addition to Ti other sulfur compounds called zwiebelanes (Zw) are produced (Block *et al.,* 1992; Ferary *et al.*, 1996). Depending on break down conditions diallyl thiosulfinate, the main Ti from garlic is transformed into 66% disulfide (DS), 14% monosulfide (S) and 9% trisulfide (TS) along with various proportions of thiosulfonate (To), vinyldithiines and ajoene (Block *et al.*, 1984).

The secondary substances of *Allium* have been extensively studied for their pesticidal effects. Available data have shown that sulphur compounds in *Allium* can be classified not only as insecticides, acaricides, nematicides, herbicides, fungicides and bactericides, but also repellents against arthropods.

The nematicidal activity of *Allium* has been reported against the root-knot *Meloidogyne incognita* and the most active compounds isolated from *Allium grayi* and *Allium fistulosum* were methylpropyl trisulfide, dipropyl disulfide (DPDS), diallyl disulfide (DADS), dipropyl Ti, methylpropenyl Ti and dipropyl thiosulfonate (Tada *et al.*, 1988). The efficacy of garlic extracts was also demonstrated. It has for instance been shown that tomato plants can be protected by dipping the roots in a 25 ppm allicin solution for 5 minutes (Gupta and Sharma, 1993).

The acaricidal activity of *Allium* has received only minor attention. However, investigators have demonstrated the repellent effect of garlic extracts on *Tetranychus urticae* mites (Boyd and Alverson, 2000). It is noteworthy that a patent (EP 0426802B1) involved the use of garlic extracts for protection of honeybees against *Varroa jacobsoni* (Ferrari, 1994).

***(ii) Sorghum-Sudan grass:*** Sorghum-Sudan grass and Sudan grass green manures have been shown to be effective against root-knot nematodes. These crops produce dhurrin, which like glucosinolates, produce a toxic compound, in this case hydrogen cyanide (HCN), when incorporated into the soil.

Sudan grass and sorghum contain the cyanoglucoside dhurrin (4 hydroxymandelonitrile β-D-glucoside) which is hydrolyzed by an endogenous β-glucosidase (dhurrinase) to produce glucose and *p*-hydroxybenzaldehyde. In plant tissue cells, dhurrin and hydrolytic enzymes are separated and they get in contact only when plant tissues are lesioned or disrupted producing the bioactive compounds (Lazzeri *et al.*, 2008).

First, select species and varieties that produce large amounts of biomass with a high concentration of glucosinolates. Generally, the concentration of glucosinolates peaks just before flowering. However, biomass continues to increase until the plants begin to dry. The time of incorporation for maximum biofumigation is not yet known.

Second, to produce ITCs, the glucosinolates must be exposed to specific enzymes, which are normally separated from the glucosinolates in the plants. This is also true for producing HCN from dhurrin in Sudan grass (Fig. 4.6). Current practice is to chop the green manure before incorporating to insure that this mixing occurs. A high-speed flail chopper, such as those used by some grass seed and asparagus growers, may be the best implement for this. Finally, ITC production is greater in wet soils than in dry soils, so if possible, irrigate following incorporation.

| Glucosinolates (Dhurrin) + Enzymes | → | Tissue Damage | → | ITCs (HCN) |
|---|---|---|---|---|

**Fig. 4.6.** Sudan grass chemicals shown in parentheses.

## 4.2. SOLARIZATION

Soil solarization is a method of killing soil pests by heat. This technique consists of covering moist soil with a plastic film during periods of intense sunshine and heat, thereby capturing radiant heat energy from the sun, causing physical, chemical and biological changes in the soil. The soil temperature is increased (by 5-15°C) to levels lethal to many soil-borne fungi, bacteria, nematodes and weeds.

Although the execution of solarization is simple, the overall mode of action can be complex, involving a combination of several interrelated processes which occur in treated soil and result in increased health, growth, yield, and quality of crop plants (Katan, 1987; Stapleton and DeVay, 1995; Stapleton, 1997).

### 4.2.1. Physical Mechanisms

Direct thermal inactivation of soil-borne pathogens and pests is the most obvious and important mechanism of the solarization process. Under suitable conditions, soil undergoing solarization is heated to temperatures which are lethal to many plant pathogens and pests. Thermal inactivation requirements have been experimentally calculated for a number of important plant pathogens and pests (Katan, 1987; Stapleton and DeVay, 1995). Although most mesophilic organisms in soil have thermal damage thresholds beginning around 39-40°C, some thermophilic and thermotolerant organisms can survive temperatures achieved in most types of solarization treatment (Stapleton and DeVay, 1995).

Because solarization is a passive solar process, soil is heated to maximal levels during the daytime, then cooled at night. The highest temperatures during solarization are achieved at or near the soil surface, and soil temperature decreases with increasing depth. Typical, diurnal maximum/minimum soil temperatures during summer solarization of open field soils in the inland valleys of California might be 50/37°C at 10 cm, and 43/38°C at 20 cm with 35/20°C air temperature flux. Solarizing soil in closed greenhouses or in containers with a limited volume of soil may produce considerably higher soil temperatures. For example, solarizing soil in 3.8 litre plastic containers resting on steel pallets under low plastic tunnels constructed using two layers of transparent "lm separated by *ca.* 23cm air space ("double tent") gave maximum/minimum soil temperatures of 75/16°C with corresponding air temperatures of 38/17°C (Stapleton *et al.*, 2000). During solarization of open fields or greenhouse floors, destruction of soil-borne pest inoculum usually is greatest near the surface and efficacy decreases with increasing depth (Stapleton, 1997). There are a number of physical factors which influence the extent of soil heating during solarization. First, solarization is dependent on high levels of solar energy, as influenced by both climate and weather. Cloud cover, cool air temperatures, and precipitation events during the treatment period will reduce solarization effciency (Chellemi *et al.*, 1997). Solarization is commercially practiced mainly in areas with Mediterranean, desert, and tropical climates which are characterized by high summer air temperatures. In order to maximize solar heating of soil, transparent plastic film is most commonly used for solarization. Transparent film allows passage of solar energy into the soil, where it is converted into longer wavelength infrared energy. This long wave energy is trapped beneath the film, creating a "greenhouse effect". Opaque black plastic, on the other hand, does not permit passage of most solar radiation. Rather, it acts as a "black body" which absorbs incoming solar energy. A small portion of the energy is conducted into soil, but most of the solar energy is lost by reradiation into the atmosphere. Nevertheless, solarization with black, or other colors of plastic mulch is sometimes practiced under special conditions (Abu-Gharbieh *et al.*, 1991; Stapleton, 1997).

The transparent film used for mulching transmits the visible spectrum of short wave length of the solar radiation but prevents the long wave infra-red radiation from the soil. The passage of light through a semitransparent body depends on the characteristics of the material that constitutes it and on the angle of the ray, i.e. the angle formed by the luminous ray and the perpendicular to the surface of the point of penetration. When both direct and diffuse solar radiation strike the plastic tarp on the soil, they are in part reflected back towards the atmosphere, in part adsorbed by the plastic material and in part penetrate the underlying soil. The short wave incident solar radiation penetrates the polyethylene sheet but the long wave radiation is prevented from soil, thus trapping the heat resulting in thermal inactivation and production of heat shock proteins and irreversible heat injury. The material used for tarping is therefore, the main element in the phase of capturing and storing the solar energy during the day, but it also acts as a barrier to dispersion of the heat energy during the night when the flux of solar energy has ceased. The transfer of energy from the soil depends on its internal conductivity which is a function of the soil type and its humidity content. The choice of the optimal characteristics of thermic films aim to have a transmittance of ≥80% in the visible spectrum and ≥25% in the long IR. To protect the quality of plastics for agricultural use the Italian Institute of Plastics has introduced "marchio di qualita" which ensures the incorporation of characteristics mentioned in Table 4.1. Hence, the PVC films are most efficient in solarization but its greenhouse effect depends on the percentage of vinyl acetate which varies between 12-14% in the films available in the market.

Table 4.1. Qualities of polyethylene for soil solarization

| Type of polyethylene | Total luminous transmittance | Greenhouse effect of the medium-long IR radiations |
|---|---|---|
| Low density polyethylene (LDPE) | 88-90% | 24-38% |
| Polyvinyl chloride (PVC) | 90% | < 15 |
| Copolymer of ethylenevinyl acetate (EVA) | 80-90% | < 25 |

Apart from solar irradiation intensity, air temperature, and plastic film colour, other factors play roles in determining the extent of soil heating via solarization. These include soil moisture and humidity at the soil/tarp interface, properties of the plastic, soil properties, colour and tilth, and wind conditions. The procedure of covering of very moist soil with plastic film to produce microaerobic or anaerobic soil conditions, but without lethal solar heating, can by itself produce varying degrees of soil disinfestation (Katan, 1987; Stapleton and DeVay, 1995).

### 4.2.2. Chemical Mechanisms

In addition to direct physical destruction of soil-borne pest inoculum, other changes to the physical soil environment occur during solarization. Among the most striking of these is the increase in concentration of soluble mineral nutrients commonly observed following treatment. For example, the concentrations of ammonium- and nitrate- nitrogen are consistently increased across a range of soil types after solarization. Results of a study in California showed that in soil types ranging from loamy sand to silty clay, $NH_4$-N and $NO_3$-N concentration in the top 15 cm soil depth increased 26-177 kg/ha (Katan, 1987; Stapleton and DeVay, 1995). Concentrations of other soluble mineral nutrients, including calcium, magnesium, phosphorus, potassium, and others also sometimes increased, but less consistently.

Increases in available mineral nutrients in soil can play a major role in the effect of solarization, leading to increased plant health and growth, and reduced fertilization requirements. Increases in some of the mineral nutrient concentrations can be attributed to decomposition of organic components of soil during treatment, while other minerals, such as potassium, may be virtually cooked off the mineral soil particles undergoing solarization. Improved mineral nutrition is also often associated with chemical soil fumigation (Chen *et al.*, 1991).

### 4.2.3. Biological Mechanisms

In addition to direct physical and chemical effects, solarization causes important biological changes in treated soils. The destruction of many mesophilic microorganisms during solarization creates a partial "biological vacuum" in which substrate and nutrients in soil are made available for recolonization following treatment (Katan, 1987; Stapleton and DeVay, 1995). Many soil-borne plant parasites and pathogens are not able to compete as successfully for those resources as other microorganisms which are adapted to surviving in the soil environment. This latter group, which includes many antagonists of plant pests, is more likely to survive solarization, or to rapidly colonize the soil substrate made available following treatment. Bacteria including *Bacillus* and *Pseudomonas* spp., fungi such as *Trichoderma*, and some free-living nematodes have been shown to be present in higher numbers that pathogens following solarization. Their enhanced presence may provide a short- or long-term shift in the biological equilibrium in solarized soils which prevents recolonization by pests, and provides a healthier environment for root and overall plant productivity (Katan, 1987; Stapleton and DeVay, 1995; Gamliel and Stapleton, 1993a).

*Chapter 5*

# METHODOLOGY

## 5.1. BIOFUMIGATION

### 5.1.1. Method of Biofumigation

- Incorporate the fresh mass into the soil. This can be done directly if the mass is coming from grown crop or with mass taken from other side and brought into the plot or field. If the mass is transported to the field, the soil should be well prepared before the incorporation. During transportation and storage of these organic materials in the field, care must be taken not to lose the gases produced from biodegradation, by covering the piles of the biofumigant with plastic until the time of application. A dose of 50 t/ha is recommended, although when problems with nematodes or fungi are very serious, 100 t/ha should be applied, a dose that can be reduced by choosing a cultivation techniques such as application in furrows. The biofumigant should be distributed uniformly, so that no concentration of pathogens will appear that could create problems for the crop. Once the biofumigant is distributed, it should be incorporated immediately into the soil.

**Fig. 5.1.** Biocidal plant incorporation into the soil.

- Water the field, if possible by sprinkling, until the soil is saturated, although watering can be done by flooding, or drip irrigation can be installed.
- Cover the soil surface tightly with a transparent plastic film for at least two weeks to retain the gases produced from the biodegradation of the organic matter. This could be the same plastic as the one used for soil solarization.
- The film is removed 3-4 weeks after and the soil slightly removed in order to permit the gases to escape from soil.
- Planting of the interested crop can be done 24 hours later.

### 5.1.2. Method of Getting Useful Chemicals Out of the Plant Tissue?

According to researchers, the most effective way is to disrupt the plant cells, by freezing or complete maceration. This will give the best biofumigation effect, but equipment to do this is not always available. A field rotavator used for chopping the plant material and incorporating it into the soil is second best. The rotavation effect can be duplicated by hand chopping the plant material before digging it into the soil but this is very labour demanding.

**Fig. 5.2.** Incorporation of plant material into soil by hand chopping.

It is also discovered that other large incorporations of green manure will suppress bacterial wilt. An example is mentioned of sweet potato leaves giving good control as well.

In Northern Australia, results in some field trials have been excellent. A paddock with a high level infection of bacterial wilt was planted with tomatoes. An untreated block of tomatoes yielded less than 2 tons/ha, while the area where brassica green manure had been applied yielded up to 20 tons/ha of tomatoes and had correspondingly lower levels of bacterial wilt.

### 5.1.3. Plowing-Down of Brassica Plants

One way to take advantage of the biopesticides produced by Brassicaceae species is to use the plants as green manures. As such, the mustard or rapeseed crop is planted within the rotation such that it is incorporated into the soil prior to seed set. Glucosinolates within the tissues are hydrolyzed by myrosinase when plant cell membranes are broken. Research indicates that soil-borne pest suppression is likely to be improved by choosing a high glucosinolate-containing variety of rapeseed or mustard and providing adequate moisture to increase allelochemical release. However, the greatest improvements in the use of Brassicaceae biofumigants to control soil-borne plant pests will be achieved by focusing on methods to increase plant cell disruption thereby maximizing glucosinolate hydrolysis and allelochemical release.

**Fig. 5.3.** Incorporation and mixing in soil of plant material using tractor-drawn implements.

## 5.2. SOLARIZATION

### 5.2.1. Method of Soil Solarization

***(i) Soil Preparation:*** Solarization is most effective when the plastic sheeting (tarp) is laid as close as possible to a smooth soil surface. Preparation of the soil begins by disking, rototilling, or turning the soil by hand to break up clods and then smoothing the soil surface. Remove any large rocks, weeds, or any other objects or debris that will raise or puncture the plastic.

***(ii) Laying the Plastic:*** Plastic sheets may be laid by hand or machine. The open edges of the plastic sheeting should be anchored to the soil by burying the edges in a shallow trench around the treated area. Plastic is laid either in complete coverage, where the entire field or area to be planted is treated, or strip coverage, where only beds or selected portions of the field are treated.

***(iii) Complete Coverage:*** In complete coverage, plastic sheeting is laid down to form a continuous surface over the entire field or area to be planted. The edges of the sheets may be joined with an ultraviolet (UV)-resistant glue or anchored by laying adjacent strips of plastic and burying both edges in soil. Anchoring the edges in the soil may be more cost effective initially than gluing the edges together but may also result in untreated soil being close to subsequently planted crops. The ends of the sheets should be held in place by burying them in the soil. If beds are formed after complete coverage, care must be taken to avoid deep tillage that could bring untreated soil to the surface. Complete coverage is recommended if the soil is heavily infested with pathogens, nematodes, or perennial weeds, since there is less chance of reinfestation by soil being moved to the plants through cultivation or furrow-applied irrigation water.

***(iv) Strip coverage:*** In strip coverage, plastic is applied in strips over preformed beds (Fig. 5.4). Strips should be a minimum of 75 cm wide; beds up to 1.5 m wide are preferred because several crop rows can be planted per bed. In some cases, strip coverage may be more practical and economical than complete coverage because less plastic is needed and it is not necessary to join the edges of the plastic sheets together. Strip coverage effectively kills most pests and eliminates the need for deep cultivation after solarization. It is especially effective against weeds, since the furrows are cultivated. With strip coverage, however, long term control of soil pathogens and nematodes may be lost because pests in the untreated soil in the rows between the strips can contaminate and reinfest treated areas.

***(v) Irrigation:*** Wet soil conducts heat better than dry soil and makes soil organisms more vulnerable to heat. The soil under the plastic sheets must be saturated to at least 70 percent of field capacity in the upper layers and moist to depths of 60 cm for soil solarization to be effective.

Soil may be irrigated either before or after the plastic sheets are laid. If the soil is irrigated beforehand, the plastic must be applied as soon as possible to avoid water loss; if heavy machinery is used to lay the plastic, however, the soil must be dry enough to avoid compaction. If the soil is to be irrigated after the plastic is laid, one or more hose or pipe outlets may be installed under one end of the plastic; drip lines may be installed before the plastic is laid; or irrigation water may be run underneath the plastic in furrows or in the tracks made by tractor wheels if the plastic sheets were applied by machine. Fields treated by strip coverage can be irrigated by drip lines on or in the bed.

**Fig. 5.4.** Plastic sheets laid by machines in strips for soil solarization.

The soil does not usually need to be irrigated again during solarization, although if the soil is very light and sandy, or if the soil moisture is less than 50 percent of field capacity, it may be necessary to irrigate a second time. This will cool the soil, but because of the increased moisture the final temperatures will be greater.

***(vi) Duration of Treatment:*** The plastic sheets should be left in place for 4 to 6 weeks to allow the soil to heat to the greatest depth possible. To control the most resistant species, leave the plastic in place for 6 weeks. Experience has shown that there is little or no need to take the temperature of the soil. The greatest concern is to solarize the soil during a period of high solar radiation with little wind or cloud cover. Soil in the Central Valley can be solarized for 4 weeks any time from late May to September. In coastal areas the best time may be August to September or May to June, transitional periods when fog or wind may be at a minimum.

***(vii) Removal of the Plastic and Planting:*** After solarization is complete, the plastic may be removed before planting. Or, the plastic may be left on the soil as a mulch for the following crop by transplanting plants through the plastic. Clear plastic may be painted white or silver to cool the soil and repel flying insect pests in the following crop. A disadvantage of leaving the plastic on the soil is that it may degrade and be difficult to clean up in the spring.

Treated soil can be planted immediately to a fall or winter crop or left fallow without the plastic until the next growing season. If the soil must be cultivated for planting, the cultivation must be shallow-less than 5 cm-to avoid moving viable weed seed to the surface.

***(viii) Greenhouses:*** Solarization in greenhouses produces significantly higher soil temperatures than solarization in fields or gardens and can therefore be more effective in cooler weather. Greenhouse solarization is extensively used in southern Europe and Japan to control diseases of strawberries, tomatoes, eggplants, cucumbers, and other intensively managed crops.

The soil surface inside the greenhouse should be leveled and irrigated before being covered with plastic sheeting. Choose a time of year with maximum solar radiation. To maximize the transmission of light it may be advisable to wash the roof of the greenhouse before treatment. Once plastic is applied, the greenhouse should be tightly closed for 4 or more weeks to contain the heat.

***(ix) Containerized Planting Media and Seedbeds:*** Soil solarization has been shown to be effective for disinfesting containerized soil and soil in cold frames. Soil temperatures should be monitored closely in this planting media to assure that temperatures are high enough to control pests. Materials can be solarized either in bags or flats covered with transparent plastic or in layers 7.5-22.5 cm wide sandwiched between two sheets of plastic. In warmer areas of California, soil inside black plastic sleeves can reach 70°C during solarization, equivalent to target temperatures for soil disinfestation by aerated steam. At these temperatures, soil is effectively solarized within a week. A double layer of plastic can increase soil temperatures by up to 50°F.

Soil temperatures can be monitored using simple soil thermometers inserted 10-15 cm into the soil mix or by using thermocouples and a digital reading logger. Temperatures can be monitored at different locations, but the duration should be lengthened to raise the temperature at the coolest location to the desired level.

***(x) Orchards and Vineyards:*** Soil solarization is most effective before or during establishment of new orchards or vineyards. It has been successfully used on a large scale to reduce Verticillium wilt symptoms in young pistachio orchards in California and has also been successfully used in vineyards and in avocado, stone fruit, citrus, and olive orchards in the state.

In the orchard or vineyard, clear plastic is either laid by hand around the bases of individual trees or vines and connected to strips laid between the rows or laid in anchored strips and glued along the tree rows. For best results, begin solarization as soon as trees are planted. Partial shading by young trees does not prevent soil heating, nor does soil solarization appear to bother most young trees during treatment. However, solarizing certain species of trees, such as herbaceous perennials, avocado, and young *Prunus* trees, with clear plastic may result in plant damage, especially when trees are young. (The Prunus trees were killed by clear plastic but not by black film).

In addition to killing soil-borne pests, solarization of orchards and vineyards can greatly reduce the amount of water needed for irrigation and increase the growth, flowering, and/or fruit set of the trees. In large commercial orchards, the cost of post-plant solarization should be compared to the benefits before making a treatment decision. Experience has shown that pests that are not eradicated by solarization may recolonize roots and soil, and pathogens and nematodes may survive in roots remaining in the soil. Periodic retreatment may be necessary.

### 5.2.2. Plastic Mulching

***(i) Clear vs. Coloured Plastic:*** Transparent or clear plastic is most effective for solarization. Black plastic, often used for mulching, does not heat the soil as well as clear plastic. It can be used for solarization but its main effect is reducing weed growth. In areas where solarization is ineffective because of low solar radiation or a heavy infestation of weeds, black plastic may combine some solarization benefit with residual weed control. It can also be used for solarizing existing crops, for example, by disinfesting soil while establishing permanent tree or vine crops.

Since soil temperatures are lower with black plastic, the treatment time must be lengthened for best results. Other colours of plastic, such as green or brown, which allow some heating of the soil but not to the degree of clear plastic, require longer treatment times. These other colours of plastic give so much less effective solarization that they should probably only be used as mulch.

***(ii) Types of Plastic:*** The thinner the plastic, the greater the heating will be. Polyethylene (PE) plastic 1 mil (0.025 mm) thick is efficient and economical but not very resistant to tearing by wind or puncture by animals. Users in windy areas should consider plastic sheets that are 1.5 to 2 mils (0.038-0.050 mm) thick. If holes or tears occur in the plastic they should be patched with clear patching tape. Users are encouraged to select plastic sheeting containing ITV inhibiting additives that prevent sheets from becoming brittle and difficult to remove from the field and extend the life of the plastic. Plastic sheets laid by hand can often be used more than once for solarization, although if the plastic is dirty or dusty reuse is less effective.

Polyethylene sheets may be modified by an additive that enables them to absorb infrared (IR) radiation and improve their capacity to retain heat. Although these are available, they have not proven to be very effective. Coloured plastic films are available that absorb light in the photosynthetic range to inhibit growth of weeds and at the same time heat the soil. These can be used for solarization but generally do not heat soil as well as transparent films. There has been considerable interest in the development of high-density or "impermeable" plastic sheeting to better contain fumigant chemicals in soil. These plastics may also improve the effects of solarization by sealing in more heat and volatile compounds. Experimental work has also been done using a sprayable polymer as a replacement for plastic sheeting. Such a material would be easy to apply and less expensive to use, but to date suitable chemicals have not been found. The use of a double layer of plastic with air space between the layers mimics the greenhouse effect and raises soil temperatures from 2° to 10°F higher than that obtained with a single layer. Using a double layer requires additional preparation time and expense but it may make soil solarization more feasible in areas with cooler climates.

***(iii) Availability:*** For small applications in gardens, UV inhibiting plastic that is 1.5 to 4 mils (0.038-0.100 mm) thick can be purchased from nursery, hardware, or lumber establishments. These are sometimes called "drop cloths" and are used to catch paint drippings. For agricultural plantings, plastic can usually be purchased in rolls from 1.8-3.6 m wide and approximately 1,200 m long. Size will vary by source (wide range of colours and sizes including solarization and IRT films).

***(iv) Disposal:*** The disposal of plastic film after solarization presents an additional expense and involves consideration of environmental pollution. At present there is no programme for recycling plastic used in soil solarization, primarily due to the relatively low amount of plastic used. A few programmes are operating in other states where a more constant supply of used agricultural plastic is available. In addition, soil adhering to used plastic makes recycling more costly. UV-treated plastics that are thicker than 4 mils (0.1 mm) may be usable for more than one season if handled carefully. Although most plastic has been put into landfills after use in solarization, some farmers store plastic at their own sites until recycling programmes can be started.

Efforts have been made to develop a plastic film for solarization that would degrade completely after use in a suitable or predictable amount of time. These "biodegradable" or "photodegradable" plastics are not currently recommended for solarization. Photodegradable plastics degrade with exposure to UV light. Although they may be effective for solarization, the timed degradation (6 to 12 weeks) has not been uniformly effective. Also, the buried part of the plastic remains in the soil until it is brought to the surface with cultivation, leaving a source of pollution in the field.

*Chapter 6*

# DISEASE MANAGEMENT

## 6.1. BIOFUMIGATION

Biofumigation is the use of volatile plant chemicals for control of soil-borne plant pathogens. These products have been shown to suppress the pathogens *Botrytis cinerea, R. solani, F. oxysporum, Didymella lycopersici,* and *Cladosporium fulvum* (Urbasch, 1984). Although the fungicidal properties of ITCs have been identified since at least 1937 (Walker *et al.*, 1937), only recently has the practice of using *Brassicas* for pest control been termed "biofumigation" (Angus *et al.*, 1994). The volatiles from several *Brassica* species suppressed the growth of the tomato pathogens *P. ultimum*, *R. solani*, and *S. rolfsii* (Charron and Sams, 1998; 1999: Harvey *et al.*, 2002).

Brassicas such as Oriental mustard have been associated in some reports with improved root health in a subsequent cash crop, such as potatoes, grown after a green incorporated brassica cover crop. Brassica cover crops can be managed as a green manure-when green residues are incorporated-and this may reduce or suppress some pathogens, including *Verticillium* in potato; *Pythium, Fusarium* and *Rhizoctonia* root rots in beans; *Pythium* in lettuce; pink root in onion; *Aphanomyces, Pythium, Rhizoctonia* and *Fusarium* root rot in peas; and cavity spot and *Fusarium* in carrot (Sanders, 2005).

The biocidal activity of *Brassicas* against fungal pathogens, nematodes, weeds, and insects is frequently attributed to ITCs from *Brassica* tissues (Delaquis and Mazza, 1995). ITCs are effective, broad-spectrum pesticides (Mithen *et al.*, 1986; Isshiki *et al.*, 1992), and substantial quantities of them can be produced for field application. Research has shown that Black mustard (*B. nigra*) and Indian mustard (*B. juncea*) produce high levels of ITC (Tollsten and Bergström, 1988) and could be utilized in a biofumigation cropping system.

Control of *V. dahliae* by *Brassica* residues has also been shown. The number of *V. dahliae* propagules in a 36-m$^2$ plot incorporated with 200 kg of chopped broccoli (*B. oleracea* var. *italica*) were lower than in control plots and were comparable to plots fumigated with methyl bromide + chloropicrin.

Since 1999 field trials have been conducted at Knoxville Experiment Station. Significant decreases in the incidence of Southern Blight of tomato were recorded, as well as increases in fruit yield, when integrating biofumigation into a sustainable production system. The research yielded important knowledge pertaining to the development of the appropriate production methods utilizing biofumigation as a management technique. Both *Brassica* cover crops and mustard seed meal incorporations have been used. The ability to amend production soils with the spreadable meal, in conjunction with its high ITC content (3-4x leaf tissue); give this enhanced biofumigation technique great potential. Problems associated with growing cover crops such as variable stands and weather complications are avoided when using the meal. As a fully organic product, pre-plant mustard meal applications could give the grower superior ability to control soil-borne pathogens.

The most serious soil-borne disease of solanaceous vegetable crops such as eggplant, tomato and potatoes in tropical environments are bacterial wilt (BW) caused by *Ralstonia solanacearum*. Biofumigation using Indian mustard (*Brassica juncea*) as green manure has been found effective in reducing the level of BW in the soil, and lessening the severity of the disease in a following solanaceous crops. A commercially available Indian mustard biofumigant green manure was shown to significantly reduce bacterial wilt in a following potato crop, resulting in spectacular yield increases (from 0.3 to 22 t/ha). The fact that such results were obtained with no purposeful selection of brassicas for high biofumigant properties indicates there may be significant scope to improve the level of suppression achieved.

The 'biofumigation' properties of brassicas as rotation, companion or green manure crops, and the use of these properties in managing the important pathogen of tropical vegetable production systems has been studied.

The effectiveness of the biocides was evaluated in commercial and small-scale farmer fields in the context of integrated disease management, using strategies that included disease resistance/tolerance, non-host crop rotation, and solarization (harnessing the sun's energy to destroy some weeds and soil pathogens).

It was demonstrated that significant suppression of the organism could be achieved using Brassica green manures, and that this arose from a combination of two separate mechanisms. The first mechanism was short-term suppression (over 2-3 days) related to the release of isothiocyanates (ITCs) from the Brassica tissues. To maximize the impact of this mechanism, strategies to increase ITC release from the plant tissues and their residence time in the soil were developed into a 'best-bet' approach. This involved: use of brassica tissues which release high concentrations of toxic ITCs (e.g. mustard); incorporating around 5 kg/m$^2$ (5% w/w) of fresh tissue into the soil; adequate maceration of the tissues at a cellular level to release the ITCs; adequate water to facilitate hydrolysis of ITCs; rapid and thorough incorporation for complete mixing of ITCs through the soil; watering or covering to retain the volatile ITCs in the soil; targeting light-textured soils with low organic matter to reduce inactivation of ITCs.

This approach could be readily adopted in highly mechanized farming systems and was very effective on light-textured soils. Field experiments in Mareeba in 2003 and 2005 showed mustard green manures reduced bacterial wilt incidence in tomatoes from 80 to 15%, with a 10-fold increase in yield (from 2.5 to 20.0 t/ha). On loam and clay soils with more organic matter in there was less evidence that ITC-related suppression was occurring.

However, a second mechanism of suppression operating over a longer period was also identified. This mechanism, occurring over a period of 3 to 4 weeks, was related to the effects on the organism of added organic amendments. This mechanism was observed on soils with higher clay and organic-matter and did not require the strict protocols listed above. Reductions of bacterial wilt incidence in potatoes (49 down to 17%) and associated yield increases (9.1 to 18 t/ha) in eight on-farm experiments were largely attributable to this process irrespective of the type of green manure used.

The scientists confirmed that the nitrogenous components of organic amendments could generate the same pattern of suppression as the green manures when added in pure forms to soil, and evidence suggests that it is related to the mobilization of specific antagonistic soil bacterial populations which are yet to be identified. More work is needed to develop strategies to fully exploit this more general mechanism of suppression.

Biofumigant green manures should not host the organism of interest. Brassica species are generally considered non-hosts of bacterial wilt and although one incidence of bacterial wilt infection was observed in a mustard line from Australia in a nursery in the Philippines. Subsequent investigations in Taiwan suggest root damage is a prerequisite. However, continual vigilance to ensure biofumigants do not become susceptible to bacterial wilt is warranted.

## 6.2. SOLARIZATION

Repeated daily heating during solarization kills many plant pathogens, nematodes, and weed seed and seedlings. The heat also weakens many organisms that can withstand solarization, making them more vulnerable to heat-resistant fungi and bacteria that act as natural enemies. Changes in the soil chemistry during solarization may also kill or weaken some soil organisms.

Soil solarization can decrease the incidence and effects of disease-causing organisms that originate in the soil. Solarization for 47-48 days in September-October in west-central Florida either reduced disease frequency or slowed disease progression of crown rot and blight caused by *Rhizoctonia* spp. and root rot caused by *Pythium* spp. when a double layer of low polyethylene mulch was used (McGovern *et al.*, 2002). In another case, solarization with a single layer of plastic mulch was more effective than methyl bromide in limiting an epidemic from *Pythium* spp. in a pepper field (Saha *et al.*, 2005). In contrast, Chellemi (2006) failed to show a difference between white plastic mulch and clear plastic mulch in the suppression of *Pythium* spp. Incidence of *Phytophthora* blight (*Phytophthora capsici*) was reduced by solarizing for 6-8 weeks (Chellemi 2006). Furthermore, densities of *Phytophthora nicotianae* and *Ralstonia solanacearum* were reduced by solarization in soil depths down to 25 cm (9.8 in) (Chellemi *et al.*, 1994). Southern blight (*Sclerotium rolfsii*) and Fusarium wilt (*Fusarium oxysporum* f. sp. *lycopersici*) can also be substantially reduced with solarization and results are comparable to a methyl bromide control (Chellemi *et al.*, 1997), but effects may be limited to the upper soil layer (Chellemi *et al.*, 1994). Jones and Overman (1986) recommended solarization as one of the methods to manage Verticillium wilt (*Verticillium albo-atrum*). Several studies have concluded that bacterial wilt caused by *R. solanacearum* is not controlled by solarization (Chellemi *et al.*, 1993, 1994, 1997).

Solarization controls populations of many important soil-borne fungal and bacterial plant pathogens, including *Verticillium dahliae*, which causes *Verticillium* wilt in many crops; certain *Fusarium* spp. that cause *Fusarium* wilt in some crops; *Phytophthora cinnamomi*, which causes *Phytophthora* root rot; *Agrobacterium tumefaciens*, which causes crown gall disease; *Clavibacter michiganensis*, which causes tomato canker; and *Streptomyces scabies*, which causes potato scab. Other fungi and bacteria are more difficult to control with solarization, such as certain high temperature fungi in the genera *Macrophomina*, *Fusarium*, and *Pythium*, and the soil-borne bacterium *Ralstonia solanacearum* (Tables 6.1 to 6.5).

Table 6.1. Fungal pathogens controlled by soil solarization.

| Scientific name | Disease caused (crop) |
|---|---|
| *Didymella lycopersici* | Didymello stem rot (tomato) |
| *Fusarium oxysporum* f. sp. *conglutinans* | Fusarium wilt (cucumber) |
| *Fusarium oxysporum* f. sp. *fragariae* | Fusarium wilt (strawberry) |
| *Fusarium oxysporum* f. sp. *lycopersici* | Fusarium wilt (tomato) |
| *Fusarium oxysporum* f. sp. *vasinfectum* | Fusarium wilt (cotton) |
| *Plasmodiophora brassicae* | Club root (cruciferae) |
| *Phoma terrestris* | Pink root (onion) |
| *Phytophthora cinnamomi* | Phytophthora root rot (many crops) |
| *Pyrenochaeta lycopersici* | Corky root (tomato) |
| *Pythium ultimum, Pythium* spp. | Seed rot or seedling disease (many crops) |
| *Pythium myrothecium* | Pod rot (peanut) |
| *Rhizoctonia solani* | Seed rot or seedling disease (many crops) |
| *Sclerotinia minor* | Lettuce drop |
| *Sclerotium cepivorum* | White rot (garlic and onions) |
| *Sclerotium rolfsii* | Southern bight (many crops) |
| *Thielaviopsis basicola* | Black root rot (many crops) |
| *Verticillium dahliae* | Verticillium wilt (many crops) |

Table 6.2. Bacterial pathogens controlled by soil solarization.

| Scientific name | Disease caused (crop) |
|---|---|
| *Agrobocterium tumefaciens* | Crown gall (many crops) |
| *Clavibacter michiganensis* | Canker (tomato) |
| *Streptomyces scabies* | Scab (potato) |

Table 6.3. Fungal pathogens unpredictably controlled by soil solarization.

| Scientific name | Disease caused (crop) |
|---|---|
| *Fusarium oxysporum* f. sp. *pini* | Fusarium wilt (pines) |
| *Macrophomina phaseolina* | Charcoal rot (many crops) |

Table 6.4. Bacterial pathogens unpredictably controlled by soil solarization.

| Scientific name | Disease caused (crop) |
|---|---|
| *Ralstonia solanacearum* | Bacterial wilt (several crops) |

Table 6.5. Some diseases controlled by soil solarization

| Disease | Crop |
|---|---|
| Verticillium wilt | Tomato, potato, eggplant, cotton, strawberry |
| Fusarium wilt | Tomato, melon, onion, cotton |
| Pink root rot | Onion |
| Southern stem rot (white mold) | Peanut |
| Rhizoctonia seedling disease (sore shin or damping off) | Potato, onion, bean |
| Crown Gall | Walnut |
| Phytophthora root rot | Ornamentals |

*Chapter 7*

# NEMATODE MANAGEMENT

## 7.1. BIOFUMIGATION

### 7.1.1. Glucosinolate-derivative compounds

Cultivation and incorporation of cover/rotation crops, especially Brassicaceae plants, occasionally suppresses soil-borne diseases, including nematodes. Utilization of volatile toxic compounds, such as isothiocyanates generated from the glucosinolates in such crops, for soil-borne disease control is generally termed biofumigation (Kirkegaard *et al.*, 1993). Isothiocyanates are known to possess broad pesticidal activity against weeds, bacteria, fungi and nematodes (Matthiessen and Kirkegaard, 2006), and are similar or relative to compounds found in commercial soil fumigants, such as metham sodium and dazomet, which release methyl isothiocyanate into the soil. Nematode suppression by biofumigation with Brassicaceae plants has been attempted in several regions worldwide, mainly with rapeseed (*Brassica napus*) and Indian mustard (*Brassica juncea*) (Mojtahedi *et al.*, 1991, 1993; Walker and Morey, 1999; Ploeg and Stapleton, 2001; Stirling and Stirling, 2003; Zasada and Ferris, 2004; Rahman and Somer, 2005).

The efficacies for soil-borne disease control were generally variable and inconsistent, a trend which seemed more salient with nematodes than with soil-borne fungi (Matthiessen and Kirkegaard, 2006). *In-vitro* assays revealed that isothiocyanates from Brassicaceae plants were nematicidal to juveniles of *M. incognita* and *Tylenchulus semipenetrans* at concentrations as low as 10*m*M (Zasada and Ferris, 2003; Lazzeri *et al.*, 2004a). However, several studies have shown that nematode control efficacy and nematode reproduction in treated soils are not correlated with glucosinolate contents in plants (Potter *et al.*, 1998; McLeod and Steele, 1999). Although glucosinolates are thought to play an important role in nematode suppression, non-glucosinolate compounds, such as other sulfur-containing compounds (Bending and Lincoln, 1999), as well as biological and physiological factors, may also be involved (Mazzola *et al.*, 2001). Although the effects of soil type on biofumigation efficacy have not been well studied, sandier soils with low organic matter content appear to allow better performance of the biofumigant.

Before selecting a plant for biofumigation purposes, the host range of the target nematode(s) needs to be checked, because most Brassicaceae used as cover crops are hosts of some important nematodes, such as *Meloidogyne* spp. (McLeod and Steele, 1999; McLeod *et al.*, 2001), and host Brassicaceae plants may increase nematode populations instead of reducing them. Biofumigant plants that are non-hosts or poor hosts, or are nematode-resistant, are strongly preferred. A study in Australia showed that *M. javanica* reproduced on *B. juncea* and *B. napus* at rates which increased the nematode population in the soil during their growth period in hot seasons; however, nematode proliferation could be prevented if these plants were grown in the winter under low temperatures (Stirling and Stirling, 2003). Use of dried pellets of glucosinolate- containing plants can solve the problem of nematode population buildup in the field, while adding a biofumigation effect (Lazzeri *et al.*, 2004b). Again, to obtain maximum control efficacy from biofumigation, several factors must be considered in addition to host status for nematodes, including: the optimum plant growth stage for soil incorporation (probably when the glucosinolate content reaches maximum levels), biomass of the plant, and adequate soil moisture for plant degradation and gas diffusion into the soil while preventing rapid gas escape from the soil to the atmosphere. Variations in glucosinolate profiles and concentrations exist among Brassicaceae crops and their developmental stages (Bellostas *et al.*, 2007).

An effective way of enhancing the nematode-control efficacy of biofumigation in small areas is to combine the incorporation of Brassicaceae plants with soil tarping using plastic film, which may prevent rapid emission of volatile nematicidal compounds from the soil to the atmosphere, and increase the soil temperature via a soil-solarization effect if performed in hot seasons. Elevation in soil temperatures due to soil solarization has been shown to improve the control efficacy of amendments with Brassicaceae crops (Gamliel and Stapleton, 1993; Stapleton and Duncan, 1998; Ploeg and Stapleton, 2001). Combinations of sublethal soil temperatures (30–38.8°C) and the biofumigation effect (toxic volatile compounds) may have a syner-gistic effect on nematode-control efficacy. Sublethal temperatures may render nematodes more sensitive to toxic compounds or to antagonistic microorganisms. Because nematode control has been obtained by soil solarization plus biofumigation with non-Brassicaceae plants, such as pepper plant residues (Piedra Buena *et al.*, 2007), this method does not appear to be specific to glucosinolate-containing plants. Other studies on the effects of fresh organic amendments and soil tarping on fungal pathogens have suggested that toxic compounds, elevated temperature, anaerobic and reducing soil conditions, and biological activity are involved in the control (Gamliel and Stapleton, 1993; Blok *et al.*, 2000).

Nitriles are another group of toxic compounds derived from glucosinolates. Simple nitriles have been shown to be involved in direct and indirect plant defense responses upon attack of *Pieris rapae* (Lepidoptera) (Mumm *et al.*, 2008). The role of nitriles in nematode suppression is not clear.

Different mustards (e.g. *Brassica juncea var. integrifolia* or *Brassica juncea var. juncea*) should be used as intercrop on infested fields. As soon as mustards are flowering they are mulched and incorporated into the soil. While incorporated plant parts are decomposing in a moist soil, nematicidal compounds of this decomposing process do kill nematodes. Two weeks after incorporating plant material into the soil a new crop can be planted or sown (it takes about two weeks for the plant material to decompose and stop releasing phytotoxic substances i.e., chemicals poisonous to plants).

Mexican marigold, also known as *Tagetes*, has been successfully used in the control of root-knot nematode in roses by a Kenyan Farmer (Report on ToT for Alternatives to the Use of Methyl Bromide for Soil Fumigation in Brazil and Kenya).

It is recommended to alternate the use of agricultural residues with green manure, especially from brassicae, using 5-8 kg/m$^2$ of green matter, although combinations of legumes and grass can be applied. In the case of the use of green manure cultivated in the same field, fast growing plants should be used to be incorporated at least 30 days after having been planted, to avoid the increase of pathogen populations. Planting brassicae after biofumigation can serve as bio-indicators of possible phytotoxicity, because the germination of these seeds is sensitive to phytotoxic substances. At the same time they are very sensitive to nematodes and permit the detection of areas in the crop where biofumigation is not effective. They act like trap plants, and like biofumigants when incorporated into the soil.

In Spain, successful application of biofumigation was achieved in strawberries, peppers, cucurbits, tomato, brassicae, cut flowers, citrus and banana. Biofumigation has also been recently applied to Swiss chard and carrot crops. The most utilized biofumigants have been goat, sheep and cow manure, and residues from rice, mushroom, brassicae and gardens.

The effectiveness of biofumigation in controlling nematodes is nearly the same as with the use of conventional pesticides. Biofumigation may also regulate viral problems by controlling vector nematodes.

### 7.1.2. Management of Root-knot Nematodes on Vegetables

#### *(i) Biofumigation*

- Use biofumigation where possible. Different mustards (e.g. *Brassica juncea* var. *integrifolia* or *B. juncea* var. *juncea*) should be used as intercrop on infested fields. As soon as mustards are flowering they are mulched and incorporated into the soil. While incorporated plant parts are decomposing in a moist soil, nematicidal compounds of this decomposing process do kill nematodes. Two weeks after incorporating plant material into the soil a new crop can be planted or sown.
- Maintain high levels of organic matter (manure and compost) in the soil.
- Incorporate neem cake powder into the soil if it is available.

#### *(ii) Crop rotation*

Rotate with onions, baby corn, sweet corn, maize, millet, sorghum, sesame, cassava or Sudan grass.

A rotation system called "Strong" is recommended for management of root-knot nematodes. The system was developed by, African Farmers' Organic Research and Training (Afforest), a NGO in Zimbabwe. It involves planting in rotation of a susceptible crop (e.g. tomatoes), followed by a tolerant crop (e.g. cabbage) and then a resistant crop (e.g. onions) before a return to a susceptible crop (e.g. tomatoes).

Use trap crops such as marigold (*Tagetes* spp.) and Indian mustard. A trap crop is a crop planted to attract a pest and is then destroyed together with the pest. Mixed cropping with marigold can also minimize root-knot nematode damage.

## 7.2. SOLARIZATION

Plant-parasitic nematodes are killed by high temperatures. Early research on *Meloidogyne javanica* (root-knot nematode) showed that movement of juveniles stopped immediately after being exposed to 50°C and there was no recovery even after returning the temperature to 25°C. Lowering the temperature resulted in longer periods required to kill juveniles; at 42°C it required 3 hours (Wallace, 1966). In Tanzania, egg masses of *M. javanica* were buried in soil (15-cm depth) and exposed to solarization. Within two to three weeks, all eggs were dead. Soil temperatures under solarization reached an average of 43°C, with maximum of 45°C (Madulu and Trudgill, 1994).

Solarization can be effective in Florida, but may be impaired by overcast skies and rainfall during the warmest summer months. In order to counteract these effects, the solarization time could be prolonged. Increasing the length of solarization will cause mortality based on the accumulation of sub-lethal, but detrimental temperatures. McGovern *et al.*, (2000) solarized soil for 41 days mainly during October in southwest Florida, which resulted in the reduction of awl (*Dolichodorus heterocephalus*) and stubby-root (*Paratrichodorus minor*) nematodes. Nematodes were exposed to average maximum temperatures of 38.4, 33.6, and 29.8°C at depths of 5, 15, and 23 cm. A recent laboratory study confirms that nematodes can be affected by an accumulation of sub-lethal temperatures (Wang and McSorley, 2008). Eggs and juveniles of *Meloidogyne incognita* were exposed to a series of temperatures from 38 to 45°C. At 44 and 45°C, juveniles were killed within one hour. For lower sub-lethal temperatures, the hours required for suppression decreased with increasing temperatures (Table 7.1). Success of solarization depends on maximum temperatures reached in the field. The higher the temperature, the shorter the duration of solarization required to kill nematodes. Temperatures over 40°C should be the goal in order to shorten the solarization period. In a 6-week solarization period (Florida: July to August), temperatures high enough to kill nematodes could be accumulated (Wang and McSorley, 2008).

Table 7.1. Hours needed to kill 100% of *Meloidogyne incognita* eggs and juveniles.

| Temperature (°C) | Hours to kill 100% | |
|---|---|---|
| | Eggs | Juveniles |
| 38 | 389.8 | -- |
| 39 | 164.5 | 47.9 |
| 40 | 32.9 | 46.2 |
| 41 | 19.7 | 17.5 |
| 42 | 13.1 | 13.8 |

Several nematodes species were negatively affected by solarization. It was effective in reducing populations of *M. incognita*, *D. heterocephalus*, *P. minor*, *Belonolaimus longicaudatus* (sting), *Criconemella* spp. (ring), and *Roytlenchulus reniformis* (reniform) (Chellemi *et al.*, 1997; McGovern *et al.*, 2002, McSorley and McGovern, 2000, McSorley and Parrado, 1986; Ozores-Hampton *et al.*, 2004). However Chellemi *et al.* (1997) and Chellemi (2006) reported that they were unable to reduce *Meloidogyne* spp. and *R. reniformis*. In some cases, solarization was able to suppress populations of *M. incognita* or *R. reniformis* initially, but numbers recovered at the end of the cropping season (McSorley and McGovern, 2000; McSorley and Parrado, 1986). In addition, when using strip solarization, it is possible that fast recolonization of plant-parasitic nematodes may occur, because row middles (occupying up to 50% of the field) will remain untreated, although this late-season recovery may not limit overall yield (Chellemi *et al.*, 1997).

Resurgence of certain nematode species may occur to higher levels than before solarization. The stubby-root nematode *P. minor* increased in numbers following three weeks of solarization (McSorley and McGovern, 2000). A possible explanation for resurgence could be that some nematode species have population reservoirs in deeper soil layers that are larger than those found in the upper soil layers. This unusual vertical distribution in soil often occurs with stubby-root nematodes. Although solarization reduces or eliminates these nematodes in the upper layers, recolonization can occur quickly by drawing upon a population pool from deeper soil layers. Similar resurgence of stubby-root nematodes has also been observed when these nematodes are managed with soil fumigation (Weingartner *et al.*, 1983).

Soil solarization can be used to control many species of nematodes (Table 7.2). However, soil solarization is not always as effective in controlling nematodes as it is in controlling fungal disease and weeds because nematodes are relatively mobile and can recolonize soil rapidly. Nematode management may therefore, require yearly treatment. Control by solarization is greatest in the upper 30 cm of the soil. Nematodes deeper in the soil profile may survive solarization (Table 7.3) and damage plants with deep root systems.

Table 7.2. Nematodes controlled by soil solarization.

| Scientific name | Common name |
|---|---|
| *Criconemella xenoplax* | Ring nematode |
| *Ditylenchus dipsaci* | Stem and bulb nematode |

| *Globodera rostochiensis* | Potato cyst nematode |
|---|---|
| *Helicotylenchus digonicus* | Spiral nematode |
| *Heterodera schachtii* | Sugar beet cyst nematode |
| *Meliodogyne hapla* | Northern root knot nematode |
| *Meliodogyne javanica* | Javanese root knot nematode |
| *Paratylenchus hamatus* | Pin nematode |
| *Pratylenchus penetrans* | Lesion nematode |
| *Pratylenchus thornei* | Lesion nematode |
| *Pratylenchus vulnus* | Lesion nematode |
| *Tylenchulus semipenetrans* | Citrus nematode |
| *Xiphinema* spp. | Dagger nematode |

Table 7.3. Nematodes unpredictably controlled by soil solarization.

| Scientific name | Common name |
|---|---|
| *Meloidogyne incognita* | Southern root knot nematode |

Nematode control by solarization is usually adequate to improve the growth of shallow-rooted, short-season plants. It is particularly useful for organic gardeners and home gardeners. Solarization may also be a beneficial addition to an integrated nematode control system. For example, excellent control of root knot nematode (*Meloidogyne incognita*) was obtained in the San Joaquin Valley by combining solarization with the application of composted chicken manure (Gamliel and Stapleton, 1993b).

*Chapter 8*

# WEED MANAGEMENT

## 8.1. BIOFUMIGATION

### 8.1.1. Glasshouse

A mixture of redroot pigweed, wild mustard and wild oat seeds were planted into potting media amended with 0.5 and 1.0 MT/ha canola, Oriental mustard or yellow mustard seed meal, along into untreated potting media, and weed seedling emergence and biomass were recorded. Yellow mustard seed meal amendment reduced wild mustard and redroot pigweed weed seedling emergence by 68% and 70%, respectively, and biomass by 68% and 80%, respectively, compared to the untreated control, but canola and Oriental mustard meal treatments had little effect on the broadleaf weeds. In contrast, Oriental mustard meal reduced wild oat emergence and biomass most effectively. Yellow mustard meal treatments affected not only weed seedling emergence, but seedling growth and mortality. Overall, Oriental and yellow mustard meal treatments had the greatest phytotoxic effect on weed emergence and biomass and have potential herbicidal uses (Brown *et al.*, 2006).

### 8.1.2. Rapeseed for Weed Control

'Dwarf Essex' rapeseed was planted on June 14, and was mowed, chopped and roto-tilled into the soil on Aug. 30. In some plots, plastic tarps were then laid immediately on the soil in order to combine the action of solarization and biofumigation. In each plot, seeds of milkweed, curly dock and foxtail were buried. These packets of seeds were retrieved in October and counted the number of seeds that germinated.

Biofumigation with rapeseed alone killed 95 percent of milkweed seeds, 68 percent of foxtail seeds and 51 percent of dock seeds. Rapeseed plots that were tarped for solarization killed 94 percent of dock and 100 percent of milkweed. However, high temperatures under tarps apparently stimulated foxtail. Twice as many foxtail seeds germinated after solarization as in the control.

### 8.1.3. Onion and Garlic Amendments for Weed Control

Results of this study showed that exposure to 39°C soil temperature was consistently deleterious to seed survival in barnyard grass (*Echinochloa crusgalli*), common purslane (*Portulaca oleracea*), London rocket (*Sisymbrium irio*), and black nightshade (*Solanum nigrum*), as compared to 23°C. Significant effects of the onion (*Allium cepa*) and garlic (*A. sativum*) amendments on weed seed viability were common, but less consistent than with the other experimental factors. No differences in weed seed viability due to soil amendment with onion versus garlic were found in the 2000 experiment, and only in barnyard grass and black nightshade in the 2001 experiment. On the other hand, seed viability differences due to amendment concentration (1% versus 3% by weight) were found in barnyard grass, black nightshade, and London rocket in the 2000 experiment, but no concentration differences were found in the 2001 experiment. In experiments, barnyard grass, common purslane, and London rocket seeds were less viable after longer incubation in the microcosms, while black nightshade was not significantly affected by exposure time.

The results indicate that the *Allium* spp. amendments, especially when combined with elevated soil temperature (solarization), may contribute to effective weed management strategies for both conventional and organic production.

## 8.2. SOLARIZATION

The use of soil solarization for weed control is in the proven, but experimental stage. We know how to make it work in the field. It has been used in the field, but there are still obstacles to its use.

Polyethylene sheeting for weed control has been widely used in agricultural and urban areas. Black polyethylene has been the dominant material where it is used as a barrier mulch. This mulch excludes light, thus prohibits photosynthesis and kills young annual weeds or even many perennial weeds if the mulch remains for a long enough period. In the early 1970s, research in Japan (Inada, 1973) demonstrated that a green polyethylene mulch that would absorb light at 450 nm wavelength, reduced weed growth, though not as effectively as black polyethylene mulch. Some of these coloured mulches are still used in greenhouses and in the field, not for soil solarization, but as a barrier for weed control.

In 1976, Katan *et al.* (1976) described the use of clear polyethylene for soil solarization (soil pasteurization) to control soil pathogens and weeds. Subsequent studies for weed control have utilized this same application technology. Various synthetic mulches have been evaluated; however, polyethylene seems to have characteristics of transmittance and durability, for greatest effectiveness.

Soil solarization for weed control is both visually dramatic and highly effective when properly conducted. Though restricted to a time interval in summer during periods of high radiation, it can effectively control species that would be a problem in the subsequent fall or spring crops. Horowitz *et al.* (1983) found in Israel that a two to four week solarization period effectively controlled annual weeds that were appreciable after one year. Rubin and Benjamin (1984) reported similar results from solarization for four to five weeks. In the field, solarization has been best adapted for the control of weeds in cool seasons, and for fall seeded crops such as onions, garlic, carrots, broccoli or other brassica crops, and lettuce. In greenhouses or in the more temperate to tropical regions, solarization can be used before planting of the spring-planted crops such as tomatoes, peppers, squash, and cucurbits. Also in the field it can be used to preplant strawberries. Other crops that have been evaluated include broad beans, potatoes, orchard trees, and vineyards. In some studies there has been no need for additional weed control measures such as hand weeding or cultivation in the annual crop following solarization. Weed control has not been effective into the second or third years following treatment, though weeds are reduced. If beds are not disturbed following treatment, weed control can be effective for two crops (Bell and Elmore, 1983).

The greatest difficulty in consistently achieving good weed control has been when the operator (farmer) cultivates after removing the mulch, thus bringing untreated soil (and seed) to the surface. Thus, much of the mulching for weed control has been on pre-formed beds, rather than flat soil and then bedding following treatment. Standifer (1984) used full-tarped beds and achieved control of annual sedges and many other annual species. Rubin and Benjamin (1984) used flat solarized soil, but did not form beds, and achieved control. Solarization can be followed by cultivating to 3 cm or 6 cm compared to an uncultivated area. This test was conducted from June to July or from August to September. No difference was found in germination of weeds at the 0 and 3 cm cultivations; however, a greater number of weeds germinated after 6 cm cultivation than at the uncultivated site. The difference was greater between cultivation depths in the August to September treatment than the more optimum June solarization period. There were no differences in control of annual grass species by cultivation at 0, 5 and 20 cm following solarization. A numerical, but not a statistical increase in germination of *Portulaca oleracea* and *Malva parviflora* occurred at 20 cm cultivation compared to the 5 cm depth. Under one- to two-week solarization experiments, Egley (1983) found increased weed emergence following 5 cm cultivation in solarized plots. Thus, it would seem that shallow cultivation would not be detrimental if the solarization was conducted under optimum conditions.

In bed solarization culture, it has been noted that more uniform control is achieved when the beds are formed in a north/south direction as compared to an east/west direction. There is always a cool north side on the east/west bed. Bed-top width solarization of 80 cm or more was required to get a uniform temperature and control in the centre of the beds where crops were to be planted.

### 8.2.1. Weed Species Controlled

Winter annual weeds that germinate during short days and cool temperatures have been effectively controlled (Table 8.1). These species seem to be very temperature sensitive and require small increments of temperature increases to achieve control. Weeds that germinate predominantly in the longer day length, warmer period of the year (summer annuals) are not as susceptible (Table 8.2). Most species, however, are controlled. In some areas of marginal solar radiation, windy or cloudy conditions, etc., some of these summer annuals may escape. Predominant species that escape include *Melilotus* sp., *Medicago* sp. and to some lesser extent *Portulaca* sp. (Porter and Merriman, 1983). Results from the literature may not give the true picture of control of some of these summer annual weeds. If solarization conditions were marginal, species would be listed as "not controlled". Results on the perennial species *Cynodon dactylon* (Bermuda grass) and *Sorghum halapense* (Johnson grass) have varied (Table 8.3). In tests at California, both species have been controlled. *Convolvulus arvensis* (field bindweed) has been controlled during the period of solarization and suppressed for two to three weeks afterward before regrowth occurred. *Cyperus esculentus* (yellow nutsedge) was only reduced by approximately 40 percent and *C. rotundus* (purple nutsedge) was partially controlled or increased following solarization.

Table 8.1. Susceptibility of winter annual weeds to soil solarization

| Weed Species | Reported | Location |
|---|---|---|
| *Anagallis coerulea* | Sensitive | Israel |
| *Avena fatua* | Moderately sensitive | Israel, California (USA) |
| *Centaurea iberica* | Sensitive | Israel |
| *Emex spinosa* | Sensitive | Israel |
| *Capsella bursa-pastoris* | Sensitive | Israel, California (USA) |
| *Lactuca scariola* | Sensitive | Israel |
| *Merurialis annua* | Sensitive | Israel |
| *Lamium amplexicaule* | Sensitive | Israel, California (USA) |
| *Poa annua* | Sensitive | California (USA), Louisiana (USA) |
| *Raphanus raphanistrum* | Sensitive | California (USA), Louisiana (USA) |
| *Sonchus oleraceae* | Sensitive | Israel, California (USA) |
| *Senecio vulgaris* | Sensitive | California (USA) |
| *Montia perfoliata* | Sensitive | California (USA) |
| *Urtica urens* | Sensitive | Israel |
| *Erodium* sp. | Sensitive | Australia |

Table 8.2. Susceptibility of summer annual weeds to soil solarization

| Weed Species | Reported | Location |
|---|---|---|
| *Conyza canadensis* | Moderately resistant | Israel |
| *Echinochloa crusgalli* | Sensitive | Louisiana (USA), California (USA), Australia |
| *Ipomoea lacunosa* | Sensitive | Mississippi (USA) |
| *Malva niceaensis* | Moderately sensitive | Israel |
| *Malva parviflora* | Sensitive | California (USA) |
| *Melilotus sulcatus* | Resistant | Israel |
| *Orobanche creuata* | Sensitive | Israel, Egypt |
| *Orobanche aegyptica* | Sensitive | Israel |
| *Portulaca oleraceae* | Sensitive-Moderately sensitive | Egypt, Israel, California (USA), Australia |
| *Digitaria sanguinalis* | Sensitive | Israel, California (USA) |
| *Sida spinosa* | Sensitive | Mississippi (USA) |
| *Tribulus terrestris* | Sensitive | Israel |

| | | |
|---|---|---|
| *Xanthium spinosum* | Sensitive | Israel |
| *Astragulus boeticus* | Moderately sensitive | Israel |
| *Solanum nigrum* | Sensitive | Egypt, California (USA), Israel, Australia |

Table 8.3. Susceptibility of perennial weed species to soil solarization

| Weed Species | Reported | Location |
|---|---|---|
| *Convolvulus arvensis* | Moderately resistant-Resistant | Israel, California (USA) |
| *Cyperus esculentusx* | Resistant | Florida (USA), California (USA) |
| *Cyperus rotundus* | Resistant | Egypt, Israel, Mississippi (USA) |
| *Cynodon dactylon* | Moderately sensitive | Israel, California (USA), Texas (USA) |
| *Sorghum halepense* | Moderately sensitive | Israel, California (USA) |

The control of seed or seedlings must take into account the seed germination depth, and both the temperature and duration of temperature. Control of many small seeded weeds such as annual bluegrass and common groundsel is more effective than large seeded weeds such as wild oat since these large seeds can germinate from depths below which thermal killing occurs.

Seed size and germination depth, however, is probably not the primary protection mechanism for weed seeds. Seeds from various weed species vary in their sensitivity to temperature (thermal death point). Common groundsel (*Senecio vulgaris*) for example is so temperature sensitive that even four hours at 40°C will reduce germination and eight hours will kill all imbibed seeds. Common purslane (*P. oleracea*), however, can be controlled with either high temperature for a short duration or lower temperature for a longer duration. Treating moist *Portulaca oleracea* seeds at 45°C for up to eight hours did not reduce germination. At 50°C, germination was reduced by 16 percent at four hours of treatment or 43 percent at eight hours. At 60°C all germination was stopped when treated for four and eight hours with 90 percent reduction at 2 hours. Barnyard grass and lambsquarters were more temperature sensitive than *P. oleracea*, but less sensitive than *S. vulgaris*. Other weeds such as *Melilotus* sp. and *Medicago* sp. seem to be very heat tolerant.

Soil solarization has been effective for the control of many weed species in various parts of the world. The treatment must be applied within the correct weather, timing and application perimeters to make the process work. However, when these factors are observed, then control is achieved. Solarization will not be an effective method of weed control in all crops or in some microclimates where it would normally be effective. This lack of control is most frequently observed under less than optimum application and weather conditions; however, it can be affected by weed species.

Treatment during periods of high radiation followed by good crop management practices can make the treatment effective and preserve the control for longer periods. Cool season crops will be most impacted; however, some summer crops can also be improved through solarization because of the decreased weed pressure in the crop seedling or new transplant stage.

Soil solarization has proved to be very effective in controlling several genera of weeds, except *Cyperus* (50-93 percent). However, the level of killing of this weed depends on soil moisture: the higher the moisture content prior to mulching, the higher the level of killing. Broomrape was controlled by solarization. It could be controlled for two successive years, in broad bean or in tomato fields, by one treatment.

In general, various weeds were also controlled, completely or partially (75-100 percent), by soil solarization, these included *Amaranthus* sp., *Portulaca* spp., *Plantago* spp., *Chenopodium murale, Vicia* sp., *Lactuca scariola, Beta vulgaris, Rumex dentatus, Cynodon dactylon, Coronopus squamatus,* and *Sisymbrium irio.*

In contrast, *Malva* spp., *Melilotus indica* and *Convolvulus arvensis* were less affected by solarization with less than 75 percent killing.

Soil solarization controls many annual and perennial weeds. While some weed species are very sensitive to soil solarization, others are moderately resistant and require optimum conditions (good soil moisture, tight-fitting plastic and high radiation) for control. Winter annual weeds seem to be especially sensitive to solarization, and control of winter annuals is often evident for more than one year following treatment. Soil solarization is especially effective in controlling weeds in fall-seeded crops such as onions, garlic, carrots, broccoli and other brassica crops, and lettuce. White sweet clover (*Melilotus alba*) is one of the few winter annuals that are poorly controlled. Although summer annual weeds are less temperature-sensitive than winter annuals, most summer annuals are relatively easily controlled by soil solarization. Control of purslane (*Portulaca oleracea*) and crabgrass (*Digitaria sanguinalis*) may be more difficult to achieve. If purslane is controlled, it is a good indicator that the soil has been adequately heated.

Solarization generally does not control perennial weeds as well as it controls annual weeds because perennials often have deeply buried underground vegetative structures such as roots and rhizomes that may resprout. Seeds of Bermuda grass (*Cynodon dactylon*), johnson grass (*Sorghum halepense*), and field bindweed (*Convolvulus arvensis*) are controlled by solarization. Rhizomes of Bermuda grass and Johnson grass may be controlled by solarization if they are not deeply buried. Solarization alone is not effective for the control of the rhizomes of field bindweed. Yellow nutsedge (*Cyperus esculentus*) is only partially controlled by soil solarization. Purple nutsedge (*Cyperus rotundus*) is not significantly affected; marginal solarization has actually induced purple nutsedge to grow.

Solarization can minimize weed emergence by causing thermal death to weed seeds. Solarization performed for 1-4 weeks reduced emergence of prickly sida (*Sida spinosa*), pigweed (*Amaranthus* spp.), morning glories (*Ipomoea* spp.), horse purslane (*Trianthema portulacastrum*), and several grass species (Egley, 1983). Solarization is also an important tool to manage nutsedge (Johnson *et al.*, 2007), which is often hard to control with regular mulches, because it grows as a rhizome (not requiring sunlight) until it encounters light, then pierces mulches with its sharp growing point and thereafter expands its leaves above the plastic film. Given sufficiently high solarization temperatures, purple and yellow nutsedges either do not emerge at all, or if they do emerge, then leaf expansion, which is stimulated by sunlight, will occur under the clear plastic, which will mostly prevent puncturing of the plastic. The plant will be exposed to the heat underneath the plastic and thermal death occurs (Chase *et al.*, 1998).

However, prolonged use of solarization can lead to a shift in weed populations when used as a long-term, single solution to weed management. In a study by Ozores-Hampton *et al.* (2004), beds that were treated with methyl bromide were dominated by pigweed (*Amaranthus retroflexus*), whereas solarized beds were primarily colonized by Bermuda grass (*Cynodon dactylon*), a perennial that is harder to control by solarization than the annual pigweed.

*Chapter 9*

# INSECT MANAGEMENT

## 9.1. BIOFUMIGATION

### 9.1.1. Masked Chafer Beetle

Masked chafer beetle larvae (*Cyclocephala* spp.) (Fig. 9.1) are a common soil-inhabiting pest of turf grasses, agronomic crops including corn and soybean, and some ornamental plants. The damage they cause is similar to damage caused by larvae of other scarabaeid pests, including the Japanese beetle (*Popillia japonica*) and European chafer (*Rhizotrogus majalis*).

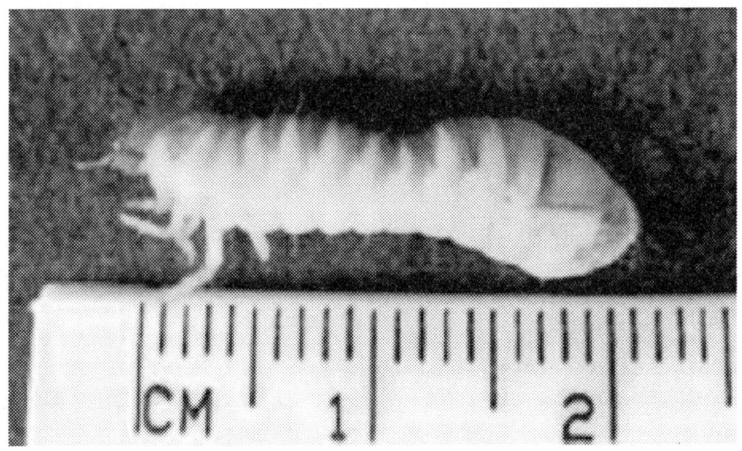


**Fig. 9.1.** Lateral view of a masked chafer beetle larva.

Different levels of fresh *B. juncea* tissue and AITC were evaluated to determine the concentration of AITC released from a cultivar of *B. juncea* and to determine the concentration required to kill masked chafer beetle larvae, as a model soil-inhabiting pest.

**Larval Mortality Experiment:** No larval mortality were observed in the control treatments, including untreated soil, soil amended with 8% tomato tissue, and the high $CO_2$ soil atmosphere (Table 9.1, Fig. 9.2). AITC was not detected in these treatments.

Mortality of grubs in the untreated control, $CO_2$ control, and tomato control differed significantly from mortality of grubs in the *B. juncea* treatments and AITC controls. AITC concentration at 0.25, 4, 8, and 24 h was positively correlated to larval mortality (Fig. 9.3). The mortality of grubs in the *B. juncea* treatments was not significantly different ($P < 0.05$) from those in the AITC controls, indicating that AITC is the predominant toxin from *B. juncea* responsible for masked chafer beetle larval mortality. At the conclusion of the experiment, the filter paper used in the AITC controls was intact, suggesting that ingestion was not vital to toxicity. This provides further evidence that the AITC vapor from both the filter paper and the *Brassica* plant material was responsible for larval mortality.

Table 9.1. Mean mortality of masked chafer beetle larvae and mean allyl isothiocyanate (AITC) concentrations over time.

| Treatment | Mortality | AITC (mg/liter)[a] | | | | |
|---|---|---|---|---|---|---|
| | | 0.25 h | 4 h[b] | 8 h | 24 h | 48 h |
| *Brassica* 1%[c] | 7* | 4.2 | 1.0 | 0.5 | 0.0 | 0.0 |
| *Brassica* 2%[c] | 33* | 7.2 | 2.9 | 1.2 | 0.6 | 0.5 |
| *Brassica* 4%[c] | 50* | 10.2 | 4.0 | 1.7 | 0.8 | 0.6 |
| *Brassica* 8%[c] | 100* | 20.3 | 11.4 | 4.9 | 2.3 | 1.7 |
| Control untreated | 0 | 0.0 | 0.0 | 0.0 | 0.0 | 0.0 |
| Control $CO_2$ | 0 | 0.0 | 0.0 | 0.0 | 0.0 | 0.0 |
| Control 8% tomato tissue[c] | 0 | 0.0 | 0.0 | 0.0 | 0.0 | 0.0 |
| Control AITC 3.9 mg/liter | 40* | 1.4 | 2.5 | - | 0.0 | - |
| Control AITC 7.5 mg/liter | 100* | 1.7 | 8.7 | - | 0.8 | - |
| Control AITC 23.7 mg/liter | 100* | 2.0 | 23.5 | - | 7.7 | - |

**a** Numbers are means of three replications.

**b** Standard errors of the means were 16.49 for larval mortality and 1.57 for AITC, respectively.

**c** Percentage is expressed as fresh tissue weight per gram of soil.

* Denotes significance at 4 h for contrasts against non-AITC controls ($P < 0.05$).

- Indicates no data were gathered at these time intervals.

Tilling a winter cover crop of *B. juncea* into the soil and covering with plastic mulch (to help retain volatile compounds) is a potential method of biofumigation (Figs. 9.4 and 9.5). Although cost may be prohibitive in standard row-cropping systems, there is little additional cost in plastic culture systems. In high-return greenhouse production systems biofumigation may prove economically feasible.

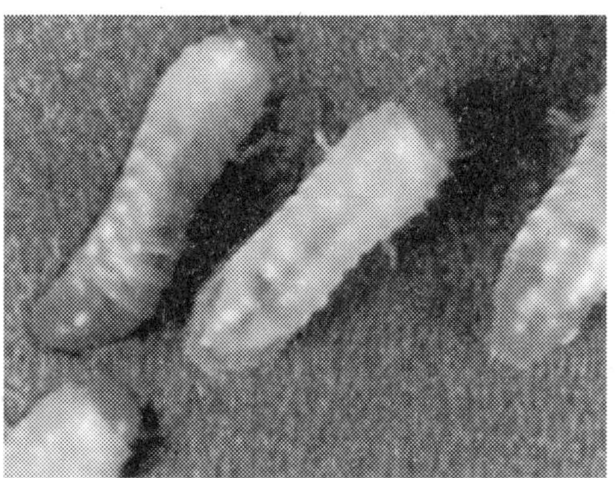

**Fig. 9.2.** Live larvae from soil amended with 8% tomato tissue.

**Fig. 9.3.** Dead larvae from soil amended with 8% *Brassica juncea* tissue

**Fig. 9.4.** *Brassica juncea* grown as a winter cover crop for biofumigation

**Fig. 9.5.** After tilling *Brassica juncea* into the soil, soil is formed into beds and covered with plastic mulch.

### 9.1.2. Termites

The termite, *Reticulitermes grassei* is more sensitive than *R. santonensis* for each disulfide. It is also noted the greater toxicity of dimethyl disulfide (DMDS) in comparison to dipropyl disulfide (DPDS) and diallyl disulfide (DADS). This result is in accordance with a previous study (Auger *et al.*, 2002) and it can be concluded that isopterans and hymenopterans are more sensitive than dipterans, lepidopterans and coleopterans (about 10 times more).

### 9.1.3. Bruchids

Dimethyl disulfide (DMDS) and dipropyl disulfide (DPDS) lead to the mortality of Coleopteran *Bruchidus atrolineatus* (Nammour *et al.*, 1989). Sulfur compounds also have an action on the Coleopteran *Calosobruchus maculatus* and its parasitoid *Dinarmus basalis* (Dugravot *et al.*, 2002).

### 9.1.4. Wireworms

The effectiveness of four new biocidal seed meals from *Brassica carinata* sel ISCI7, *Eruca sativa* cv. Nemat, *Barbarea verna* sel. ISCI100 and *Sinapis alba* cv. Pira applied broadcast and in furrow was evaluated using wireworms (*Agriotes brevis, A. sordidus, A. ustulatus*) put, at different times, in vials or pots filled with sandy soil and planted with winter wheat or maize. The first two biocidal meals caused a very high larval mortality and prevented wireworms from damaging crop seedlings. The best results were obtained with broadcast applications. Insecticide activity of volatiles produced by the glucosinolate enzymatic hydrolysis disappeared in 2-3 days. This makes the treatment particularly safe from an environmental point of view, but opens some practical problems to set up an effective procedure to cause high wireworms mortality in open field conditions. It can hypothesized that biocidal seed meals can give a high wireworm's control effect only if all the following conditions occur concurrently:

- Homogeneous broadcast biocidal seed meal application.
- Effective and prompt incorporation into the soil.
- Suitable soil temperature and humidity conditions.
- Treatment with biocidal meal when there is the presence of most of the wireworms.
- Population in the upper part of the soil.

All the information about the biology and behaviour of different *Agriotes* species has to be carefully taken into consideration to optimize application procedure. In Italy the most suitable conditions should be usually found in early spring and early autumn with bare, warm and humid soils, just before and after crop growing season. In order to maximize wireworm population control at least 10-15 days with soil temperature over 18-20°C after cold or dry conditions should pass so that also the larvae in little mobile larval stages (pre-moulting, moulting and hardening stages) in the deeper soil layers can turn into the feeding stage and move upwards looking for food.

According to the biology of the different *Agriotes* species, it may be also implemented for effective population management; for example applying biocidal seed meals in the very suitable period of presence of first instar larvae can cause a significant population reduction in next years (Furlan *et al.,* 2004).

### 9.1.5. Black vine weevil

Black vine weevil (*Otiorhynchus sulcatus*) larvae cause significant root damage and have major economical impact in the horticultural industry. There are no effective pesticides or pest management strategies available for control of larvae in field or nursery landscape situations.

Pacific Gold (*Brassica juncea*) or Ida Gold (*Sinapis alba*) seed meal at 3, 6, and 12% quantities were evaluated against Black vine weevil. Pacific Gold meal treatments resulted in 100% weevil mortality while Ida Gold meal had no effect on weevil mortality. In Petri-dish studies relatively small amounts of Pacific Gold meal was effective at killing black vine weevil larvae. Soil amended with seed meal of Pacific Gold and Ida Gold, however, resulted in phytotoxicity and stunted nursery tree growth in both glasshouse and field studies. Glasshouse phytotoxicity studies on nursery trees showed varying results. Ida Gold meal showed phytotoxicity on cedar growth while Pacific Gold proved toxic to dogwood tree growth. With cherry and fir Pacific Gold treatments resulted in higher growth compared to the control. Phytotoxicity of Pacific Gold meal also resulted in stunted rhododendron and ash trees compared to pure bark mulching. In conclusion, Pacific Gold seed meal has very high potential as a bio-pesticide to be used in controlling black vine weevil larvae in horticultural tree production although phytotoxic effects need to be considered (Seamons *et al.*, 2004).

### 9.1.6. Aphids

Toba *et al.* (1977) demonstrated that wheat, *Triticum aesativum* delayed the frequency and severity of aphid transmitted non-persistent viruses in cantaloupe by controlling the aphid vector.

Hooks *et al.* (1998) planted zucchini (*Cucurbita pepo*) between rows containing buckwheat (*Fagopyrum esculentum*) and yellow mustard (*Sinapis alba*) and found that densities of *Aphis gossypii* and *Bemisia argentifolii* and the conveyance of aphid-transmitted diseases and squash silver leaf on zucchini were reduced in these living mulch treatments when compared with monoculture plantings. Additional research by Frank (2004) and Frank and Liburd (2005) have shown that living mulch plants containing buckwheat and white clover have higher numbers of natural enemies than synthetic mulch or bare-ground plots.

### 9.1.7. Citrus California Red Scale

The mortality of adult red scale, *Aonidiella aurantii* on detached orange fruits treated with a mineral oil emulsion activated by Biofence technology was 100%, significantly greater than mineral oil alone (47.5%). With the aim of improving biodegradability and lowering environmental impact of the formulation mineral oils were substituted by vegetable one with no statistical differences. Even this time, vegetable oil alone determined a mortality reached (42.5%) significantly lower than the treatment with new technology (97.5%). A clear "dose-effect" of liquid formulation was demonstrated. Linear regression between values of mortality and doses showed $R^2$ = 0.936 and F = 29.01 on mineral oil, and $R^2$ = 0.944 and F = 33.9, on vegetable oil (Rongai *et al.*, 2006).

### 9.1.8. Fungus Gnat on House Plants

Two rates of *Brassica juncea*, 'Pacific Gold,' seed meal (1.7 g and 4.2 g), no meal, 4.2 g *B. napus* 'Athena' canola seed meal (control), and 10ml/L Gnatrol(standard) were applied to four common houseplant species in 10-cm diameter pots by spreading the meal on the surface of the potting medium. Two week post treatment, fungus gnat larval mortality was significantly higher with the 4.2 g *B. juncea* seed meal treatment than the other treatments. Plants showed significant phytotoxic post-treatment effects with the *B. juncea* seed meal treatments at both doses. *Brassica juncea* seed meal is toxic to fungus gnat larvae, but it may show signs of phytotoxicity in some plant species, suggesting that all target plant species should be evaluated prior to use of this product (McCaffrey *et al.*, 2006).

### 9.1.9. Cowpea Pests

Oparaeke *et al.* (2000) evaluated the extracts of garlic (*Allium sativum*) bulb and African nutmeg (*Monodora myristica*) seeds both applied at 10% in field trials during 1997-98 against insect pest control on cowpea. They reported that all treatments exhibited significant protection from *Maruca vitrata* and pod sucking Hemiptera damage compared with untreated control. *A. sativum* and *M. myristica* treated plots recorded 37.3 and 35.0 and 36.1 and 42.7% pod damage in 1997 and 1998, respectively compared to 67.8 and 98.95 pod damage in the untreated control. *A. sativum* and *M. myristica* treated plots recorded 24.6 and 13.0% and 30.0 and 17.2% higher grain yields in the 1997 and 1998 seasons, respectively compared to untreated control.

### 9.1.10. Potato Pests

The field study on effect of intercropping on population dynamics of major insect pests and vectors of potato (cutworm, various defoliators, epilachna beetle, aphid, whitefly and other pests as well as damage caused by the pests) revealed that all the treatments viz., Potato + recommended pesticide schedule of the area, Potato + Onion in alternative rows (1:1), and Potato + Garlic in alternative rows (1:1) were statistically superior and significant over control.

## 9.2. SOLARIZATION

Stapleton and DeVay (1995) reported that the greatest use of solarization as a pest management tool is in greenhouses, organic farms, and backyard gardens. It may be possible to combine solarization with reduced dosages of chemical fumigants to obtain good insect control.

### 9.2.1. Melon Thrips, *Thrips palmi*

In parts of Japan, hundreds of hectares of greenhouses that produce cucumber or sweet pepper are treated by solarization for 7 days following crop removal (Horiuchi 1991). The main target pest for this treatment is the melon thrips, *Thrips palmi*. Heating the soil is not the objective in this case, but rather raising the air temperature to heat the crop residue and the thrips that remain on the plants. All plants are pulled from the soil and left in the greenhouse.

### 9.2.2. Pepper Weevil, *Anthonomus eugeni*

Costello and Gillespie (1993) found that maintaining greenhouse temperatures at 25°C for 10 days after removing all plant residues killed adults of the pepper weevil, *Anthonomus eugeni*. A combination of maintaining 20°C and deploying large numbers of yellow sticky traps to capture adults was also used successfully.

*Chapter - 10*

# SOURCE OF NUTRIENTS

## 10.1. BIOFUMIGATION

### 10.1.1. Increased Potentially Minerizable Nitrogen (PMN)

Soil amended with *Brassica* meals: *Brassica carinata* meal, thermally deactivated *B. carinata* meal, non-biocidal sunflower (*Helianthus annuus* L.) defatted meal, *Brassica juncea* green manure (GM), *B. carinata* meal + *B. juncea* GM, Inorganic N, Inorganic N + *B. carinata* meal and *B. carinata* meal + *B. juncea* GM were evaluated for potentially mineralizable N (PMN). At time t 1 (N content in soil 28 days after incubation) the soil amended with *B. carinata* + fertilizer N showed the highest PMN; soil green manured with *B. juncea* was not significantly different from the control, whereas the PMN for the other treatments was higher than the control, without any significant difference between treatments. At time t 2 (N content in soil 91 days after incubation) the PMN decreased as follows: *B. carinata* meal + *B. juncea* GM + Fertilizer N > *B. carinata* meal + Fertilizer N > *B. carinata* meal + *B. juncea* GM and Fertilizer N > *B. carinata* meal, *B. carinata* meal deactivated, *B. juncea* and Sunflower meal > Control. The lack of significant differences between biocidal and thermally deactivated *Brassica* meals shows that the PMN was not affected by the presence of biocidal substances. Green manuring with *B. juncea* produced a delayed, but remarkable increase of PMN. The simultaneous supply of inorganic N and organic matter (whether meal or GM) gave rise to higher PMN values (Marchetti *et al.*, 2008)

### 10.1.2. Improved Nutrient Cycling/Availability of Nutrients

Mustard green manures can improve nutrient cycling by taking up nutrients that might otherwise be lost to leaching. When a green manure is returned to the soil, much of the nitrogen that the mustard takes up, whether from fertilizer or residual from the previous crop, will become available to the following crops. This is also true for other nutrients. As the saying goes, "feed the soil and the soil will feed the plants." A rule of thumb for green manures incorporated the same season as the following crop is that 50% of the nitrogen in the green manure will be available. The actual amount depends on the plant composition, soil temperature, soil moisture, and losses due to leaching.

Leaching of nutrients released during decomposition is a risk with mustard green manures because they are incorporated in the fall. This risk may be reduced when the mustard green manure is incorporated with a large amount of wheat straw, which will tend to tie up any available nitrogen in the soil. To do this, farmers must direct seed the mustard through standing wheat stubble. If, however, the wheat straw is incorporated before mustard planting, it will tie up nitrogen and more fertilizer will be required to grow the mustard.

Spring incorporation would lessen the risk of leaching, but farmers prefer to incorporate in the fall for several reasons. First, they do not want to have to deal with a green manure in the spring when there are many other things to do. Second, there is less risk of damage from any soil-borne pathogens that might have been stimulated by the green manure. Finally, because mustards will not always survive winters in this region, a spring incorporation of dead plants would not give them the green manure effects that are important for pest control.

Green manures can also increase nutrient availability through weathering of soil mineral components. This weathering may be caused by the production of acids by microorganisms during the decomposition of the green manure.

### 10.1.3. Nutrient Scavenging

Brassicas can provide greater than 80% soil coverage when used as a winter cover crop (Haramoto and Gallandt, 2004). Depending on location, planting date and soil fertility, they produce up to 8,000 lb. biomass/acre. Because of their fast fall growth, brassicas are well-suited to capture soil nitrogen (N) remaining after crop harvest. The amount of nitrogen captured is mainly related to biomass accumulation and the amount of N available in the soil profile.

Because they immobilize less nitrogen than some cereal cover crops, much of the N taken up can become available for uptake by main crops in early to late spring. Brassicas can root to depths of 6 feet or more, scavenging nutrients from below the rooting depth of most crops. To maximize biomass production and nutrient scavenging in the fall, brassicas must be planted earlier than winter cereal cover crops in most regions, making them more difficult to fit into grain production rotations.

### 10.1.4. Effect on Crop Plants

***(i) Rice:*** Winter rape (*Brassica napus*) is one of the important oil crops in China, and due to its high biomass; can also be used as green manure. Glucosinolates (GSLs) are rich in rape plant, and hydrolysis by the myrosinase enzyme yields toxic isothiocyanates (ITCs) and nitriles, which can be used as biofumigants in controlling pests, such as nematodes, fungi and bacteria. A long term field experiment conducted since 2003 has investigated the effects winter rape green manure on soil fertility, rice yield and quality, and plant senescence. The winter rape and rice cultivars were "gaoyou 605" and "No6 guodao", respectively. The results in year 2007 showed that returning rape as green manure had the following impacts: Increased soil fertility - the content of soil organic matter, total N, available P, available K in the treatment were increased by 7%, 23%, 39% and 15% above the control, respectively. The activity of the soil proteinase, catalase and phosphatase was increased by 38%, 11% and 12% above the control and postponed the senescence of rice. During the grain filling periods, the content of the soluble protein, the activity of POD and the activity of SOD was 7%, 14% and 41% higher than that of the control, respectively. The yellow leaf ratio of control was 37%, while in the treatment it was 27%. The above treatment also increased the rice biomass accumulation, yield and quality. Compared with control, the biomass during the heading period, LAI and yield were increased by 15%, 11% and 38%, respectively. The grain milling and appearance quality traits were also improved (Jian *et al.*, 2008).

***(ii) Wheat:*** There is some evidence that canola provides more N for following cereals (38-40). Possible reasons are the breakdown of microbial biomass during biofumigation and that brassica residues break down more rapidly than wheat residues.

Besides evaluating the break-crop benefit and the mechanism for disease control, research on brassica-wheat sequences has led to insights into the water and N relations of subsequent crops. Averaged over 8 comparisons, dry land wheat growing after canola extracted 23 mm more soil water than wheat growing after wheat, leading to subsoil drying to matric potentials down to –5 MPa in comparison to values of –3 MPa after wheat. Over 26 comparisons wheat after canola contained 17 kg N/ha more than wheat after wheat (Fig. 10.1). While the additional N may have been due in part to more soil N mineralized after canola, it was also associated with less mineral N remaining in the soil at the time of wheat maturity. The main reason for the additional extraction of soil water and mineral N was greater demand by healthy plants and continued uptake by healthy wheat roots.

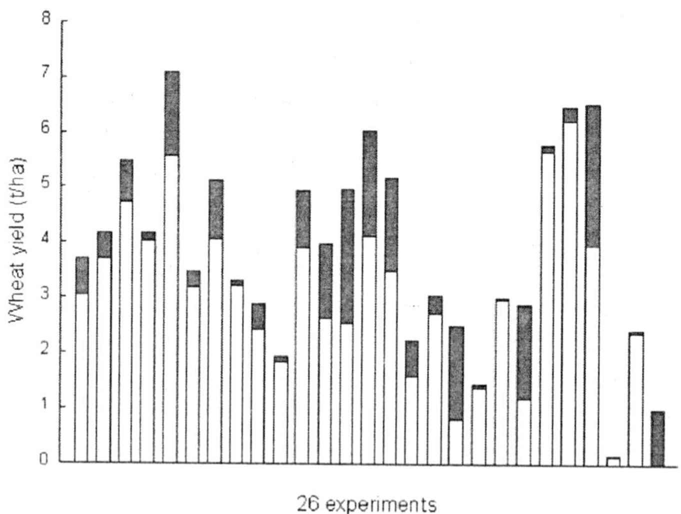


**Fig. 10.1.** Yield of wheat growing after wheat (open bars) and the increase (green bars) or decrease (red bars) of yield for wheat growing after canola in 26 experiments in southern Australia

Field experiments with break crops in southern New South Wales in the 1980s suggested that wheat growing after Indian mustard (*Brassica juncea*) yielded more than wheat growing after canola (*B. napus*), which in turn out yielded wheat (*Triticum aestivum*) growing after non-Brassica crops. The average yield of wheat growing after canola was 19% greater than wheat growing after wheat, and the yields after Indian mustard and Linola were respectively 20 and 19% greater. The average N uptake of wheat after wheat was 95 kg/ha, compared to 112, 113 and 120 kg/ha for wheat after canola, Indian mustard and Linola respectively. The two-crop N uptake by canola-wheat and mustard-wheat were 17 and 14% respectively greater than by wheat-wheat (Table 10.1).

Table 10.1. Mean yield, N uptake and gross margins (Coefficients of variation are shown in brackets).

| First crop | Wheat | Canola | Mustard | Linola |
|---|---|---|---|---|
| First crop yield (t/ha) | 3.63 (48) | 1.91 (60) | 1.76 (62) | 1.41 (56) |
| Following wheat yield (t/ha) | 3.29 (48) | 3.91 (46) | 3.96 (43) | 3.91* (44) |
| First crop N uptake (kg/ha) | 92 (53) | 107 (64) | 99 (63) | 76 (61) |
| Following wheat N uptake (kg/ha) | 95 (44) | 112 (36) | 113 (33) | 109* (54) |

| | | | | |
|---|---|---|---|---|
| Two-year crop N uptake (kg/ha) | 187 (39) | 219 (46) | 213 (44) | 185* (54) |
| Two-year gross margin ($A/ha) | 552 (84) | 699 (86) | 634 (89) | 515* (71) |

***(iii) Carrot:*** A field study was initiated to investigate the influence of *Brassica juncea, B. napus,* and *Sinapis alba* meals on soil and carrot plant growth parameters. Among 2 MT/ha rates and the unamended control, N uptake and carrot yields were not different and across all treatments, averaged 117 kg N/ha and 37 t/ha, respectively. Increases in plant-available N ($NO_3-$ and $NH_4+$) in soil, from pre-amendment to harvest, were 85, 66, 36, and 15 kg N/ha for *B. napus, B. juncea, S. alba* and the control, respectively (Snyder *et al.*, 2006).

### *(iv) Tomato*

**(a) Incubator simulation of soil solarization:** Concentrations of soil $NH_4$-N, $NO_3$-N, P, K, Ca, and Mg following four-week incubation are shown in Table 10.2 and compared with reference values obtained after six weeks of field solarization (Stapleton *et al.*, 1985). Simulated solarization generally resulted in increased concentrations of mineral nutrients similar to those seen after field solarization in the loamy sand and silty clay soils.

Table 10.2. Comparative effect of simulated vs. field solarization on concentrations of soluble mineral nutrients in two soils

| Soil Type Treatment[1] | $NH_4$-N (?g/g) | $NO_3$-N (?g/g) | $NH_4$-+$NO_3$-N (?g/g) | P (?g/g) | K (?g/g) | Ca+Mg (me/l) |
|---|---|---|---|---|---|---|
| Loamy Sand | | | | | | |
| Heated (moist soil) | 0.1a[2] | 15.0a | 15.1a | 7.4a | 90a | 6.3a |
| Control (moist soil)[3] | 0.1a | 7.5b | 7.6b | 7.7a | 96ab | 4.0b |
| Control (dry soil) | 1.5b | 0.9c | 2.4c | 6.3 | 100ab | 2.5c |
| Solarized (6 wk)[4] | 5.9c | 11.1ab | 17.0a | 7.4a | 108b | 4.3b |
| Silty Clay | 33.0b | 35.7a | 68.7b | 20.7a | 448a | 7.2a |
| Heated (moist soil) | | | | | | |
| Control (moist soil) | 0.2c | 15.1b | 15.4c | 16.5a | 444a | 3.1b |
| Control (dry soil) | 7.4c | 7.9b | 15.3c | 13.0a | 426a | 2.8b |
| Solarized (6 wk) | 73.5a | 33.9a | 107.4a | 19.0a | 428a | 7.2a |

1. Soil moistened to get field capacity and incubated in polyethelene bags for four weeks. Incubator diurnal heating regime: high 45°C; low 28°C.
2. Values within columns followed by different letters are different at (P<0.05) according to Duncan's Multiple Range Test.

3. Soil prepared as for heated samples but incubated for four weeks in darkness at ambient air temp. (ca. 20-24°C).

4. Reference values from concurrent field experiment.

**(b) Effect of treatments on soil N and pH levels**: Pre- and post-treatment mean values for Kjeldahl-N, $NH_4$-N, $NO_3$-N, and soil pH are shown in Table 10.3. No consistent differences in Kjeldahl-N values were found among treatments. Fertilizer application, with or without solarization, normally resulted in increased $NH_4$-N and $NO_3$-N levels, and lower soil pH. Fertilized soil which was then solarized normally had higher N values than non-solarized plots.

Table 10.3. Effect of ammonia-based fertilizers and/or solarization on soil N and pH levels[1]

| Treatment | Loam soil | | | | | Fine sandy loam soil | | | | |
|---|---|---|---|---|---|---|---|---|---|---|
| | Kjeldahl- N (μ g/g) | $NH_4$ - N (μ g/g) | $NO_3$- N (μ g/g) | $NH_4$+ $NO_3$-N (μ g/g) | pH | Kjeldahl-N (μ g/g) | $NH_4$- N (μ g/g) | $NO_3$- NH (μ g/g) | $NH_4$+ $NO_3$-N (μ g/g) | pH |
| Pre-Treatment | 810 | 8 | 24 | 32 | 7.0 | 510 | 8 | 12 | 20 | 7.1 |
| Post-Treatment Control | 1027 | 6 | 32 | 38 | 7.0 | 608 | 5 | 14 | 19 | 7.1 |
| Solarization | 945 | 5 | 50 | 65 | 7.1 | 595 | 8 | 41 | 49 | 6.7 |
| Aqua ammonia | 1070 | 5 | 26 | 31 | 7.1 | 788 | 5 | 22 | 27 | 6.9 |
| Aqua ammonia + Solarization | 1027 | 30 | 38 | 68 | 7.1 | 735 | 11 | 31 | 42 | 6.9 |
| Urea | 945 | 7 | 28 | 35 | 7.0 | 735 | 10 | 53 | 63 | 6.7 |
| Urea + Solarization | 1027 | 19 | 57 | 76 | 6.8 | 773 | 31 | 104 | 135 | 6.7 |
| Ammonium sulphate | 1138 | 58 | 74 | 132 | 6.6 | 788 | 29 | 53 | 82 | 6.6 |
| Ammonium sulphate+ Solarization | 1095 | 64 | 39 | 103 | 6 8 | 735 | 14 | 39 | 53 | 6.7 |
| Ammonium phosphate | 985 | 17 | 35 | 52 | 6.9 | 713 | 28 | 39 | 67 | 6.8 |
| Ammonium phosphate + Solarization | 1027 | 56 | 42 | 98 | 6.8 | 595 | 28 | 34 | 62 | 6.9 |

Fertilizer were applied at 305 kg N/ha, and soil was solarized for 3.7 weeks.

## 10.2. SOLARIZATION

### 10.2.1. Increased Availability of Nutrients

In general, availability of some nutrients can be expected to increase with solarization, because the heat generated under clear plastic will encourage accelerated decomposition of organic matter. In a field study in southwest Florida, soil nutrients were not affected by solarization compared to conventional treatment with methyl bromide (Ozores-Hampton *et al.*, 2004). Solarization coupled with compost increased soil nutrient levels more than treatments with methyl bromide or solarization alone, both of which were combined with inorganic fertilizer. Seman-Varner *et al.* (2008) measured nutrient concentration in the soil and plant tissue of an okra crop following different durations of solarization. While soil potassium (K) and manganese (Mn) were higher following solarization, Copper (Cu) and zinc (Zn) were lower. In addition, soil pH was slightly decreased by solarization. Soil phosphorus (P), magnesium (Mg), calcium (Ca), and iron (Fe) were not affected by solarization. Nutrients supplied to the crop were exclusively provided by chopped cowpea hay. Okra tissue concentrations of K, nitrogen (N), Mg, and Mn were higher when grown on solarized plots. In contrast, concentrations of P and Zn were lowered by solarization.

The increased availability of mineral nutrients following soil solarization are particularly those tied up in organic fraction, such as $NH_4$-N, $NO_3$-N, P, Ca, and Mg, as a result of the death of the microbiota. Extractable P, K, and Ca, Mg sometimes have been found in greater amounts after soil solarization. The liberation of N compounds (vapour and liquid) is a component of the mode of action, increased concentration of reduced N would then nitrify after termination of soil solarization to provide $NO_3$ for increased crop growth.

Soils covered with plastic film mulches usually retain a higher level of soluble minerals. Constant moisture content, higher temperature and better aeration of the soil all tend to favour higher microbial populations in the soil thus ensuring more complete nitrification.

Plastic mulch prevents leaching of nutrients, particularly nitrogen. The dominant advantage of using polyethylene is that it aids in the retention of nutrients within the root zone, thereby permitting more efficient nutrient utilization by the crop.

### 10.2.2. Decomposition of Organic Matter

Solarization increases the levels of available mineral nutrients in soils by breaking down soluble organic matter and increasing bioavailability. Soil solarization also speeds up the breakdown of organic material in the soil, often resulting in the added benefit of release of soluble nutrients such as nitrogen ($N03^-$, $NH4^+$), calcium ($Ca^{++}$), magnesium ($Mg^{++}$), potassium ($K^+$), and fulvic acid, making them more available to plants.

### 10.2.3. Increased Plant Growth and Yield

Solarization increased okra biomass; the longer the duration of solarization, the greater the increase in okra biomass (based on comparison among 2-, 4-, and 6-week solarization periods). The positive yield response indicated that solarization did not impair organic matter decomposition and subsequent release of plant nutrients. Both of these studies (Ozores-Hampton *et al.*, 2004; Seman-Varner *et al.*, 2008) suggest that solarization does not interfere much with beneficial soil organisms that decompose organic matter.

Plants often grow faster and produce both higher and better quality yields when grown in solarized soil. This can be attributed the increase in soluble nutrients, and relatively greater proportions of helpful soil microorganisms.

Solarization may also result in an increased growth response (as evidenced by increased trunk diameters) and yield in orchard trees, by increasing the availability of plant nutrients and changes the soil microflora to favour biological pest control.

Improved growth of rice and wheat plants was observed following solarization. Soil inorganic nitrogen levels were consistently higher as was extractable manganese.

### 10.2.4. Increase in Beneficial Soil Organisms

Solarization may also result in increased relative populations of beneficial organisms (i.e., rhizosphere bacteria such as *Bacillus* spp. and *Pseudomonas* spp., actinomycetes, *Trichoderma* spp., and mycorrhizal fungi).

Many beneficial soil organisms are able to either survive solarization or recolonize the soil very quickly afterwards. Important among these beneficials are the mycorrhizal fungi and fungi and bacteria that parasitize plant pathogens and aid plant growth. The increased populations of these beneficials can make solarized soils more resistant to pathogens than nonsolarized or fumigated soil. Earthworms are generally thought to burrow deeper in soil to escape the heat.

*Chapter 11*

# IMPROVED SOIL QUALITY BY BIOFUMIGATION

## 11.1. INCREASED INFILTRATION RATE

Soil quality is measured by specific attributes such as infiltration, soil respiration, bulk density, and aggregation; optimal levels of these attributes contribute to soil health. A healthy soil is one that will support crop growth without soil degradation or harm to the surrounding environment.

Infiltration is the process of water entering the soil. The rate of infiltration is important in irrigated agriculture, especially where center pivot irrigation is used. In these systems, the water application rate at the outside of the circles is often higher than the infiltration rate of the soil. This can result in runoff and ponding in low areas, which can lead to increased incidence of root-rot diseases. Where fertilizers and pesticides are being applied through the water (chemigation), high infiltration rates allow the chemicals to stay where they fall. Good infiltration is also correlated with good aeration, which also can reduce soil-borne disease problems.

Green manures can increase infiltration rates, but the effect is not direct. It begins, as with many green manure effects, with the addition of organic matter to the soil. As soil microorganisms digest this organic matter, they produce various substances called soil glues. These glues bond soil particles together into stable aggregates, or soil "crumbs," which allow water to move more quickly into the soil.

Table 11.1. Average infiltration rates after consecutive 1-inch applications of ponded water.

| Point in rotation | Average infiltration rates* (in/min) | | |
|---|---|---|---|
| | 1st inch | 2nd inch | 3rd inch |
| September 3, 1999<br>After wheat harvest (MGM)† | 1.39[a] | 0.48[a] | § |
| After wheat harvest (no MGM) | 0.13[b] | 0.18[b] | |

| | | | |
|---|---|---|---|
| November 2, 2000<br>After potato harvest (MGM) | 0.20[a] | 0.19[a] | 0.16 |
| After sugar beet harvest (no MGM) | 0.39[a] | 0.05[b] | |
| March 7, 2001<br>Potatoes/Winter (MGM) | 0.57[a] | 0.10[a] | § |
| Sugar beets/Winter (no MGM) | 0.06[b] | 0.05[b] | |
| March 5, 2002<br>Potatoes/Winter (MGM) | 0.14[b] | 0.09[a] | 0.08 |
| Fallow/Winter (no MGM) | 0.10[b] | 0.05[b] | |

* Mean separation (within date and inch applied) by PLSD at 0.01 level. † MGM - Rotation with mustard green manures. [a,b] Figures with the same letter are not statistically different. § Third inch not measured on neighbouring field due to time constraints.

Infiltration rates have been measured on adjacent fields having similar soil textures. One was managed under the Gies cropping system (Table 11.1). The other was in a rotation more typical of the Columbia Basin, with no green manures.

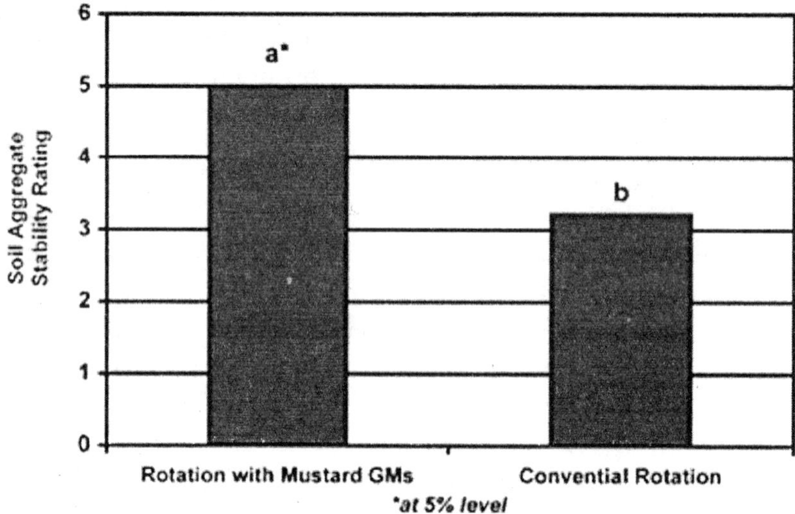


**Fig. 11.1.** Stability rating for aggregates immersed in water (slake method), Spring 2002

Infiltration rates were generally much greater under the Gies cropping system. One exception was after potato harvest on the Gies field and sugar beet harvest on the adjacent field (2000). Infiltration rates on both fields were measured on soils that had been fluffed up during harvest. In this condition, the infiltration rate for the first inch of applied water was lower in the Gies field than in the adjacent field.

However, the situation was reversed when a second inch of water was applied. It is believed that the aggregates in the field not receiving green manures were not stable in water. After the first inch of water was applied, they broke down and sealed the soil. Infiltration in the Gies soil was stable, even when a third inch of water was applied. This difference in aggregate stability of the soils was confirmed in later measurements (Fig. 11.1).

## 11.2. REDUCED WIND EROSION

Another benefit of increased aggregation due to green manures is reduced wind erosion. Farmers in the Columbia Basin have observed this in fields receiving green manures. Research to confirm their observations are on-going.

## 11.3. REDUCED SOIL COMPACTION

Other important benefits of green manures such as reduced soil crusting, improved infiltration, increased organic matter and the subsequent increase in beneficial microflora and microfauna are often overlooked. Soil improvement following a *Brassica* may produce longer lasting effects on disease suppression and crop health, as well as soil 'health'. Field studies were conducted in an attempt to quantify these beneficial properties. Soil penetration resistance is a measure of soil compaction, which increases over time with cultivation. Generally, at field capacity, growth of plant roots is not affected at <1000 kPa; while seedling emergence will be retarded and root growth restricted at 1000-2000 kPa; and root growth is impeded other than through cracks, channels and points of weakness at >2000 kPa.

The effects of biofumigation (mulching and subsequent irrigation) with five incorporated cover crops (*Sinapis alba* 'Ida Gold,' *Brassica juncea* 'Pacific Gold,' *Brassica napus* x *Brassica campestris* 'BQ mulch' *Vicia faba*, and *Triticum aestivum* x *Secale cereale*) on Romaine lettuce were evaluated. Generally, in the topsoil, penetration resistance was lower at 15-210 mm in ground previously sown with the brassicas, compared to fallow, indicating that plants or residues reduce topsoil compaction. Different brassicas performed equally well. Poor drainage from soil compaction will create a favourable environment for *Sclerotinia minor* lettuce drop disease.

In the subsoil (255-410 mm), crop residue incorporated by rotary hoe to about 200 mm is less likely to be important. However, *Brassica* tap roots may assist in breaking up deep compacted soil. Reduced subsoil compaction was recorded following Nemfix, BQ Graze and BQ

Mulch, but not Fumus. The difference may be related to their higher densities from better crop establishment. The results indicate that optimum plant density and tap roots combine to break up compacted subsoil. Lower soil compaction was also evident in a divided field with either BQ Mulch or ryegrass grown over winter, followed by potatoes in spring. After harvest, lower penetration resistance occurred in the area previously planted with BQ Mulch. In another example, compaction following two brassicas and pyrethrum in three adjacent blocks was compared. The perennial pyrethrum was terminated after three years, while the brassicas were sown in winter and rotary hoed in spring. Lower soil compaction was evident in the BQ Mulch and Fumus blocks, compared to pyrethrum.

*Chapter - 12*

# COMMERCIAL PRODUCTS FOR BIOFUMIGATION

## 12.1. CALIENTE MUSTARDS

### Green Manure and Biofumigation

Caliente Mustards have been bred specifically for biofumigation and green manuring (Fig. 12.1). The naturally occurring biofumigant gas (ITC) is produced by the plants when the crop is chopped, incorporating this compound and the green material into soil results in many benefits including improved soil structure, health and fertility; suppression of various soil-borne diseases and pests and increased soil microbial activity.

Fig. 12.1. Caliente Mustards

### What is Green Manuring?

The incorporation of green plant tissue into the soil is known as 'green manuring', this plant tissue can be from any source, and even weeds can count as green manure. It is an economical and practical method to improve soil structure and restore productivity to unused or overworked soil.

## Benefits of Green Manuring

- Improves soil fertility and structure.
- Adds nutrients and organic matter.
- Improves soil aeration.
- Increases water infiltration and holding capacity.
- Reduced soil erosion from wind and water.
- Increases soil biodiversity by stimulating the growth of beneficial microbes and other soil organisms.
- Ideal for organic crop rotations and conventional systems where Integrated Crop Management (ICM) and reduced risk inputs are favoured.

## What is Biofumigation?

Biofumigation refers to the suppression of various soil-borne pests and diseases by naturally occurring compounds. Caliente Mustard produces isothiocyanate (ITC) a natural gas released from all brassicaceous plant tissues. The gas is produced when plant cells are damaged (by crushing or chopping) and compounds called glucosinolates (present in all brassicas) come into contact with an enzyme (myrosinase) in the presence of water.

It is believed that this mechanism was originally developed by the plant as a defense against sucking and chewing insects, but over time many of the glucosinolates (the compounds that make certain brassicas 'hot' e.g. radishes) have been bred out of brassica crops to make them more palatable for human and animal consumption e.g. cabbage and cauliflower, or to produce better quality oil e.g. oil seed rape.

To create the best biofumigation effect, Caliente Mustard plants must be chopped as finely as possible before immediately incorporating into soil, simply ploughing in the crop will not give the same effect.

The following diagram simplifies the chemical reaction that takes place in a plant cell during biofumigation (Fig. 12.2):

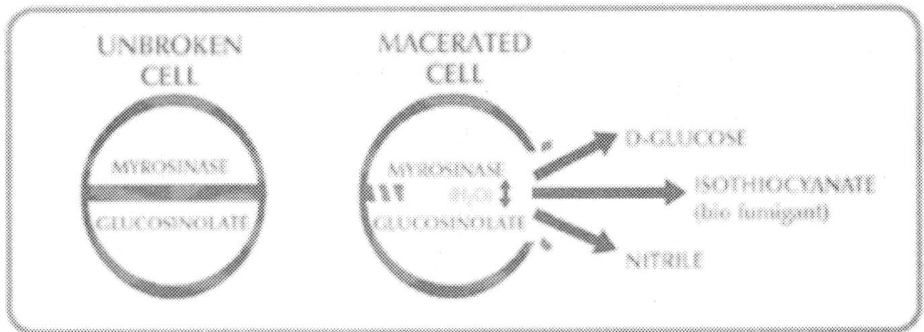


Fig. 12.2. The biofumigation process.

## How is Caliente Mustard different?

Caliente Mustards are bred specifically for biofumigation, they contain very high levels of the correct glucosinolates (there are many different types, and some do not produce lots of ITC), and high levels of the myrosinase enzyme, in addition they have the potential to grow into large plants producing high levels of biomass and therefore more green tissue to produce ITC when chopped.

Fig. 12.3. Growing of Caliente Mustard and incorporating in soil.

Caliente Mustards have been developed over many years, with new blends and varieties developed for performing specific functions, and for growing in different climates and situations. They are not simple green manure or cover crops and should not be confused with fodder mustard which is commonly available as a green manure and will not generate the same biofumigation effect.

When considering disease suppression, it is important to note that the biofumigation aspect is not the only important factor. It is a combination of the biofumigation plus the incorporation of green material into the soil which encourages beneficial microbe multiplication (Fig. 12.3). These beneficial soil microbes play an important part in 'out-competing' pathogenic microbes for space in the soil profile, helping to keep soil disease levels down.

## What are the Benefits of Caliente Mustard?

There are a number of benefits common to all crop types, soil types, climate and situation:

- Improvements to soil structure, fertility and biodiversity from the green manure effect.
- Provides a good source of nutrition to following crops by recycling nutrients like nitrogen.
- Encourages improved root systems and penetration (of following crop).
- Measurable yield increases in following crops.
- Weed suppression, mainly of soft seeded annuals, most beneficial to crops following soon after Caliente Mustard incorporation.

## Benefits to Potato and Root Crops

- Suppression of a range of soil-borne diseases, including *Verticillium* wilt, *Rhizoctonia* spp., *Pythium* spp., *Fusarium* spp., and *Sclerotinia* spp.
- Suppression of a range of nematodes including some cyst, root-knot and free living species.
- May improve skin finish.
- May deter wireworm populations.

## Benefits to Vegetable and Salad Crops

- Suppression of a range of soil-borne diseases including *Pythium* spp., *Sclerotinia* spp., *Fusarium* spp., *Rhizoctonia* spp., and others.

- Suppression of a range of nematode species.
- Offers significant reduction in weeding costs, in terms of subsequent herbicide applications, and/or labour for hand/ mechanical weeding. This is a particularly important benefit for organic systems.

## Benefits to Soft and Cane Fruit Crops

- Suppression of a range of soil-borne diseases including *Verticillium* wilt, *Fusarium* spp., and others.
- Suppression of a range of nematode species.

## Benefits to Other Crops/Situations

Caliente Mustards and Nemat are suitable for a wide range of other applications and benefits have been seen in the following crops/situation:

- Before replanting top fruit.
- In forestry and tree production.
- In protected cropping (soil grown) under glass and plastic.
- In ornamental plant and hardy nursery stock production, especially field grown plants where crop rotation is lacking and soil disease is present.
- In cut flower production.
- In bulb production.
- In viticulture (wine and table grape production) especially where nematodes are problematic.

## Caliente Mustard Products:

Caliente Brand Mustard 119

- Excellent all round variety with a proven track record in many cropping situations.
- Can be sown in spring or late summer for quick crops, or mid-autumn for overwintered crops in milder areas of the UK.
- Best variety for organic systems (Fig. 12.4).

## Caliente Brand Mustard 99

- Hottest variety, giving the best biofumigant action due to its very high levels of glucosinolates (30% more than variety 119).

- However it does require good growing conditions including additional fertilizer and irrigation in dry periods.

Fig. 12.4. Caliente Brand Mustard 119.

### Caliente Brand Mustard 61

- Large leaved variety producing high levels of biomass under ideal conditions.
- Longer season and slowest to flower out of the varieties.
- Requires warmth so suitable for mid-summer UK growing and needs irrigation where soil moisture is low.
- Best variety for warmer climates (Spain, Portugal etc.).

### Nemat (*Eruca sativa*)

- Not mustard but a white flowered rocket with biofumigant properties.
- Trap crop for various nematodes, including some root knot and cyst species. Plant roots contain highest levels of glucosinolates.

- Dense foliage produces good biomass although crop is shorter in height than the mustards.
- Mowing before flowering can extend the crops growing (and trapping) period.
- Tolerant to a range of temperatures, frost tolerant, and relatively drought tolerant once established.

**Fig. 12.5.** Nemat- a white flowered rocket with biofumigant properties.

Caliente Mustards and Nemat are bred by the ISCI Institute, Bologna, Italy and all varieties are protected by plant breeder's rights.

## 12.2. MUSTCLEAN MUSTARDS

### Biofumigation

The CSIRO developed the "mustard bomb" effect. This is where the mustard plant is chopped or macerated at the stage when about 30% is flowering. The plant material is immediately incorporated into the soil and irrigated. With the presence of water natural biofumigant gas is released through the soil. Mustclean mustards are widely used in pest and disease management systems for the suppression of various soil borne diseases and pests.

**Varieties**

Several blends are sold with different characteristics.

**Mustclean New Generation 666 (*Brassica juncea*)**

- Original line noted for its very high levels of 2-phenylethylglucosinolates released through its roots.
- Grows up to ten feet tall in mid season and is suited to heavy clay soils where the mustard bomb process is less effective.

**Mustclean Rex (Brassica juncea/Sinapis alba)**

- Mustclean Rex is a blend of two new varieties.
- The *Sinapis alba* line is derived from wild species collected in Israel and is extremely hot.
- The plant tissue contains very high levels of glucosinolates.
- It produces less biomass thus the addition of the second line.
- The *Brassica juncea* line is mid season, tall growing producing good biomass and glucosinolates.

## 12.3. ORGANIC BIOFUMIGATION MIX 1

This mix is made up of Organic Juan Triticale and Pacific Gold Mustard. The combination is good for weed suppression, cycling nutrients and adding organic matter to the soil. Pacific Gold Mustard has very high levels of a glucosinolate (2 Propenyl) which can help to reduce disease causing organisms when managed for its biofumigation potential.

- Certified Organic by C.C.O.F.
- 95% Organic Juan Triticale + 5% Pacific Gold Mustard
- Plant at 100 pounds per acre.
- Plant this mix as you would a grain crop (the mustard will flow out with the triticale) but at a shallow depth of about 1/2 inch with a firm finished seed bed.
- For home gardens 5 lbs. per 1000 sq. ft.

## 12.4. SEEDCO COMMERCIAL PRODUCTS

Seedco released the following specialized cover crop products - Biomax (green manuring), Weedcheck (Fodder Radish), Rebound (annual swards), Bt Assist (insectory habitat mix), Nemfix (soil fumigation of nematodes) and Permagreen (perennial swards) under the brand name "VinEco". These VinEco products were widely promoted to the wine industry through seminars, grower meetings, field days, printed brochures, magazine and journal articles, conference presentations and direct grower contact. The products were enthusiastically taken up by industry with sales outstripping supply. Feedback for the industry was very positive and grape production responses were very promising with major benefits being seen in subsequent years (Robertson and Grant, 2008).

## 12.5. BIOFUM BLENDS (MIXTURES FOR BIOFUMIGATION)

With Brassicaceae like *Raphanus sativus* (oil seed radish), *Sinapis alba* (white mustard), *Brassica carinata* (Ethiopian mustard) and *Brassica napus* (fodder rape) it is possible to reduce and control soil-borne diseases. This can be achieved thanks to the resistances of these plants or with biofumigation.

Under the brand name BioFum Blends, special selected varieties are developed and marketed for a maximum biofumigation effect. This means that these blends contain varieties with the highest glucosinolate level.

*Chapter 13*

# HORTICULTURAL CROPS

## 13.1. IMPORTANCE OF HORTICULTURAL CROPS

Agriculture is the mainstay of the Indian economy, and horticulture is a crucial component thereof. Horticulture development has been accorded high priority in the XI$^{th}$ Five Year Plan. The impact of enhanced investment in horticulture has been highly encouraging in terms of vastly improved quality production and export potential-an increase from 96.1 million MT in 1990-91 to 275.4 million MT during 2007-08.

India is the second largest producer of fruits next only to China contributing 12.0% of the total world production. Total production of fruits during 2007-08 has been estimated at 63.503 million MT from 5.775 million ha (Bijay Kumar, 2009) (Table 13.1). India is the largest producer of mango, banana, and papaya. India produces 41.0% of the world mangoes, 29.0% of the bananas and 30.0% of papayas. In grapes, India has recorded the highest productivity per unit area in the world (26 MT/ha). The per capita consumption of fruits has increased from 40 to 85g/day within the last one decade.

Table 13.1. Area, production and productivity of horticultural crops in India (2007-08)

| Horticultural crops | Area ('000 ha) | Production ('000 MT) | Productivity (MT/ha) | % Share of hortil. crops |
|---|---|---|---|---|
| Vegetables | 7803 | 125887 | 16.1 | 60.84 |
| Fruits | 5775 | 63503 | 11.0 | 30.69 |
| Plantation crops | 3226 | 12045 | 3.7 | 5.82 |
| Spices | 2603 | 4103 | 1.6 | 1.98 |
| Flowers | 161 | 870 | 5.4 | 0.42 |
| Aromatic & Medicinal plants | 386 | 325 | 0.8 | 0.16 |
| Nuts | 132 | 177 | 1.3 | 0.09 |
| Mushrooms | --- | 37 | --- | --- |

| Honey | --- | 65 | --- | --- |
|---|---|---|---|---|
| Total | 20087 | 207012 | 10.3 | |

India is the second largest producer of vegetables next only to China contributing 13.0% of the total world production. Vegetables occupy an area of 7.803 million ha with a production of 125.887 million MT (Bijay Kumar, 2009). India produces 18.0% of the world onions. In vegetables, India occupies the first position in the production of peas and okra, second in onion, brinjal, cabbage, cauliflower and third in potato and tomato. The per capita consumption of vegetables has increased from 96 to 175 g/day within the last one decade.

Floriculture is estimated to cover an area of 161,000 ha with production of 870,000 MT of loose flowers. The area under cut flowers has increased significantly in recent years and the production has reached 4341 million numbers.

India has emerged as the largest producer of spices, tea, cashew and areca nut. Coconuts sustain an area of 1.940 million ha with a production of 108.94 million MT. Areca nut is cultivated in 0.386 million ha with a production of 0.476 million MT. Cashew nuts are grown in an area of 0.868 million ha with a production of 0.665 million MT. The annual production of spices is 4.103 million MT from 2.603 million ha.

Horticulture has emerged as an important sector for diversification of agriculture. Presently, horticulture has established its credibility in improving income through increased productivity, generating employment and in enhancing exports besides providing protective food, household nutritional security, precious raw material for various processing and ancillary industries and finally in enhancing the aesthetic aspects of life. The focused attention on investment in horticulture during the last decade has been rewarding in terms of increased production and productivity of horticultural crops with manifold export potential. Horticulture has played a significant role in women empowerment, providing employment opportunities through mushroom cultivation, floriculture, processing, nursery raising, vegetable seed production etc.

## 13.2. EXPORT OF HORTICULTURAL PRODUCE

Horticulture in India is fast emerging as a major commercial venture, because of higher remuneration per unit area and the realization that consumption of fruits and vegetables is essential for health and nutrition. In the last one decade, export potential of horticultural crops has significantly increased attracting even multinationals into floriculture, processing and value added products.

The present annual growth rate of horticulture is more than 6.5% which contributes 24.5% from the mere 8% of area to the GDP and 54.55% to the export earnings in the agriculture sector (Tables 13.2 and 13.3).

Table 13.2. Export of horticultural produce from India (2007-08) (Bijay Kumar, 2009)

| Horticultural product | Export quantity (MT) | Value (Rs. Crores) | % Share of value |
|---|---|---|---|
| Floriculture & seeds | | | |
| Floriculture | 42016.60 | 64983.50 | 13 |
| Fruit & vegetable seeds | 8031.09 | 12146.68 | 2 |
| Total | 1654282.66 | 159332.63 | |
| Fruits & vegetables | | | |
| Fresh onions | 1378373.17 | 116330.57 | 25 |
| Other fresh vegetables | 275909.49 | 43002.06 | 9 |
| Walnuts | 5062.06 | 11803.06 | 2 |
| Fresh mangoes | 79060.88 | 14193.95 | 3 |
| Fresh grapes | 85897.79 | 30192.45 | 6 |
| Other fresh fruits | 159569.71 | 25643.64 | 5 |
| Total | 549949.22 | 173163.18 | |
| Processed fruits & vegetables | | | |
| Dried & preserved vegetables | 118787.46 | 42406.20 | 9 |
| Mango pulp | 156835.52 | 50582.79 | 10 |
| Pickle & Chutneys | 145216.00 | 29359.48 | 6 |
| Other processed fruits and vegetables | 129110.24 | 50814.71 | 10 |
| Grand Total | 2754181.00 | 505659.00 | |

Table 13.3. Export of Horticultural produce from India (2007-08) (Bijay Kumar, 2009)

| Horticultural produce | Quantity in MT | Value in lakh Rs. |
|---|---|---|
| Apple | 32665 | 3330 |
| Banana | 16662 | 2607 |
| Oranges | 29261 | 2726 |
| Grapes | 96723 | 31706 |

| | | |
|---|---|---|
| Guava | 2495 | 417 |
| Litchi | 161 | 63 |
| Mango | 54350 | 12741 |
| Papaya | 10879 | 1123 |
| Pineapple | 4194 | 339 |
| Sapota | 2150 | 503 |
| Brinjal | 338 | 191 |
| Cauliflower | 70 | 2 |
| Onion | 100860648 | 103577 |
| Peas | 814 | 270 |
| Tomato | 134845 | 15290 |
| Potato | 78450 | 4142 |
| Sweet potato | 346 | 35 |

The major destinations for export of different horticultural products and potential international markets for fresh and processed products are presented in Table 13.4.

Table 13.4. Major destinations for export of different horticultural product (Bijay Kumar, 2009)

| Horticultural crops | Major destinations for export |
|---|---|
| **Fruits** | |
| Fresh grapes | UK, UAE, The Netherlands, Bangla Desh, Germany, Sri Lanka, Norway, Belgium, Nepal, Oman, Russian Republic |
| Pomegranate | UAE, Saudi Arabia, Sri Lanka, Kuwait |
| Fresh guava | Saudi Arabia, Netherlands, UAE, Kuwait, Yaman Arab Republic, The Netherlands, Nepal, Sudan, Canada, Oman, Japan |
| Fresh Gooseberries | UAE, Qatar |
| Fresh mango | UAE, UK, Bangla Desh, Nepal, Saudi Arabia, Kuwait, USA, Bahrain, Singapore, Germany |
| Fresh apple | Bangla Desh, Nepal, China, USA, Saudi Arabia, Germany, UAE |
| Fresh Banana | UAE, Saudi Arabia, Bahrain, Kuwait, Qatar, Nepal, Oman, Maldives, UK, Bangla Desh |
| Fresh orange | Bangla Desh, Nepal, USA, Canada, The Netherlands, Oman, Spain, China, UAE, Malaysia |
| Fresh litchi | Nepal, UAE, UK, Bahrain, Bahamas, Canada, Oman |
| Fresh papaya | UAE, Saudi Arabia, Qatar, Bahrain, France, Nepal, |

| | |
|---|---|
| | Kuwait, Bahamas, Germany, The Netherlands |
| Fresh pineapple | Nepal, UAE, Maldives, Saudi Arabia, Spain, Pakistan, Kuwait, USA, Bangla Desh, Armenia |
| Fresh sapota | UAE, UK, Bahrain, Saudi Arabia, Qatar, Canada, Bahamas, USA, Oman, Kuwait |
| Mango pulp | UK, Canada, Bangle Desh, Bahrain, Netherlands, Japan, Saudi Arabia |
| Mango slices | UK, Saudi Arabia, Japan, USA |
| Dried grapes | UK, Malaysia, Netherlands, UAE |
| Mango pickles | Saudi Arabia, UAE, UK, Canada |
| Mango chutney | Oman, USA |
| Guava jam & jelly | UK, Germany, Hong Kong, USA, Netherlands, Saudi Arabia |
| Dried tamarind | Egypt, UAE, Pakistan, Saudi Arabia, UK |
| Dried figs | Nepal, Bangla Desh |
| **Vegetables** | |
| Fresh brinjal | UK, France Saudi Arabia, Germany, Hong Kong, Portugal, Sweden, Canada, Belgium, UAE |
| Fresh cauliflower | Nepal, UAE, Maldives, Canada |
| Fresh onion | UAE, Singapore, Sri Lanka, Bangla Desh, Malaysia, Nepal, Bahrain, Saudi Arabia, Mauritius, Oman |
| Fresh peas | Saudi Arabia, UAE, Kuwait, USA, Qatar, Nepal, Comoros, Oman, Singapore, Liberia |
| Fresh tomato | Pakistan, UAE, Bangle Desh, Nepal, Maldives, Taiwan, China, Oman, Kuwait, Tanzania Republic |
| Fresh potato | Nepal, Mauritius, Sri Lanka, Maldives, Seychelles, UAE, Iran, Bangladesh, Pakistan, Malaysia |
| Fresh sweet potato | UAE, Maldives, Israel, France, Kuwait, Oman, Bahrain, Canada, Qatar |
| Gherkin | France, USA, Spain, Belgium |
| Dried vegetables | UAE, Sri Lanka, USA, UK, Singapore |
| Dehydrated onion | Germany, Romania, Sri Lanka, USA, Malaysia, Netherlands, UK |
| Vegetable seeds | USA, The Netherlands, France, Italy, Japan, Pakistan |
| **Flowers** | |
| Flowers | USA, Japan, The Netherlands, UK, Italy |

Horticulture today is not merely a means of diversification but forms an integral part of food and nutritional security, as also an essential ingredient of economic security. Adoption of horticulture both by small and marginal farmers brought prosperity in many regions of the country of which Maharashtra, Karnataka and Andhra Pradesh are the prime examples. A Golden Revolution through significant production increases in horticultural crops is already on the horizon.

Since India is now a signatory member of WTO, it will require its produce and products to be competitive both in the domestic and export markets. This would call for the use of improved horticultural practices such as use of HYV/F1 hybrids, recommended agro-techniques and plant protection measures, hi-tech intervention and scientific post-harvest management.

After the success of 'Green Revolution' in agriculture, India is now moving towards the **'Golden Revolution'** comprising fruits, vegetables and flowers. Commercialization of horticulture along with post harvest management, processing and marketing assume great significance and add value not only in the context of farmers' income, but also its contribution to the agriculture GDP. Commercialization of horticulture is largely dependent upon use of appropriate technology and market oriented production.

*Chapter - 14*

# FRUIT CROPS

Fruits are amongst the most cherished bounties of nature. They are the major sources of Vitamins 'A' and 'C' and minerals like calcium and iron. Fresh fruits have low calorific values and are low in fats. A large variety of fruits are grown in India. Of these, banana, citrus, pineapple, papaya, sapota, jack fruit and avocado (tropical fruits); mango, grape, guava, litchi and loquat (subtropical fruits); apple, pear, peach, plum, apricot, almond and walnut (temperate fruits) and amla, ber, pomegranate, annona, fig and phalsa (arid zone fruits) are important. The leading fruit growing states are Maharashtra, Andhra Pradesh, Tamil Nadu, Kerala, Gujarat, Karnataka, Bihar, Uttar Pradesh, West Bengal and Orissa.

Table 14.1. Major fruit producing countries of the world (2007-08)(Bijay Kumar, 2009)

| Country | Area (in '000 ha) | Production (in '000 MT) | Productivity (in MT/ha) | % share of world production |
|---|---|---|---|---|
| China | 9587 | 94418 | 10 | 19 |
| India | 5775 | 63503 | 11 | 12 |
| Brazil | 1777 | 36818 | 21 | 7 |
| USA | 1168 | 24962 | 21 | 5 |
| Italy | 1246 | 17891 | 14 | 3 |
| Spain | 1835 | 15293 | 8 | 3 |
| Indonesia | 846 | 11615 | 14 | 2 |
| Mexico | 1100 | 15041 | 14 | 3 |
| Iran | 1256 | 12102 | 10 | 2 |
| Turkey | 1049 | 12390 | 12 | 2 |
| Others | 22841 | 208036 | 9 | 42 |
| Total (World) | 48481 | 512070 | 13.09 | |

India is the second largest producer of fruits next only to China contributing 12.0% of the total world production. The present area under fruit crops is estimated to be about 5.775 million hectares with an estimated production of 63.503 million MT (Tables 14.1 to14.3). Fruit crops and nuts contribute 31.81% share to the total production of horticultural crops. The nut crops are grown in 0.1209 million hectares with an annual production of 0.1335 million MT (Table 14.4). India is the largest producer of mango, banana and papaya. India produces 41.0% of the world mangoes, 29.0% of the bananas and 30.0% of world papaya. In grapes, India has recorded the highest productivity per unit area (26.0 MT/ha) in the world. The per capita consumption of fruits has increased from 40 to 85 g/day within the last one decade.

Table 14.2. Crop-wise area and production of fruit crops in India (2007-08)

| Fruit Crops | Area ('000 ha) | Production ('000 MT) | Productivity (MT/ha) | % share of India production |
|---|---|---|---|---|
| Apple | 264 | 2002 | 7.6 | 3.2 |
| Banana | 647 | 23205 | 35.9 | 36.5 |
| Citrus | 843 | 7574 | 9.0 | 11.9 |
| Grapes | 64 | 1677 | 26.0 | 2.6 |
| Guava | 178 | 1975 | 11.1 | 3.1 |
| Litchi | 69 | 418 | 6.0 | 0.7 |
| Mango | 2205 | 13792 | 6.3 | 21.7 |
| Papaya | 80 | 2686 | 33.4 | 4.2 |
| Pineapple | 80 | 1216 | 15.1 | 1.9 |
| Pomegranate | 122 | 858 | 7.0 | 1.4 |
| Sapota | 150 | 1238 | 8.2 | 1.9 |
| Others | 1071 | 6862 | 6.4 | 10.8 |
| Total | 5775 | 63503 | 11.0 | |

Table 14.3. State-wise area, production and productivity of fruit crops (2007-08) (Bijay Kumar, 2009)

| State | Area ('000 ha) | Production ('000 MT) | Productivity (MT/ha) | % share of India production |
|---|---|---|---|---|
| Maharashtra | 1432.3 | 11047.6 | 7.7 | 18 |
| Andhra Pradesh | 853.0 | 10772.3 | 12.6 | 17 |
| Tamil Nadu | 292.3 | 7530.0 | 25.8 | 12 |
| Gujarat | 306.9 | 5849.7 | 19.1 | 9 |

| | | | | |
|---|---|---|---|---|
| Karnataka | 279.0 | 4588.0 | 16.4 | 7 |
| Uttar Pradesh | 316.0 | 3757.5 | 11.9 | 6 |
| Bihar | 286.2 | 3252.4 | 11.4 | 5 |
| West Bengal | 194.3 | 2766.7 | 14.2 | 4 |
| Kerala | 323.3 | 2579.8 | 8.0 | 4 |
| Jammu & Kashmir | 194.9 | 1435.9 | 7.4 | 2 |
| Assam | 122.7 | 1410.8 | 11.5 | 2 |
| Orissa | 265.3 | 1275.1 | 4.8 | 2 |
| Madhya Pradesh | 46.7 | 1237.0 | 26.5 | 2 |
| Punjab | 61.6 | 1055.4 | 17.1 | 2 |
| Others | 800.5 | 3924.8 | 4.7 | 8 |
| Total | 5775 | 62503 | 11.0 | |

India has been exporting fruits and fruit products since several decades. However, in the recent past export promotional activities received greater attention. Mango and grapes are the main items in fresh fruit export and some other fruits like pomegranate, sapota, banana, citrus, litchi, apple and strawberry also entered fresh fruit export market in limited quantities.

Table 14.4. Crop-wise area, production and productivity of nut crops in India (2007-08) (Bijay Kumar, 2009)

| Nut crop | Area ('000 ha) | Production ('000 MT) | Productivity (MT /ha) |
|---|---|---|---|
| Almond & walnut | 132 | 177 | 1.3 |

## 14.1. BANANA, *MUSA* SPP.

### 14.1.1. Diseases

#### (i) Bacterial Wilt (Moko Disease), *Ralstonia solanacearum*

**(a) Symptoms:** The disease is characterized by rapid wilting and collapse of leaves (Fig. 14.1), premature ripening of fruits, discolouration of the vascular strands and wilting and blackening of suckers. In Cavendish cv., the lower leaves develop a yellowish tinge which quickly spread to all the other leaves. Sometimes second or third leaf or both the leaves collapse near the junction of petiole and lamina. In severe cases, plants collapse before the emergence of bunches. The presence of yellow fingers in otherwise green stem often indicates the occurrence of moko disease. Internally, a firm dry rot is found in the fruit. Fruit stalks also show discoloured vascular strands.

**Fig. 14.1.** Moko wilt of banana (Courtesy, K. Srinivasan, TAFE, Chennai).

Initially one of the youngest three leaves turns pale-green or yellow in color and breaks down at the petiole and the pseudostem. Later all the other leaves collapse around the pseudostem. An infected finger or fruit shows dry and rotted pulp that is colored brown or black, and the presence of bacterial discharges.

### (b) Management

**Green Manuring:** In recent years Australian banana growers have started growing brassica green manures for disease control.

**Antagonistic and Cover Crops:** Growing cover crops like kudzu (*Pueraria phaseoloides*) and sorghum is helpful in reducing the soil inoculum. Use of antagonistic crops such as garlic and sorghum helps in reducing the disease incidence.

Growing of radish, mustard and broccoli have reduced bacterial wilt significantly (50-60%) in most of the experiments, though researchers believe the treatment is more effective on sandy soil than on heavy clay soil.

**Crop Rotation:** Rotation with maize or rice for more than 2 years is also effective.

### (ii) Panama Wilt, *Fusarium oxysporum* f. sp. *cubense*

Panama wilt is one of the most devastating diseases of banana in the world.

**(a) Symptoms:** The entry of the fungus is facilitated by root damage caused by the nematodes (*Radopholus similis, Meloidogyne incognita*, etc.). The fungus blocks the vascular system and causes wilting (Fig. 14.2). The infected plants show characteristic yellowing of leaf blades developing as a band along the margin and spreading towards midrib. The leaf wilts and the petiole buckles. The leaf hangs between the pseudo stem while the middle of lamina is still green. All leaves eventually collapse, whereas the petioles join the pseudo stem and die. Often the emerging heart leaf gets affected. Four to six weeks after the appearance of first symptoms, only the pseudo stem with dead leaves hanging round it remains. Young and old plants show dwarfing or stunting. When an affected rhizome is cut transversely, the disease is seen localized in the vascular strands. Individual strand appears yellow. Red or brown dots or streaks are also seen. The cut stem smells like rotten fish. The suckers growing out of diseased corms wilt and eventually the whole mat dies.

**Fig. 14.2.** Panama wilt affected banana plant.

### (b) Management

**Biofumigation and Bioagents:** Application of bioagents such as *Trichoderma viride* and *Pseudomonas fluorescens* in the soil and incorporation of cassava residue at 10 MT/ha plus rice bran with 21 days fallowing before planting found to check the occurrence of the disease effectively.

### 14.1.2. Nematodes

#### (i) Burrowing Nematode, *Radopholus similis*

*R. similis* causes 'rhizome rot' or 'toppling' or 'black head' disease of banana and is becoming a serious problem.

**(a) Symptoms:** Wounding of banana roots by the burrowing nematode usually induce reddish-brown cortical lesions which are diagnostic of the disease (Fig. 14.3). These lesions are clearly seen when an affected root is split longitudinally and examined immediately. Root and rhizome necrosis is manifested by varying degrees of retarded growth, leaf yellowing and falling of mature plants.

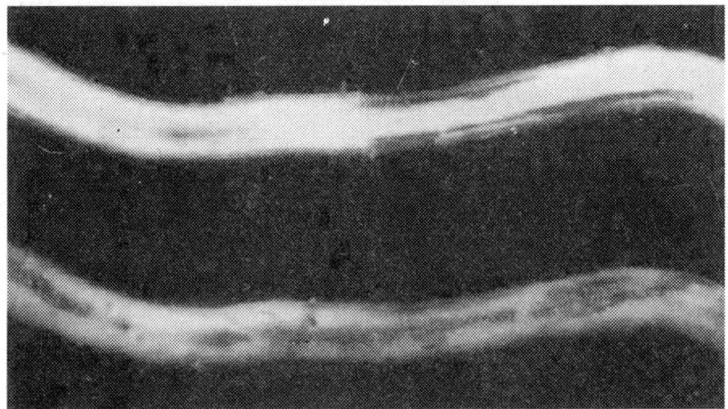

**Fig. 14.3.** Banana roots infected with *Radopholus similis*. Upper – Longitudinally cut root; Lower – Complete root (Courtesy: Union Carbide Agril. Products Co. Inc., 1986).

With the increase in nematode population, feeding roots are invaded and destroyed as fast as they are formed. The resulting setback in the uptake of plant nutrients leads to debility of the plant and production of smaller fruits. The lesioning of the primary roots together with the girdling and death of these anchor roots makes the plant prone to 'tip over' by wind action (Fig. 14.4).

#### (b) Management

**Cover Cropping:** After destroying an infested plantation, planting of cover crops like *Panicum maximum* var. *trichoglume* and *Phaseolus atropurpureus* did not support *R. similis* population. Planting of sugarcane or pangola grass (*Digitaria decumbens*) following the destruction of bananas eliminated *R. similis* after 10 weeks (Loos, 1961).

**Crop Rotation:** Rotation of banana with paddy, sugarcane, green gram, cotton, pangola grass, sorghum, tobacco or cassava suppressed the nematode population and increased the yield and controlled *R. similis* (Rajendran *et al.*, 1979). Crop rotation with rice in wet lands and with vegetables, cotton and cereals in garden lands helps to reduce the nematode population. Rice or green gram grown after banana suppressed the population of *R. similis* (Rajendran *et al.*, 1979).

**Fig. 14.4.** Premature fall of banana plants due to infection of *Radopholus similis*. Front – Toppled plants from non-treated plots; Back – Treated plants.

**Intercropping:** The nematode population both in soil and roots were considerably reduced when banana was intercropped with nematode antagonistic plants such as marigold. *Crotalaria juncea* was most effective in reduction of soil nematodes and *Coriandrum sativum* was more effective in reduction of root nematodes. *C. juncea* gave maximum bunch weight (16.7 kg) recording 51% more than control (11.03 kg) followed by *Tagetes erecta* (16.2 kg), *Sesamum indica* (15.87 kg), *Acorus calamus* (15.67 kg) and carbofuran (15.23 kg) (Charles and Venkitesan, 1993).

Intercropping with sunnhemp, coriander, marigold, radish or lucerne reduced nematode population and increased yield (Vadivelu *et al.*, 1987). Intercropping sunnhemp in banana field and ploughing the green manure 45 days after sowing was found effective in reducing *R. similis* population by 38.4%, followed by marigold and cowpea which recorded 29.0 and 22.3% reduction, respectively (Shanthi, 2003). Banana intercropped with *Crotalaria* reduced *R. similis* population in root and increased plant growth and yield (Charles *et al.*, 1985). Lakshmana Murthy (1983) reported that when banana was intercropped with papaya and marigold, *R. similis* was not encountered. Intercropping with *Tagetes, Crotalaria* and radish significantly reduced the *R. similis* population.

**Mulching:** Covering soil (mulching) with black polythene at 20 per cent moisture depletion recorded highest yield with low population of *R. similis* (Vadivelu *et al.*, 1987). Covering soil with black polyethylene, sugarcane trash and banana trash are reported to reduce *R. similis* in banana roots (Bhattacharya and Rao, 1984).

**Organic Amendments:** A significant reduction in the nematode population and increased yield in plants treated with press mud (15 t/ha) or neem cake (1.5 t/ha) was recorded. Application of press mud is economical and eco-friendly.

**Integrated Method:** Integration of paring of banana suckers, dipping in 0.5% monocrotophos solution for 45 min along with intercropping with marigold or sunnhemp gave higher fruit yield (62.838 and 61.816 MT/ha, respectively) and benefit cost ratio (1.70 and 1.28, respectively) (Patil *et al.*, 1999) (Table 14.5).

Table 14.5. Effect of paring of suckers, dipping in insecticide solution and intercropping on nematode population and yield of banana

| Treatment | Nema popn in 200 ml soil | Nema popn in 5g roots | Yield (MT/ha) | Benefit: cost ratio |
|---|---|---|---|---|
| Untreated control | 454 | 29 | 58.439 | --- |
| Paring + dipping in 0.5% monocrotophos | 283 | 18 | 60.927 | 1.20 |
| Paring + dipping in 0.5% monocrotophos + incorporation of sunnhemp at 10g/m$^2$ | 252 | 14 | 61.816 | 1.28 |
| Paring + dipping in 0.5% monocrotophos + incorporation of cowpea at 10g/m$^2$ | 274 | 17 | 60.483 | 1.04 |
| Paring + dipping in 0.5% monocrotophos + incorporation of marigold at 2g/m$^2$ | 239 | 15 | 62.838 | 1.70 |
| Carbofuran at 1.25g a.i./plant | 240 | 14 | 61.860 | 1.58 |

### (ii) Lesion Nematode, *Pratylenchus coffeae*

**(a) Economic Importance and Losses:** *P. coffeae* has been reported to cause decline of banana plantations. The lesion nematode causes most damage to banana in association with *F. o.* f. sp. *cubense* (Vadivelu *et al.*, 1987). In India, crop loss caused by the root-lesion nematode in banana cv. Nendran was reported to be 45%.

**(b) Symptoms:** The symptoms caused by *P. coffeae* are very similar to those caused by *R. similis*. Both *R. similis* and *P. coffeae* are associated with the blackhead toppling diseases of bananas.

**(c) Management**

**Intercropping:** A significant reduction in *Pratylenchus* sp. population (85%) was observed in the banana field where *Tagetes erecta* was grown as intercrop. The yield of the plants significantly increased (12 kg/plant) when intercropped with *Tagetes* spp. compared with the untreated control (7 kg/plant) (Sundararaju *et al.*, 2002). Intercropping with sunnhemp in banana field was found effective in reducing *P. coffeae* population by 38.4%, followed by marigold and cowpea which recorded a 29.0 and 22.3% reduction, respectively. Rice or green gram after banana suppressed the population of *P. coffeae*.

**Mulching:** Covering soil with black polyethylene, sugarcane trash and banana trash are reported to reduce *Pratylenchus* sp. in banana roots (Bhattacharya and Rao, 1984).

**Organic Amendments:** Application of distillery sludge at 2.5 kg + vermicompost 1 kg + neem cake at 1 kg + poultry manure at 2.5 kg/plant at 3, 5, and 7 months after planting significantly reduced *P. coffeae* population. It was on par with distillery sludge at 2.5 kg + neem cake at 1 kg treatment.

**(iii) Spiral Nematode, *Helicotylenchus multicinctus***

The spiral nematode is considered to be the important nematode pest in banana causing severe economic loss.

**(a) Economic Importance and Losses:** The spiral nematode causes serious decline of banana yield to the tune of 34 to 56% with delayed flowering. It is responsible for 33.83 per cent loss in yield, 55.88 per cent loss in number of fruits per bunch and delayed fruiting by 134 days (Vadivelu *et al.*, 1987).

**(b) Symptoms:** The spiral nematode incites discrete, relatively shallow, necrotic lesions on banana roots. *H. multicinctus* causes extensive root necrosis, die-back and disfunction leading eventually to debility of the entire plant.

**(c) Management**

**Crop Rotation:** Rice or green gram after banana suppressed the population of *H. multicinctus*.

**Intercropping:** Intercropping with sunnhemp in banana field was found effective in reducing *H. multicinctus* population by 38.4%, followed by marigold and cowpea which recorded a 29.0 and 22.3% reduction, respectively.

**Organic Amendments:** Application of distillery sludge at 2.5 kg + vermicompost 1 kg + neem cake at 1 kg + poultry manure at 2.5 kg/plant at 3, 5, and 7 months after planting significantly reduced *H. multicinctus* population. It was on par with distillery sludge at 2.5 kg + neem cake at 1 kg treatment.

**Organic Amendments and Bioagents:** Significant reduction in population of *H. multicinctus* was observed in banana plants treated with *Pseudomonas fluorescens* at 2g/*Glomus mosseae* at 25g along with press mud at 3 kg/plant. These treatments also enhanced the plant height, pseudo stem girth, number of leaves, leaf area and bunch weight (Jonathan and Cannayane, 2002).

### (iv) Root-knot Nematodes, *Meloidogyne incognita, M. javanica*

The root-knot nematodes, *M. incognita* and *M. javanica* attack bananas. In 1893, Cobb made one of the first associations of a root-knot species with bananas. Root-knot nematodes have not been reported to cause yield reductions in bananas, but may interact with other nematodes or soil pests. *M. incognita* caused 30.90% loss in fruit yield of banana (Jonathan and Rajendran, 2000).

**(a) Symptoms:** The infected plants were stunted, having small chlorotic leaves and galled roots (Fig. 14.5). Galling of the plant roots is most commonly found in areas previously planted with sugarcane. The galls vary in size and occur at the tips as well as in other areas along the root. Roots with galled tips cease to grow and sometimes develop secondary roots above the gall. Swollen female nematodes were found inside the galled and sometimes non-galled roots.

### (b) Management

**Biofumigation:** Bello *et al.* (2004) obtained promising levels of control of *M. incognita* and *M. javanica* in banana plantations in the Canary Islands with biofumigation using urban waste and urea, or manure and banana residues.

**Organic Amendments:** Application of distillery sludge at 2.5 kg + vermicompost 1 kg + neem cake at 1 kg + poultry manure at 2.5 kg/plant at 3, 5, and 7 months after planting significantly reduced *M. incognita* population. It was on par with distillery sludge at 2.5 kg + neem cake at 1 kg treatment.

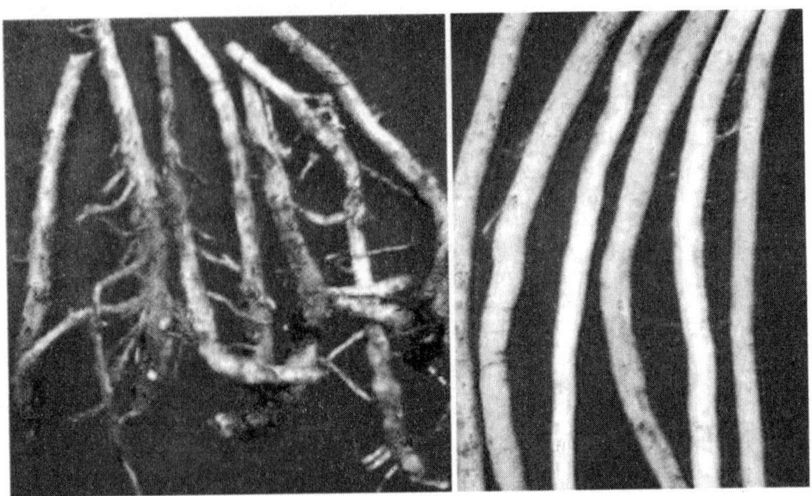

**Fig. 14.5.** Root-knot nematode on banana. Left – Infected; Right – Healthy (Courtesy: NRC on Banana, Trichy).

### (v) Cyst Nematode, *Heterodera oryzicola*

The cyst nematode is an important nematode found on banana cv. Nendran in Kerala.

**(a) Economic Importance and Losses:** *H. oryzicola* has been reported to cause considerable damage to banana in certain pockets of Kerala. An initial inoculum level of 100 and 1000 viable cysts per plant could reduce bunch weight by 20.5 to 56.6 per cent. The nematode infection was observed to affect the quality of fruits by increasing acidity and reducing total sugars (Charles, 1989).

*H. oryzicola* attacks the minute tertiary feeder roots and not observed to attack the fleshy roots and rhizome portions. The affected feeder roots turn black in colour and detach easily (Charles, 1989).

**(b) Symptoms:** The cyst nematode attacks thin lateral roots and the root weight is considerably reduced (62-87%). The damage caused to the root tissue may suppress the flow of food materials to the various parts of the plant resulting in general decline, stunting and premature defoliation.

#### (c) Management

**Organic Amendments:** Application of distillery sludge at 2.5 kg + vermicompost 1 kg + neem cake at 1 kg + poultry manure at 2.5 kg/plant at 3, 5, and 7 months after planting significantly reduced *H. oryzicola* population. It was on par with distillery sludge at 2.5 kg + neem cake at 1 kg treatment (Sundararaju *et al.*, 2002).

### (vi) Reniform Nematode, *Rotylenchulus reniformis*

The reniform nematode, *R. reniformis* attack bananas.

**(a) Economic Importance:** In field plots where *R. reniformis* was the predominant parasitic nematode, yield reduction of 19.3 tonnes per ha was observed.

**(b) Symptoms:** Nematodes feed on secondary or tertiary roots of bananas and destroy feeder roots. Necrotic lesions are produced in the area of the roots around the female head. The posterior portion of the female body and the egg mass are easily seen with a hand lens. High populations of *R. reniformis* cause severe necrosis and destruction of feeder roots.

### (c) Management

**(a) Crop Rotation:** Crop rotation with Pangola grass gives effective control of the reniform nematode on bananas.

### 14.1.3. Weeds

The major monocot weeds of banana fields are *Cynodon dactylon, Cyperus rotundus, Digitaria marginata* and *Eleusine aegyptium*, and dicot weeds are *Euphorbia* spp., *Polygonum plebejum, Portulaca oleracea* and *Mimosa pudica*.

***(i) Intercropping:*** With closer spacing of 1.6 m x 1.6 m, only one intercrop of radish followed by short duration legume like green gram is possible during the initial 3-5 months after planting, the most remunerative combination being green gram-ginger for northern zone giving a net profit of Rs. 7312 and field beans-squashes for southern zone giving Rs. 3330 from intercrops. Intercropping with cowpea drastically reduced weed growth with corresponding higher yield in Robusta banana. For Basrai banana spaced at 1.8 m x 1.8 m in medium black soil of Gujarat, the most economic intercrop was turmeric (Table 14.6). Some farmers take early cauliflower, cabbage, yam and elephant foot yam in early growth period of ratoon crop which are harvested before flowering of banana. Onion was found to be the best intercrop in Orissa. Intercropping of banana with green gram, black gram, soybean, groundnut or cowpea had no significant adverse effect, whereas intercrops gave additional returns, black gram being the best intercrop. In Kerala, intercropping of banana with vegetables such as brinjal, colocasia, yam, dioscorea, chillies and okra in wet-land is common.

Table 14.6. Intercropping schedules and profit earned in banana fields.

| Intercrops | Sowing time | Harvesting time | Crop duration (days) | Net income ( Rs./ha) |
|---|---|---|---|---|
| **Assam, Bihar, West Bengal, Uttar Pradesh and Gujarat** | | | | |
| Toria (*Brassica compestris* cv. Pusa Toria) | September | December | 70 | 1248 |
| Radish | October | November | 35 | 1200 |
| Cucumber/Pumpkin | June | September | 80 | 2200 |
| Turmeric | August | March | 240 | 6300 |
| Ginger | August | March | 240 | 6400 |
| Coriander | September | December | 95 | 3340 |
| Groundnut | June | October | 100 | 1025 |
| Bottle gourd | October | January | 100 | 2800 |
| Okra (Pusa Savani) | June | September | 105 | 1050 |
| Maize (Composite) | June | September | 102 | 1221 |
| **Andhra Pradesh, Maharashtra, Karnataka, Tamil Nadu and Kerala** | | | | |
| Field bean | September | December | 100 | 1131 |
| French bean (Contender) | September | December | 90 | 2300 |
| Groundnut | June | October | 100 | 1025 |
| French bean (Contender) | October | January | 90 | 2300 |
| Green pea | September | December | 90 | 1800 |
| Radish | October | November | 35 | 1200 |
| Pumpkin | September | December | 85 | 2200 |
| Field bean | June | September | 100 | 1131 |
| Squash | September | December | 45 | 1200 |
| Radish | November | January | 45 | 1200 |

Intercropping with ginger, cowpea, colocasia and tapioca in various varieties of banana have been reported (Mathews *et al.*, 1987). These intercrops did not have any adverse effect on the growth and yield of banana.

***(ii) Mulching:*** Mulching appears to be another effective method of weed control. Generally organic mulches like dried oak leaves and inorganic mulches like black polythene were most effective.

***(iii) Cover Cropping:*** Growing cover crops to suppress weeds is also another way of controlling weeds in this crop (Chacko and Reddy, 1981; Chundawat *et al.*, 1984).

## 14.2. CITRUS, *CITRUS* SPP.

### 14.2.1. Diseases

#### (i) *Phytophthora* Root Rot, *Phytophthora nicotianae* var. *parasitica*

**(a) Symptoms:** Profuse gumming on the surface of the attacked bark is the main symptom (Fig. 14.6). When gumming occurs on the stem, droplets of gum trickle down the stem. The bark gradually turns brown to dark brown and develops longitudinal cracks. A thin layer of wood tissue is also affected. When gumming starts close to the soil, the disease spreads to the main roots and then around the base of the trunk. As a result of severe gumming, the bark becomes completely rotten and the tree dies owing to girdling effect. The trees usually blossom heavily and die after the fruits mature. In such cases, the disease is called as foot rot or collar rot. The pathogen produces symptoms of decline through rotting of the rootlets, girdling of the trunk and dropping of the blighted leaves. The fruits lying on the ground are liable to invasion by the pathogen and develop brown rot.

Leaf fall and fruit rot phase of the disease is severe on mandarin oranges in heavy rainfall areas of south India. Quick shedding of leaves is the earliest symptom. The infection starts as water-soaked lesions at the leaf base. By the time the lesions extend to the whole leaf, the leaf drop off. The infection may spread to young twigs and fruits of all stages. The affected fruits show water-soaked patches on rind and subsequently drop off and rot.

#### (b) Management

**Biofumigation:** Biofumigation with yellow and oriental mustard cover crops reduced citrus nematode and *Phytophthora* root rot by 90%. The beneficial fungi such as *Trichoderma* have been found at much higher levels in the mustard treated soils. This does not occur when the soil is fumigated with chemicals. Mustard cover crops are now common place in the Salinas Valley, California.

**Organic Amendments and Bioagents:** *Trichoderma harzianum* has been advocated to have a potent antagonistic action against

*Phytophthora* root rot of Coorg mandarin when applied along with coffee waste, poultry manure and FYM (Sawant and Sawant, 1989).

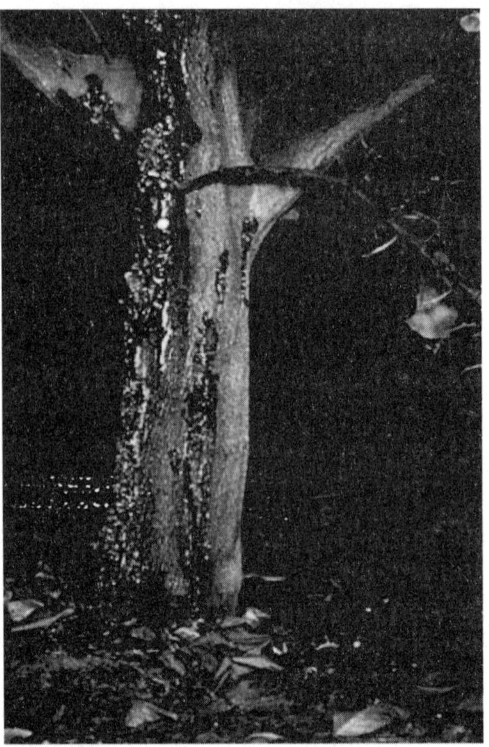

**Fig. 14.6.** Gummosis on main stem of Nagpur mandarin tree.

### 14.2.2. Nematodes

#### (i) Citrus Nematode, *Tylenchulus semipenetrans*

**(a) Symptoms:** 'Slow decline' of citrus is a diseased condition of trees with symptoms similar to those caused by drought and malnutrition. Affected trees exhibit reduced vigour (Fig. 2.34), chlorosis and falling of leaves, twig dieback and consequently, reduced fruit production (Prasad and Chawla, 1965). This decline of the tree is gradual and persists until the crop is so small that tree maintenance may become uneconomical.

As the nematode feeds on roots (Fig. 14.7) and reproduces, a large proportion of the feeder roots of citrus trees, particularly in the upper soil layers, is inactivated or destroyed, the uptake of water and minerals from the soil is reduced and the symptoms appear in the above ground tree parts. Heavily infested roots are darker in colour with branch rootlets shortened, swollen and irregular in appearance than in normal ones (Chona *et al.*, 1965).

Soil particles usually cling tightly, even after washing, to the gelatinous egg masses which cover the protruding part of the nematode body (Fig. 14.8). In heavily infested roots, the cortex separates readily from the vesicular stele.

**Fig. 14.7.** Citrus root infected with *Tylenchulus semipenetrans.*

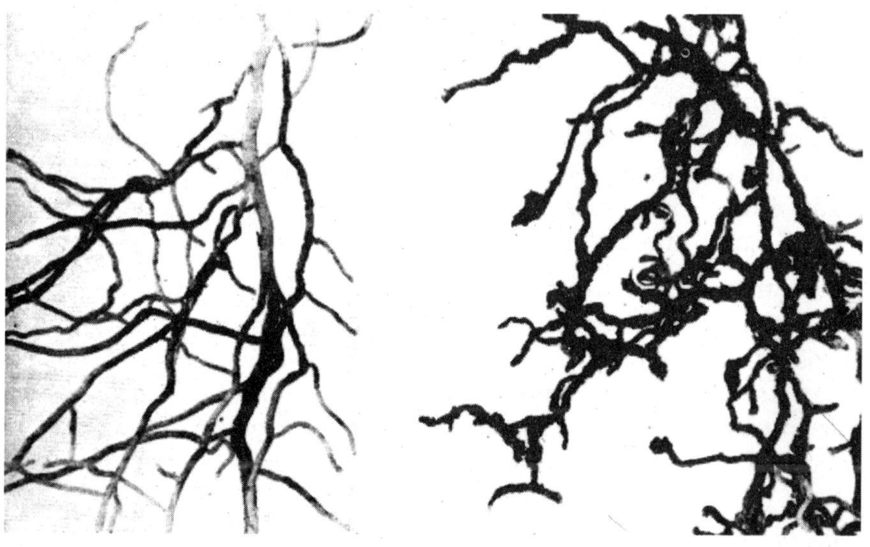

**Fig. 14.8.** Citrus root infected with *Tylenchulus semipenetrans.* Left – Healthy, Right – Infected.

## (b) Management

**Intercropping:** Interculture of acid lime with marigold and mustard markedly reduced the rate of multiplication of *T. semipenetrans* to 10.03 and 8.59, respectively (Mani, 1988). Interculture of onion, garlic or marigold in the citrus orchards not only reduces the nematode population but also provides additional income.

**Biofumigation:** Castor (*Ricinus communis*), Aak (*Calotropis procera*) and Datura (*Datura strumonium*) chopped leaves applied at

20 and 40 g/kg soil in 9 sq. m. area around tree trunks can result in reduction of nematode population and increase in yield (Baghel, 1995).

Bello *et al.* (2004) obtained promising levels of control of the citrus nematode *T. semipenetrans* in an uprooted orange orchard in Spain with biofumigation using urban waste and urea, or manure and banana residues.

**Biofumigation and Solarization:** The effect of tissue mulching and irrigation on survival of the citrus nematode, *T. semipenetrans* was measured. Oriental mustard (*Brassica juncea*), yellow mustard (*Sinapis alba)*, faba bean (*Vicia faba*) and a mixture of cereals (rye and triticale) were chopped with a mower at late flowering and sprinkler-irrigated with 30 mm of water within 0.5 hour. Any organic material reduced citrus nematode survival. Nematode suppression was 92% greater after oriental and yellow mustard (except yellow mustard/ plastic) than after cereal or legume, indicating mustard allelochemicals possibly suppressing nematodes.

Biofumigation with yellow and oriental mustard cover crops reduced citrus nematode and *Phytophtora* root rot by 90%. The beneficial fungi such as *Trichoderma* have been found at much higher levels in the mustard treated soils. This does not occur when the soil is fumigated with chemicals. Mustard cover crops are now common place in the Salinas Valley, California.

**Organic Amendments and Bioagents:** Management of the citrus nematode, *T. semipenetrans* based on the application of neem cake/castor cake each at 10 g per plant along with an antagonistic fungus *Paecilomyces lilacinus* at 4 g (grown on paddy seeds) per plant was found to be extremely effective in reducing the citrus nematode population with a consequent increase in the growth of acid lime seedlings (Parvatha Reddy and Khan, 1991).

*Trichoderma harzianum* in combination with neem/karanj/castor oil cake was effective in increasing the growth of acid lime seedlings and reducing the citrus nematode population both in soil and roots. The parasitization of citrus nematode females with *T. harzianum* increased in the presence of the oil cakes (Parvatha Reddy *et al.*, 1996).

### (ii) Root-knot Nematodes, *Meloidogyne javanica, M. indica*

Whitehead (1968) reported *M. indica* on citrus from India (Fig. 14.9). Mani (1986) reported that *M. javanica* caused severe crop loss of acid lime in Andhra Pradesh and it was pathogenic to acid lime and sweet orange (Fig. 14.10). It is confined to only coastal Andhra Pradesh.

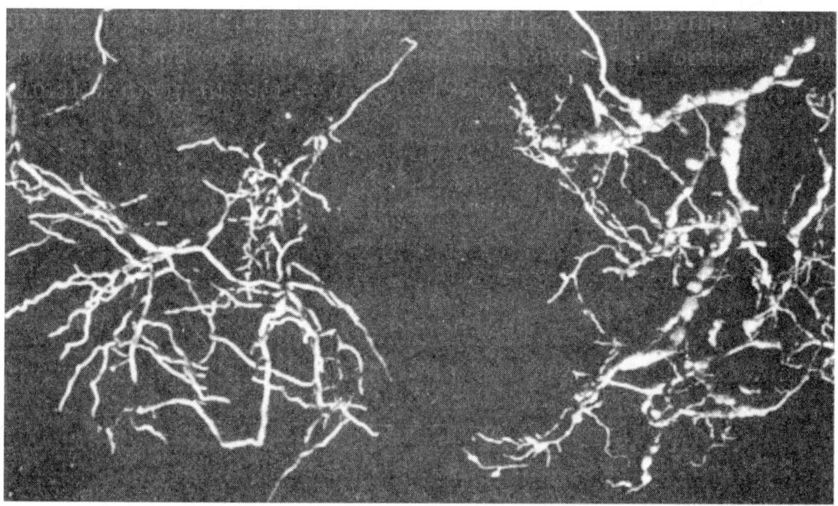

**Fig. 14.9.** Acid lime roots infected with *Meloidogyne indica*. Left – Healthy roots; Right – Infected roots.

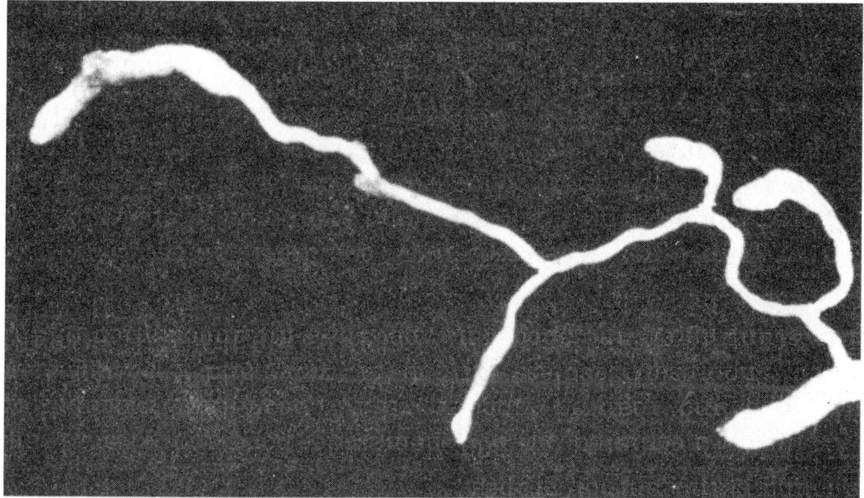

**Fig. 14.10.** Tomato roots infected with *Meloidogyne javanica*.

**(a) Symptoms:** *M. javanica* infected trees are poor in vigour, unthrifty in appearance and show severe stunted growth. They fail to flower and produce fruits even after several years of planting. The roots have conspicuous galls on pioneer and fibrous roots. In advanced stage, large cavities can be observed in place of galls. Egg masses can be seen as thin films spread over the root surface. Nematode infestation gets aggravated if vegetables like okra, brinjal, cucurbits, tomato and tobacco are grown as intercrops in orchards or as rotational crops in nurseries (Mani, 1986).

### (b) Management

**Intercropping:** Growing sunnhemp in tree basin reduced the population of root-knot nematodes. *Crotalaria* roots exert a toxic effect on *Meloidogyne* population. *Meloidogyne* larvae freely entered roots of *Crotalaria* but failed to survive. The possibility of growing *Crotalaria* in tree basin for the control of root-knot nematodes infecting citrus needs further investigation.

**Crop Rotation:** Once the life of the citrus orchard comes to an end, it is highly preferable to grow cereals and millets for at least 2-3 seasons before initiating new orchard in the same field. This will help in reducing the populations of *Meloidogyne* species and *T. semipenetrans* which are common in citrus orchards.

### 14.2.3. Weeds

The predominant monocot weeds of citrus fields are *Cynodon dactylon, Cyperus* spp., *Digitaria marginata, Eleusine indica, Setaria* spp. and *Imperata cylindrica, and* dicot weeds are *Amaranthus caturus, Bidens pilosa, Lagasca mollis, Acanthospermum hispidium, Euphorbia* spp., *Borrreria articularis* and *Evolvulus alsinoides.*

**(i) Intercropping:** Growing cover crops to reduce the density of weeds in the inter-rows is popular. Intercropping also brings more money to the grower and improves the health of the trees (Singh *et al.*, 1961). For summer season, vegetables like tinda, bottle gourd, bitter gourd, onion, chillies, green gram, cowpea and cotton are recommended for growing in 3 to 4 years old citrus orchards. In winter season, vegetables like pea, turnip, cauliflower, carrot, radish and chickpea are grown. Bajwa *et al.* (1978) grew even wheat and maize as intercrops in citrus orchards. Intercropping of citrus with green manure crops is helpful in maintaining organic matter content and moisture in soil besides improving N content to a great extent (Aiyappa *et al.*, 1970). Growing of vegetables and fodder crops like bean, carrot, tomato, berseem and senji (*Melilotus parviflora*) in orchards near cities, whereas in those located away from the market, onion, potato, chilli, pulses and kohl rabi for *rabi* season and bottle gourd, bitter gourd and okra during *kharif* season were recommended. In soils deficient in organic matter and N content, berseem, senji, chickpea, pea, cluster bean and cowpea were suggested. Cost-benefit analysis of some intercropping systems revealed that potato was more profitable giving Rs. 42,000 net profit/ha followed by cabbage (Rs. 20,000/ha) and pea (Rs. 16,5000/ha) compared with no intercropping (Rs. 12,516/ha).

***(ii) Cover Cropping:*** Cover cropping with leguminous forage crops like *Trifolium alexandrium* and *Phaseolus trilobus* was found highly profitable as short duration crops. For longer periods, lucerne was of some value. *Stylosanthes* cover crop had no harmful effect on kinnow and lemon trees and gave 21.2 t/ha fodder.

***(iii) Mulching:*** Mulching with alkathene films, wheat bran, rice husk or paddy straw also helps in keeping the weeds under check.

## 14.3. PAPAYA, *CARICA PAPAYA*

### 14.3.1. Weeds

The important monocot weeds that occur in papaya field are *Cynodon dactylon, Cyperus rotundus*, *Eleusine indica, Chloris barbata* and *Digitaria marginata* and dicot weeds are *Parthenium hysterophorus, Lagasca mollis, Croton bonplandianum, Euphorbia geniculata, Phyllanthus niruri* and *Evolvulus alsinoides.*

***(i) Mulching:*** The old dried leaves and other organic materials are used for mulching in the basins of papaya.

***(ii) Cover Cropping*:** Growing cover crops in papaya orchards is another way of controlling weeds.

***(iii) Intercropping*:** Low growing vegetables like chillies, onion, cabbage, cauliflower, radish and tomatoes can be grown as intercrops for about 6 months with advantage (Cheema and Dani).

## 14.4. PINEAPPLE, *ANANAS COMOSUS*

### 14.4.1. Nematodes

### (i) Root-knot Nematode, *Meloidogyne incognita*

**(a) Economic Importance and Losses:** *M. incognita* has been found on pineapple plants in various parts of the country. Losses for the first crop were greater than 40% due to the attack of root-knot nematodes.

**(b) Symptoms:** The root growth is greatly slowed down. The nematode also reduces plant weight, leaf weight and fruit weight by 47.5%. The root infested simultaneously by several larvae stops the growth and may produce many secondary roots.

### (c) Management

**Crop Rotation:** Crop rotation is important to avoid build-up of root knot and other nematodes that contribute to large crop losses.

Some crops usually included in rotation with pineapples are groundnut and rice.

Yield increases have been observed when planting was done in a field, which was previously had pigeonpea, owing to decrease in infestation by *M. incognita*. Crop rotation with pangola grass resulted in elimination of *M. incognita* after 1 year.

Planting of pineapple after pangola grass reduced the population of *M. incognita, Criconemoides* spp. and *Helicotylenchus* spp. and increased the fruit yield in the planted crop (23.5 MT/acre as against 15.1 MT/acre in control) and first ratoon crop (18.4 MT/acre as against 9.3 MT/acre in control) (Ayala *et al.*, 1967).

**Green Manuring:** To prepare the land used for pineapple production, green manure plants like cowpea, can be grown and incorporated into the soil prior to planting pineapples. To be effective, green manure crops known to reduce or eliminate root knot nematode infestation should be planted between pineapple crops.

**Biofumigation:** The addition of chopped pineapple material to soil at 125-375 tonnes/ha significantly reduced galling caused by root-knot nematode.

### (ii) Reniform Nematode, *Rotylenchulus reniformis*

The reniform nematode is a major problem in pineapple throughout the world. Heavy infestations combined with moisture stress can result in complete ratoon failures.

**(a) Economic Importance and Losses:** The reniform nematode can reduce the pineapple plant populations by more than 83% in the fields.

**(b) Symptoms:** Leaves of infected plants are less erect than those of healthy plants, are reddish in colour and show poor growth. Heavy infestations may result in plant collapse and death. Reniform nematode infection inhibits secondary root formation and root systems are poorly developed. Improper management of reniform populations typically leads to ratoon crop failures.

#### (c) Management

**Crop Rotation:** Rotation of pineapple with French marigold (*Tagetes patula*), Rhodes grass (*Chloris guyana*), sunnhemp (*Crotalaria juncea*), *Desmodium unicatum*, sugarcane or Pangola grass were effective in controlling *R. reniformis*. Sugarcane is generally considered a non-host for *R. reniformis*, so planting pineapple into sugarcane soils may decrease problems with the nematode, provided that weed hosts are not present.

Planting of pineapple after pangola grass reduced the population of *R. reniformis* and increased the fruit yield in the planted crop (23.5 MT/acre as against 15.1 MT/acre in control) and first ratoon crop (18.4 MT/acre as against 9.3 MT/acre in control) (Ayala *et al.*, 1967).

### 14.4.2. Weeds

The predominant weeds are Cynodon dactylon, Cyperus spp. and Digitaria marginata among monocots, and Lagasca mollis, Portulaca oleracea and Mollugo pentaphylla among dicots.

***(i) Mulching:*** Mulching with black polythene is a very effective method of weed control. Use of black polythene 150 gauge is recommended as it helps to maintain high soil temperature, retain moisture and controls weeds.

***(ii) Cover Cropping:*** A ground cover of sunnhemp (*Crotalaria juncea*) was planted in the organic production block at Maui Pineapple. Sunnhemp was planted either in rows or broadcast and incorporated with light disking. Weeds were allowed to grow in an area adjacent to the sunnhemp planting and represented the current intercrop practice referred to as a weedy fallow (WF). The sunn hemp was harvested for seed and the residues were incorporated. Pineapple was planted and weed counts were recorded.

Dry weight accumulation of weeds was reduced by 67% in broadcast seeded sunnhemp and 24% where sunnhemp was seeded in rows. These observations indicated that incorporated sunnhemp residues might be able to inhibit weed growth. Allelopathic effects may be responsible for reduced weed growth. It appears that sunnhemp and Dwarf Essex are promising cover crops for weed suppression during the pineapple inter crop cycle.

## 14.5. SAPOTA, *ACHRAS SAPOTA*

### 14.5.1. Weeds

The predominant weeds of sapota fields are *Setaria glauca, Digitaria marginata, Andropogan* sp., *Heteropogon contortus* and *Cymbopogan caesius* among monocots, and *Bidens pilosa, Blumea* sp. and *Oxalis corniculata* among dicots.

***(i) Intercropping:*** Intercropping with banana, papaya, pineapple, cocoa and vegetables like cowpea, French beans, tomato, brinjal, cabbage, cauliflower, cucurbits and peas in inter rows and turmeric and ginger in basins may also be taken up for the first 6 to 10 years to suppress weeds.

## 14.6. LITCHI, *LITCHI CHINENSIS*

### 14.6.1. Weeds

***(i) Intercropping:*** Growing of intercrops like legumes during pre-bearing stage would bring more income to the grower besides improving the health of trees and suppression of weeds. In Bihar, intercrops like turmeric, ginger, pointed gourd, sweet potato, tomato, radish, cabbage, turnip and brinjal have been suggested. In young litchi orchards, papaya and banana are suggested as fillers; cucurbits like pumpkin, cucumber, ridge gourd, bitter gourd and leguminous crops like green gram, black gram and cowpea during summer; and radish, beet root, turnip, cauliflower and carrot for *rabi* season are recommended.

## 14.7. MANGO, *MANGIFERA INDICA*

### 14.7.1. Weeds

The major monocot weeds that occur in mango orchards are *Cynodon dactylon* and *Cyperus rotundus*, and dicot weeds are *Bidens pilosa, Tridax procumbens* and *Phyllanthus maderaspatensis*.

***(i) Intercropping:*** Cover cropping with sunnhemp in light soils and sesbania in heavy soils is a practice followed in mango orchards to suppress the growth of weeds, to bring additional income to the grower until the trees begin to bear and to improve the health of trees if the intercrops grown are of right type (Singh *et al.*, 1961). The recommended intercrops in mango orchards for summer season are bottle gourd, bitter gourd, onion, chillies, cowpea, black gram and green gram. For winter season, the intercrops suggested are peas, turnip, cauliflower, carrot, radish and gram.

## 14.8. GRAPEVINE, *VITIS VINIFERA*

### 14.8.1. Diseases

#### (i) Gray Mold, *Botrytis cinerea*

The disease occurs in the entire grape growing regions of the world. It is one of the principal causes of post-harvest spoilage in storage. The pathogen is capable of growing at low temperatures.

**(a) Symptoms:** In early stage, tissue just beneath the surface of fruit is infected loosening the skin from the flesh. The affected area turns light brown. The fungal infection advances into the inner flesh resulting in a soft watery mass of decayed tissue. Under moist atmosphere, the fungus sporulates on the surface of the fruit and the typical powdery grey mold stage becomes evident. Infected fruits shrivel and turn dark brown. The disease starts in midseason and continues to develop until harvest in the absence of rain. The fungus infects stigma and style and becomes latent in the necrotic stigma and style tissues at the stylar end of the berry. In a compact infected bunch, fruits inside may split during growth. Infection of such fruits results in bunch rot. Even a single field infected berry may cause 'nest rot' in transit and storage.

**(b) Management**

**Biofumigation with *Muscodor albus*:** Biofumigation with the fungus *Muscodor albus* controlled post-harvest gray mold of table grapes caused by the fungus *Botrytis cinerea*. Post-harvest biofumigation of table grapes with *M. albus* is a flexible approach to manage gray mold because it is compatible with the various phases of the handling process, cold storage, packaging, and shipment of the fruit. *M. albus* was never in direct contact with grapes during biofumigation. It was effective when 10 or 20g of *M. albus* grain formulation per kilogram of grapes was placed inside clamshell containers, as well as within the liner outside the perforated grape cluster bags. It was slightly more effective at ambient room temperature than at 0.5°C. The treatment can be applied by simply placing active *M. albus* formulations within packages of grapes. Biofumigation with the natural volatiles from *M. albus* may be promising to develop into a product for commercial use to prolong the post-harvest life of table grapes instead of sulfur dioxide gas-generating pads now used for this purpose.

### 14.8.2. Nematodes

**(i) Root-knot Nematodes, *Meloidogyne spp.***

Two species of root-knot nematodes, *M. javanica* and *M. incognita* are recognized as the major pests of grapes causing economic damage. In India, *M. javanica* is most prevalent in northern part of country (Baghel *et al*., 1980) and *M. incognita* in southern part (Darekar and Patil, 1985). *M. incognita* is also reported from some parts of Haryana. *M. incognita* was responsible for 55% loss in fruit yield of grapes

(Rajagopalan and Naganathan, 1977), while *M. javanica* caused 53% loss in yield (Baghel and Bhatti, 1983).

**(a) Symptoms:** The root-knot nematode infestation is not manifested by any typical above-ground symptoms. Patches of poorly branched vines with scant foliage, pale and small leaves, and poor bearing are the indications of root-knot nematode damage. In young plants, premature decline, weak vegetative growth is commonly associated with nematode attack. The visibly unthrifty growth is generally attributed to moisture stress, low fertility, nutritional deficiency and other adverse conditions. However, the confirmation of nematode attack is possible by assaying soil and root samples. The root system shows typical localized swellings particularly on feeder roots and young secondary roots and females may be found on internodal trunk just below the ground level. *Meloidogyne incognita* has been reported to stimulate the production of many new fine rootlets above the site of nematode infection resulting in "hairy root" condition. Depending upon the variety of grape, *M. javanica* forms galls of varying size and shape and distorts the normal appearance of roots.

### (b) Management

**Intercropping:** *Tagetes patula* used as an intercrop reduced 39-44% nematode population and significantly increased vine yield. Under field conditions, 27.0 to 38.7% soil population and 30.1 to 35.7% root population of nematode on grapevine was reduced when intercropped with marigold, asparagus or sunnhemp. Marigold was found most effective (Baghel and Gupta, 1986). Intercropping with marigold, asparagus or sunnhemp gave effective control and resulted in increased fruit yield. Cultivation of onion or garlic in the basin area brings additional remuneration besides reducing the nematode population.

### (ii) Dagger Nematode, *Xiphinema index*

Dagger nematodes are ectoparasites and feed on succulent tissues of young roots. *X. index, X. diversicudatum* and *X. americanum* have been found pathogenic on grapes.

*X. index* inoculated at 500 nematode per vine altered auxin relationship and reduced vine growth. Concomitant inoculation of grape cv. Thompson Seedless with 500 *X. index* and *P. vulnus* caused greater stunting of vine shoot and root than did inoculation of either nematodes alone. Individually *P. vulnus* caused greater stunting than *X. index*.

**(a) Symptoms:** Dagger nematodes feeding on grape roots result in terminal swelling, cessation of root elongation and distortion due to malformation of rootlets (Fig. 14.11). Discolouration, decay of roots and death of growing points of feeder roots occur at later stage.

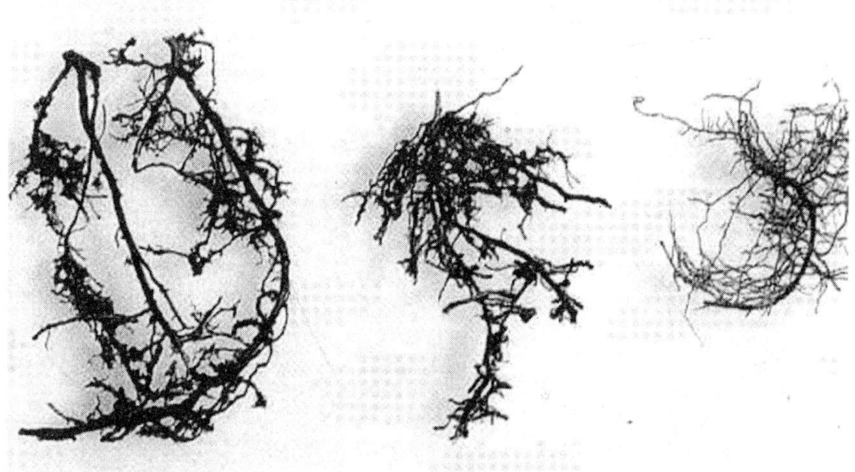

**Fig. 14.11.** Dagger nematode (*Xiphinema index*) damage to grapevine roots. Left– Infected, Right – Healthy. (Courtesy: Union Carbide Agril. Products Co. Inc., 1986).

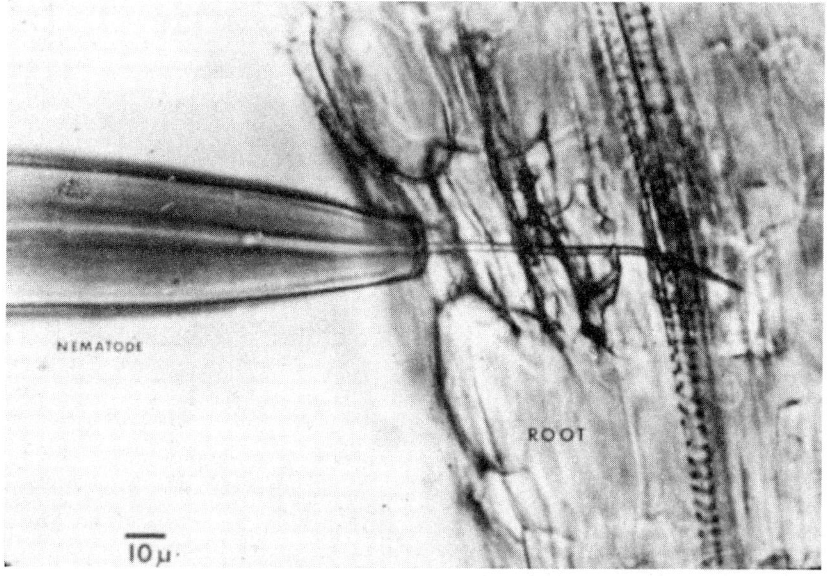


**Fig. 14.12.** *Xiphinema index* feeding on grapevine root.

Fan leaf is caused by a virus that also causes yellow mosaic and vein banding. All three are transmitted by the nematode *X. index*.

Leaf symptoms that resemble a fan are very conspicuous on Mission and French Colombard (Fig. 14.13). Infected vines have shortened and more irregular internodes. Lateral sprout development, double nodes, and stem fasciations cause a bushy appearance. Many berries shatter, and others do not develop beyond shot size. The disease affects berry set in grapes and can result in yield reductions of up to 80%, making it a highly important disease from an economic perspective.

**Fig. 14.13.** Grapevine infected with fan leaf virus disease.

**(b) Transmission:** *X. index* transmits the fan leaf yellow mosaic vein banding virus (GVFV-GVYMV) disease of grape. Leftover roots of dead and old plants in vineyards act as source of viral inoculum of dagger nematode year after year. *X. americanum* has been found transmitting tobacco mosaic virus, grape yellow vine virus, peach rosette mosaic virus and tomato ring spot virus.

**(c) Management**

**Cover Crops:** Results from vineyard with a high incidence of grapevine-fanleaf virus and its vector nematode species *Xiphinema index* (Bello *et al.*, 2004), suggested that biofumigation may be useful to reduce the fallow period necessary to eliminate vector nematodes in uprooted vineyards.

In the fall, the seed bed is planted with Brassica crops high in glucosinolate. In the spring, Brassicas are green manured 4 to 6 inches deep in the soil. A follow-up irrigation cycle speeds up the Brassica decomposition and aids microbes in the decomposition process. Gases released from the plant material can kill nematodes, weed seeds, and other disease organisms.

Vineyard test results are yielding healthier vines and fewer dagger nematodes. The holistic farming plan is showing good vine growth, low or no presence of dagger nematodes, and noticeably improved soil structure. Biofumigation is elevating soil health while providing the vine with nutrients for improved pest tolerance and resistance.

**Crop Rotation:** In France, to save grapevines from attack by *X. index* and subsequently infection by grapevine fan leaf virus, a 7-year rotation with cereal and alfalfa crop has been recommended (Lamberti, 1981).

### 14.8.3. Weeds

The common weeds present are *Polygonum plebjum, Euphorbia geniculata, Amaranthus viridis, Portulaca oleracea, Oxalis* sp. and *Mollugo pentaphylla* among dicots, and *Cynodon dactylon, Cyperus rotundus, Digitaria marginata, Eleusine indica* and *Setaria glauca* among monocots.

***(i) Mulching:*** Mulching with *Sarkanda*, wheat and rice straw, wheat bhusa, sugarcane trash, dried leaves, saw dust and black polythene film (Muthuswamy and Thamburaj, 1964) have been advocated for conserving soil moisture, for fostering rapid growth of plants and suppression of weeds (Dhillon and Bajwa, 1976; Teotia and Srivastava, 1971). *Sarkanda* was found to be the best mulch for grape in Punjab and black polythene in Tamil Nadu.

***(ii) Cover Cropping:*** Cover cropping is another way of reducing weed growth in vineyards. The cover crops grown in grape orchards are cowpea, French beans, cucurbits, sunnhemp and green gram.

***(iii) Intercropping:*** Intercropping with some vegetables has been suggested during the first 2 years after planting to keep weeds under check.

## 4.9. MULBERRY, *MORUS* SPP.

### 14.9.1. Nematodes

#### (i) Root-knot Nematodes, *Meloidogyne* spp.

Govindaiah *et al.* (1991) reported 11.8% loss in herbage yield of mulberry due to *M. incognita.*

### (a) Management

**Intercropping:** Intercropping of mulberry with marigold at wider spacing of 30 to 45 cm controlled *M. incognita*, increased leaf moisture and leaf yield of mulberry (Govindaiah *et al.*, 1990).

## 14.10. GAUVA, *PSIDIUM GUAJAVA*

### 14.10.1. Weeds

The problematic weeds of guava fields are *Cynodon dactylon* and *Cyperus rotundus* among monocots, and *Bidens pilosa, Tridax procumbens, Acanthospermum hispidium* and *Lagasca mollis* among dicots.

***(i) Intercropping:*** Intercropping with cauliflower, pea, French bean and senji in *rabi* season; cowpea, cluster bean, black gram and green gram during *kharif* season and cucurbits during summer have been recommended. Okra, onion, garlic, turmeric, cabbage, cauliflower, chilli and leafy vegetables can also be grown as intercrops in guava orchards. Under dryland conditions, *Stylosanthes* cover crop gave additional fodder yield of 21.2 t/ha.

***(ii) Mulching:*** The dried leaves are usually mulched into the soil in the basins soon after one or two rains. The growing of green manure crops like cowpea, cluster bean, sunnhemp and dhaincha during rainy season and clean cultivation during the rest of the year is recommended.

Mulching with black polythene and sugarcane trash reduced weed intensity of guava + amla cropping system. Maximum reduction in weed population (14.5/m$^2$) and their dry weight (33.8 g/m$^2$) was noticed in black polythene mulching (Table 14.7). Among the organic mulches, the minimum weed population was recorded in sugarcane trash (29.4/m$^2$) followed by paddy straw (30.8/m$^2$). The reduction in weed population was due to elimination of light intensity and regulation of soil temperature.

Table 14.7. Effect of various mulches on weed population under guava + amla cropping system.

| Mulching material | No. of weeds/m$^2$ | Dry wt. of weeds ( g/m$^2$) |
|---|---|---|
| Black polythene | 14.5 | 33.8 |
| Paddy straw | 30.8 | 64.8 |
| Sugarcane trash | 29.4 | 59.3 |
| Dried lemon grass | 116.8 | 89.3 |
| Unmulched | 101.7 | 310.2 |

## 14.11. APPLE, *PYRUS MALUS*

### 14.11.1. Diseases

#### (i) Nursery Diseases (White Root Rot, Collar Rot, Hairy Root and Crown Gall)

#### (a) Management

**Soil Solarization, Organic Amendments and Bioagents:** Soil solarization in combination with organic amendments and biocontrol agents in general gave good control of soil-borne diseases of apple in nursery. However, cent percent control of all diseases was achieved in plots where *T. viride* or *T. harzianum* or *Bacillus* sp. was incorporated in combination with organic amendments (deodar needles, neem cake). Collar rot pathogen was highly sensitive to the biocontrol agents and organic amendments both in solarized and unsolarized soil. The incidence of hairy root and crown gall was also generally reduced under these treatments (Table 14.8).

Table 14.8. Effect of soil solarization, organic amendments and biocontrol agents against soil-borne diseases in apple nursery.

| Treatment | White root rot (%) | Collar rot (%) | Hairy root (%) | Crown gall (%) | Average disease (%) |
|---|---|---|---|---|---|
| SS | 1.15 | 0.00 | 2.25 | 4.00 | 1.85 |
| SS + NC | 0.00 | 0.00 | 4.15 | 1.50 | 1.41 |
| SS + DN | 3.75 | 0.00 | 2.00 | 0.50 | 1.56 |
| SS + Tv | 0.00 | 0.00 | 0.00 | 2.50 | 0.62 |
| SS + Th | 0.00 | 0.00 | 0.00 | 0.00 | 0.00 |
| SS + Ba | 0.00 | 0.00 | 0.00 | 0.00 | 0.00 |
| SS + NC + Tv | 0.00 | 0.00 | 0.00 | 0.00 | 0.00 |
| SS + NC + Th | 0.00 | 0.00 | 0.00 | 0.00 | 0.00 |
| SS + DN + Tv | 2.38 | 0.00 | 1.75 | 0.00 | 1.03 |
| SS + DN + Th | 0.00 | 0.00 | 0.00 | 0.00 | 0.00 |
| SS + NC + DN + Tv | 0.00 | 0.00 | 0.00 | 0.00 | 0.00 |
| SS + NC + DN + Th | 0.00 | 0.00 | 0.00 | 0.00 | 0.00 |
| US + NC | 4.50 | 1.00 | 7.00 | 3.00 | 3.87 |
| US + DN | 7.75 | 0.75 | 9.45 | 6.55 | 6.12 |
| US+ Tv | 1.50 | 0.00 | 6.75 | 3.95 | 3.05 |
| US+ Th | 1.43 | 0.00 | 5.85 | 4.15 | 2.85 |

| | | | | | |
|---|---|---|---|---|---|
| US + Ba | 2.01 | 0.00 | 3.15 | 5.25 | 2.60 |
| US + NC + Tv | 0.00 | 0.00 | 0.50 | 0.00 | 0.12 |
| US + NC + Th | 0.00 | 0.00 | 0.00 | 1.50 | 0.37 |
| US + DN + Tv | 2.36 | 0.00 | 3.75 | 1.75 | 1.96 |
| US + DN + Th | 2.27 | 0.00 | 4.25 | 2.15 | 1.09 |
| US + NC + DN + Tv | 0.00 | 0.00 | 2.05 | 0.05 | 0.52 |
| US + NC + DN + Th | 0.00 | 0.00 | 1.00 | 0.00 | 0.25 |
| US | 22.82 | 2.05 | 16.00 | 8.00 | 12.21 |

SS = Soil Solarization, NC = Neem Cake, DN = Deodar Needles, Tv = *Trichoderma viride*, Th = *Trichoderma harzianum*, Ba = *Bacillus* sp., US = Unsterilized Soil.

### (ii) Root Rot, *Rhizoctonia solani*

#### (a) Management

**Biofumigation:** A low glucosinolate content (21.8 micromoles/g) rape seed meal (RSM) (*Brassica napus*) applied to orchard soils altered communities of both pathogenic and saprophytic soil micro-organisms. RSM amendment reduced infection by native and introduced isolates of *Rhizoctonia* spp. from apple roots. Root infection by *R. solani* AG-5 was also suppressed in split-root assays where a portion of the root system was cultivated in RSM-amended soils and the remainder grown in the presence of the pathogen but lacking RSM. *R. solani* hyphal growth was not inhibited by RSM amendment. Thus, glucosinolate hydrolysis products did not appear to have a significant role in the suppression of *Rhizoctonia* spp. obtained via RSM amendment. RSM amendment elevated populations of *Pythium* spp. and of ammonia-oxidizing bacteria that release nitric oxide but suppressed fluorescent pseudomonad numbers. *Streptomyces* spp. soil populations increased significantly in response to RSM but not soybean meal (SM) amendment. The vast majority of *Streptomyces* spp. recovered from the apple rhizosphere produced nitric oxide and possessed a nitric oxide synthase homolog. The transformations in the bacterial community structure are associated with the observed control of *Rhizoctonia* root rot, with nitric oxide (NO) production by soil bacteria potentially having a role in the induction of plant systemic resistance (Cohen *et al.*, 2005).

### (iii) Replant disease

Synergistic interactions of fungi, bacteria and nematodes were found responsible for causing apple replant problem.

### (a) Management

**Biofumigation:** Field trials were established to evaluate the efficacy of brassicaceae seed meal amendments for control of the soil-borne pathogen/parasite complex inciting apple replant disease, and capacity of such treatments to enhance tree growth and yield. Multiple seed meals were evaluated with and without a post-plant mefenoxam drench. Although *Brassica juncea* seed meal did not stimulate root infection by resident *Pythium* spp., application of mefenoxam to such treated soils nevertheless stimulated tree growth, apparently resulting from the suppression of root infection by *Phytophthora*. To date, tree growth in *B. juncea* seed meal/mefenoxam treated soils has been superior to all other treatments, including pre-plant soil fumigation (Mazzola and Abi Ghanem, 2006).

### (iv) White Root Rot, *Dematophora necatrix*

**(a) Symptoms:** This is a soil-borne disease which is most important as the fungus can cause death of plants. The disease appears on underground plant parts and causes complete rotting of the roots. The fine roots are attacked first which are completely devoured and infection spreads to main root through secondary root system. Lateral roots turn dark brown and become infected with white flocculent fungus during monsoon months. Bronzing of leaves, stunted growth and size are important above-ground symptoms. Root rot affected trees are usually associated with a heavy blossom and fruiting next year. However, in succeeding years, few leaves emerge and much of the immature fruits fail to reach maturity. Infected trees often persist for 2-3 years depending upon infestation of fungus.

### (b) Management

**Soil Solarization:** Soil solarization by polyethylene tarping improved the control of white root rot of apple.

**Soil Solarization and Bioagents:** Use of antagonists like *T. viride*, *T. harzianum*, *Enterobacter aerogenes* and *Bacillus subtilis* along with soil solarization have been found to protect the plants from root rot infection.

### (v) Collar Rot, *Phytophthora cactorum*

The disease is also known as crown rot and trunk canker is universally present in all the apple growing regions of the world. On susceptible varieties, it causes extensive losses even resulting in death of apple trees within a few years.

**(a) Symptoms:** The above-ground symptoms are often confused with white root rot. The infection starts from the collar region and spreads to the under-ground parts. Bark at the soil level becomes slimy and rots resulting in cankered areas. The attacked trees are recognized by chlorotic foliage with red colouration of veins and margins.

**(b) Management**

**Soil Solarization and Bioagents:** Use of antagonists like *T. viride*, *T. harzianum*, *Enterobacter aerogenes* and *Bacillus subtilis* along with soil solarization have been found to protect the plants from collar rot infection.

### 14.11.2. Nematodes

**(i) Lesion Nematode, *Pratylenchus spp.***

**(a) Management**

**Organic Amendments:** Rape seed meal (RSM) amendment reduced recovery of *Pratylenchus* spp. from apple roots. Suppression of *Pratylenchus* was attained to an equivalent extent by amending soils with either RSM or soybean meal (SM) when applied to provide a similar N content. RSM amendment elevated populations of *Pythium* spp. and of ammonia-oxidizing bacteria that release nitric oxide but suppressed fluorescent pseudomonad numbers. *Streptomyces* spp. soil populations increased significantly in response to RSM but not SM amendment. The vast majority of *Streptomyces* spp. recovered from the apple rhizosphere produced nitric oxide and possessed a nitric oxide synthase homolog. The transformations in the bacterial community structure are associated with the observed control of *Pratylenchus* spp., with nitric oxide production by soil bacteria potentially having a role in the induction of plant systemic resistance (Cohen *et al.*, 2005).

### 14.11.3. Weeds

*Rosa moschata, R. eglantaria, Rubus* spp. and *Barberis* spp. are some of the predominant weeds of apple orchards.

***(i) Mulching:*** The use of oak leaves and black polythene mulch proved as effective as chemical treatments for the management of weeds. Oak leaves were the best for suppressing weeds. Increase in yield up to 40% was also realized in apple. Mulching with peat controlled weeds effectively. Apple trees growing in straw mulched strips of 1.5 m width made more growth and yielded twice the weight of fruit from the control plots (Halloway and White, 1968).

***(ii) Intercropping:*** Growing intercrops in rows between the trees and in the basins of apple trees suppressed weeds and supplemented the income. The suitable intercrops like tomato, cabbage, beans, strawberry, early potato, mustard, wheat, oat and to some extent barley can be taken during summer and rainy season.

***(iii) Cover cropping:*** Cover cropping can be done during rainy season with sod or leguminous crops.

## 14.12. STONE FRUITS

### 14.12.1. Diseases

#### (i) Brown Rot, *Monilinia laxa*

**(a) Symptoms:** Symptoms of the disease are blighting of blossom and leaves, canker production on woody tissues and rotting of fruits. Blossom blight is the first symptom during spring and attacked parts turn grey to dark brown. The fungus spreads through peduncle and reach branches causing twig blight. Stem cankers usually develop from blighted twigs or fruit spurs. Fruit rot is the most destructive phase of the disease. Rotted fruits either fall down or hang as firm mummies. Conidia and ascospores produced on the mummified fruits and canker spots serve as a source of primary infection (Fig. 14.14).

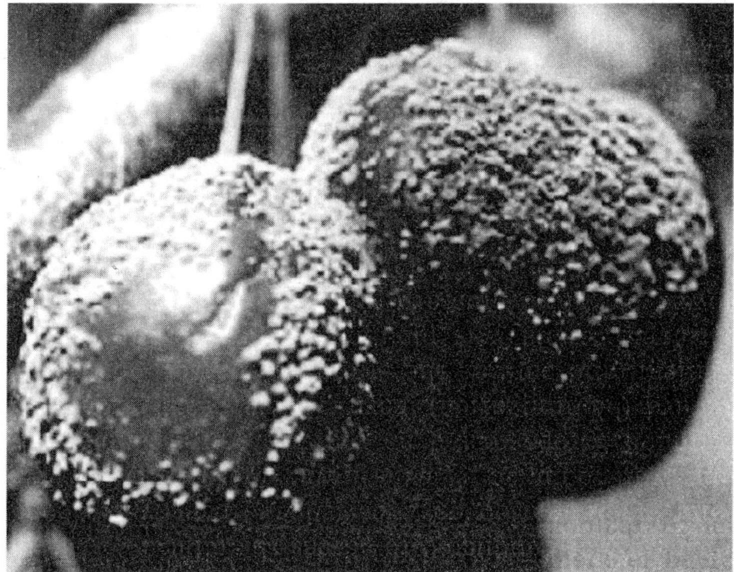

**Fig. 41.14.** Brown rot of cherry fruits.

**(b) Management**

**Biofumigation:** Allyl-isothiocyanate (AITC) from sinigrin (*Brassica carinata*), butenyl-isothiocyanate from gluconapin (*B. rapa*), 2-phenylethyl isothiocyanate from gluconasturtin (*Barbarea verna*) and 4-methyl-thiobutyl isothiocyanate from glucoerucin (*Eruca sativa*) were tested in vapour phase on inoculated stone fruits. Allyl-isothiocyanate (AITC) treatments were effective against *Monilinia laxa* on stone fruits, treated in the same conditions but at a reduced concentration. Also butenyl isothiocyanate vapours controlled the brown rot produced by *M. laxa* although with a higher initial concentration. At these concentrations the pathogens control was obtained without any phytotoxic effects on fruits. These results have led to study the best practical large scale application, using Brassicaceae meals as starting material to produce the active isothiocyanates for fruit biofumigation (Bernardi *et al.*, 2004).

## 14.13. PEACH, *PRUNUS PERSICA*

### 14.13.1. Nematodes

#### (i) Root-knot Nematode, *Meloidogyne* spp.

**(a) Symptoms:** Above-ground symptoms on peach plants infected with *Meloidogyne* spp. include-poorly developed foliage, lack of vigour, die-back of twigs and yield reduction. In periods of draught, infected trees wilt more quickly than non-infected ones. The peach grove infested with root-knot nematode is usually composed of unevenly sized trees which are due to uneven distribution of nematodes and their population variation within one field.

The root system of nematode infested trees has poorly developed secondary and tertiary branches bearing galls of varying sizes (Fig. 14.15). When many juveniles penetrate a root, root tip is devitalized and elongation of the root ceases resulting in the production of many fine tertiary branches forming brush-like appearance. Nematode damage to roots create 'suit effect' in which plant photosynthates are diverted downwards, making the nutrients available for root repair, thus impairing the tree from desired energy for the production of fruit resulting into yield loss.

**(b) Management**

**Biofumigation:** Bello *et al.* (2004) obtained promising levels of control of *M. incognita* in peach orchards with biofumigation using urban waste and urea, or manure and banana residues.

**Fig. 14.15.** Root-knot nematode on peach.

## (ii) Peach-Tree Short-Life, *Mesocriconema xenoplax*

*M. xenoplax* is ectoparasitic and economically important nematode which is one of the factors in bacterial canker and peach tree short life (PTSL) disease. The estimated losses that occurred in peach orchards suffering from PTSL in Georgia were 30-70% after five years. *M. xenoplax* was responsible for 33% loss in fruit yield of peach (Anon, 1990).

**(a) Symptoms:** Parasitism of *M. xenoplax* causes chlorosis, leaf drop, stunting and loss of vigour on top of peach trees. Peach roots showed lesions, pits and fewer feeder roots (Fig. 14.16). The roots are predisposed to bacterial canker caused by secondary infection of *Pseudomonas syringae*.

### (b) Management

**Soil Solarization:** A 37-100% reduction in population of *M. xenoplax* and other nematodes has been achieved by soil solarization in South Africa (Barberecheck and Broembsen, 1986).

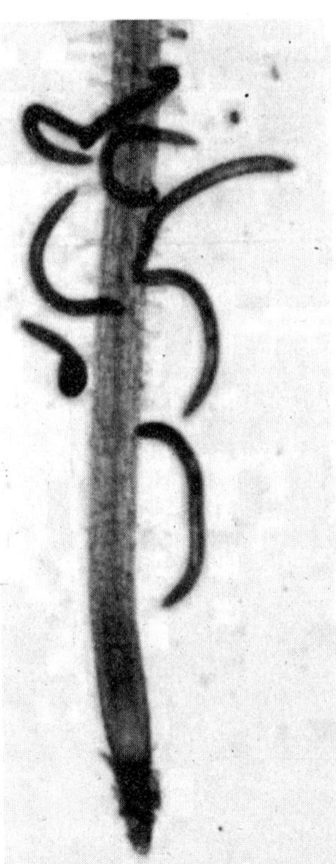

**Fig. 14.16.** Peach roots infected with *Mesocriconema xenoplax.*

**Crop Rotation:** Pre-plant crop rotation with wheat has been shown to be comparable to methyl bromide fumigation in suppressing the ring nematode (*M. xenoplax*) population prior to establishing a peach orchard.

Coastal Bermuda grass and small grains, especially wheat, are poor to non-hosts for ring nematode. Several consecutive years of growing coastal Bermuda grass or wheat on an anticipated peach orchard site may possibly reduce the necessity for using a preplant fumigant.

**Biofumigation and Solarization:** Four months (January 1999) after establishing the sorghum biofumigant and methyl bromide plots and prior to planting peach trees, ring nematode (*M. xenoplax*) populations were greatest ($P < 0.05$) in the unfumigated soil than in sorghum + plastic, sorghum w/o plastic, and methyl bromide fumigated plots. In September 1999 (12 months after incorporating the sorghum as a green manure), no differences in nematode populations were detected among the unfumigated and two sorghum treatment plots. However, nematode populations were still suppressed ($P < 0.05$) in the methyl bromide plots. At 24 months after methyl bromide application (September 2000), the nematode population density in fumigated soil did not differ from the other treatment plots.

Ring nematode populations continued to increase in subsequent sampling dates. Peach trees developed typical peach-tree short-life (PTSL) symptoms and died during this experiment. In 2001 and 2002, percentage of PTSL tree death was greater in unfumigated w/o plastic (29% and 54%) followed by sorghum + plastic (12% and 46%), sorghum w/o plastic (8% and 50%), unfumigated soil + plastic (4% and 29%), and methyl bromide (4% and 29%), respectively. No differences in trunk diameter were detected among any of the treatments. Sorghum as a green manure with and without plastic did suppress the population of *M. xenoplax* in the early stages of this experiment, but suppression did not last as long as preplant methyl bromide fumigation (i.e., 12 vs. 24 months, respectively) (Nyczepir and Rodriguez-Kabana, 2004).

### 14.13.3. Weeds

***(i) Intercropping:*** Intercropping orchards with green gram, chickpea, pea and onion reduced the density of weeds both in basins and inter rows (Jauhari and Srivastava, 1972).

After intercropping for 2 years with soybean and pineapple, the N content in peach leaves was higher. Intercropping of peach orchards with short duration crops like cowpea and sesame up to 4 years, and the long duration crops like turmeric could be grown even beyond 6 years without any adverse effects on the productivity of peach trees and an income of Rs. 10,000 could be obtained.

***(ii) Cover Cropping or Green Manuring:*** A suitable cover or green manure crop may be sown in rainy season after the fruits are picked and ploughed under during winter. Peach trees growing in straw mulched strips, 3 m wide, grew at a faster rate than the controls under permanent pasture and produced twice as many fruits of larger size in their fourth and fifth seasons. Peach under cover cropping had better physiological condition of the soil, increase in the organic matter content of the soil and improved water uptake by peach trees.

## 14.14. PEAR, *PYRUS COMMUNIS*

### 14.14.1. Diseases

#### (i) Blue Mold, *Penicillium expansum*

#### (a) Management

**Biofumigation:** Allyl-isothiocyanate (AITC) from sinigrin (*Brassica carinata*), butenyl-isothiocyanate from gluconapin (*B. rapa*), 2-

phenylethyl isothiocyanate from gluconasturtin (*Barbarea verna*) and 4-methyl-thiobutyl isothiocyanate from glucoerucin (*Eruca sativa*) were tested in vapour phase on inoculated pears. *P. expansum* was controlled by exposing pears for 8 hr in AITC-enriched atmosphere. The treatments were more effective when realized 24 hr and 48 hr after pear inoculation. AITC treatments controlled a thiabendazole-resistant strain of *P. expansum*, reducing the incidence of blue mould by 90% (Bernardi *et al.*, 2004).

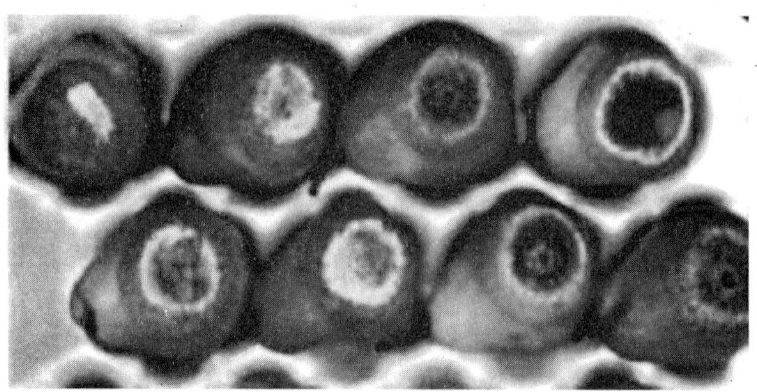

**Fig. 14.17.** Blue mold rot on pear

### 14.14.2. Weeds

***(i) Mulching:*** An opaque plastic mulch suppressed weeds, conserved moisture, increased soil temperature in cool weather and reduced them in warm weather and increased the yield in pear. Black polythene mulch controlled annual weeds well.

## 14.15. APRICOT, *PRUNUS ARMENIACA*

### 14.15.1. Weeds

***(i) Cover Cropping*:** Cover cropping the inter rows with mustard and wheat reduced the density of weeds in apricot.

## 14.16. STRAWBERRY, *FRAGARIA* SPP.

### 14.16.1. Diseases

#### (i) Root Rot, *Phytophthora cinnamoni*

#### (a) Management

**Biofumigation and Solarization:** The use of biofumigation with *Brassica carinata* and soil solarization for reducing populations of *Phytophthora* spp. in the soil and enhancing strawberry production was evaluated. Treatments were soil solarization (S), biofumigation + solarization (B+S), and the untreated control (C). B+S increased plant growth (foliar surface), fruit weight, and strawberry yield the most each year. Plant growth differences were observed relative to S and C, with foliar surface ($cm^2$) of B + S, S and C of 502, 414 and 351, and 435, 346 and 228 in January 2006 and 2007, respectively. Furthermore, S increased plant growth and strawberry yield relative to C. Fruit weight (g/fruit) of B + S, S and C was 25, 22 and 17 and 23, 20 and 17 in 2006 and 2007, respectively. In addition, both treatments reduced *Phytophthora* soil population relative to C. The current work contributes to the development and optimization of biofumigation with *Brassica* and soil solarization as alternatives to the traditional use of chemicals in strawberry production (Porras *et al.*, 2009).

**Biofumigation and Chemicals:** Maximum disease control was achieved in plots which were mulched with polythene and sprayed with ridomil MZ (74.47%) followed by paddy straw mulch and ridomil MZ sprays (64.92%). The fruit yield of strawberry also increased significantly in combined treatment of polythene mulch and ridomil MZ (97.57%) followed by paddy straw mulch and ridomil MZ (90.04%) (Table 14.9) (Bharadwaj and Sharma, 2000).

Table 14.9. Effect of fungicide alone and in combination with mulch on incidence of fruit rot and yield of strawberry

| Treatment | Disease incidence (%) | Fruit rot control (%) | Fruit yield (kg/plot) | % increase in fruit yield |
|---|---|---|---|---|
| Polythene mulch | 18.43 | 54.36 | 1.600 | 71.99 |
| Ridomil MZ | 24.67 | 38.92 | 1.258 | 35.23 |
| Paddy straw mulch | 28.02 | 30.63 | 1.056 | 13.51 |
| Polythene mulch + Ridomil MZ | 10.31 | 74.47 | 1.838 | 97.57 |
| Paddy straw mulch + Ridomil MZ | 14.17 | 64.92 | 1.768 | 90.04 |
| Control | 40.39 | --- | 0.930 | --- |
| CD (P = 0.05) | 5.025 | 14.263 | --- | --- |

### 14.16.2. Nematodes

### (i) Lesion Nematode, *Pratylenchus penetrans* and Wilt, *Verticillium dahliae* Disease Complex

**(a) Symptoms:** Infested plants are stunted, pale green, have reduced yields and eventually die. Root symptoms appear as distinct brown lesions in early or light infestation. The entire root system eventually becomes black and necrotic as populations increase and secondary organisms invade (Fig. 14.18).

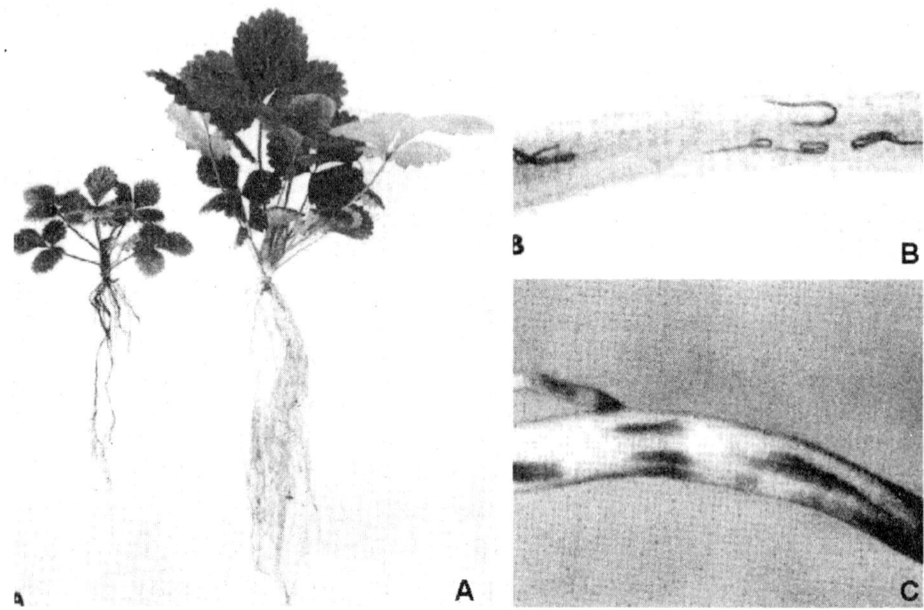


**Fig. 14.18.** The lesion nematode (*Pratylenchus penetrans*) on strawberry. A-Left, infected; Right, healthy. B-Top, Lesion nematodes inside the root; Bottom, Lesions on the root.

### (b) Management

**Crop Rotation and Biofumigation:** From 2004 and 2006, a field experiment was conducted to assess the efficacy of a 2-year crop rotation and biofumigation on *Pratylenchus penetrans* densities and Verticillium wilt damage (*Verticillium dahliae)* in strawberry in Quebec, Canada. Crops tested as rotation were: canola followed by white mustard (biofumigation), corn, forage pearl millet cvs. CFPM 101 and Tifleaf, and oats as the standard control. For the biofumigation treatment, canola was ploughed under at late bloom (end of July) followed by white mustard which was ploughed at full bloom (October).

Nematode population densities were significantly lower after biofumigation with both pearl millet cultivars than after corn and oats. In 2006, both pearl millets and biofumigation significantly reduced by an average of 21% the incidence of Verticillium wilt and allowed a 54% average increase in the development of strawberry plants cv. Jewel.

**Biofumigation and Solarization:** In 2007, a field trial was set out to investigate the potential of improving the efficacy of biofumigation by incorporating poultry manure (15 t/ha) at canola-crop ploughing time, with and without the addition of a plastic mulch. Plastic mulch significantly improved biofumigation with a 95% reduction in the number of *P. penetrans* densities when compared to canola ploughed with poultry manure but no plastic mulch (Bélair and Coulombe, 2008).

### (ii) Root-knot Nematode, *Meloidogyne hapla*

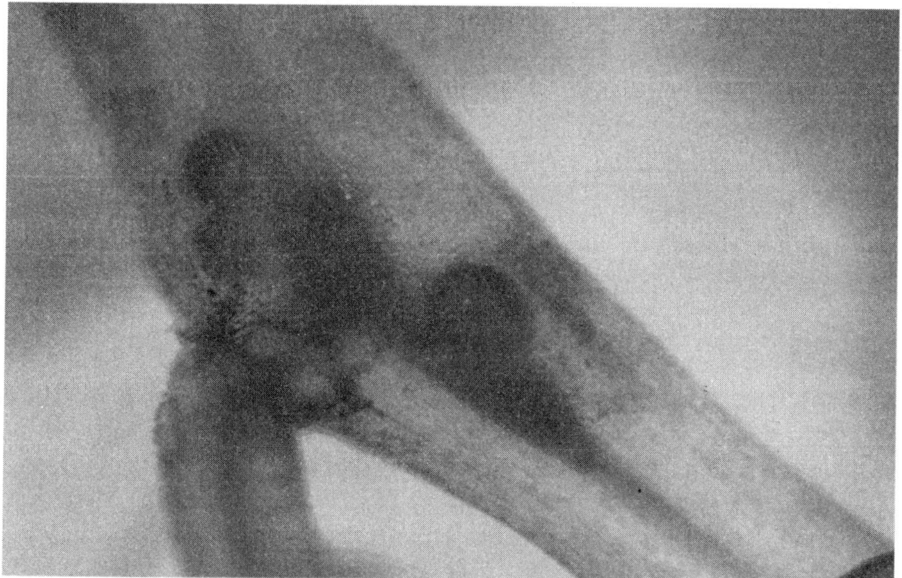

**Fig. 14.19.** *Meloidogyne hapla* females inside strawberry roots.

### (a) Management

**Solarization:** Solarization was tested in a soil with *Meloidogyne, Pratylenchus. Helicotylenchus, Tylenchorhynchus* and *Hoplolaimus* spp. (Table 14.10). In general, combinations of solarization and herbicides gave good control of nematodes and healthy plants, compared with non-solarized plots.

**Crop Rotation:** Rape and mustard plantings in rotation with strawberries have checked the increase of some nematodes (Brown and Morra, 1997).

**Biofumigation and Solarization:** Solarization + MB/Telopic or MB, were effective in controlling root-knot nematodes compared to solarization and biofumigation (Table 14.11)

Table 14.10. Effect of solarization for control of nematodes on strawberry at different localities

| Localities | Solarized | | | Non-solarized | | |
|---|---|---|---|---|---|---|
| | Av. no. of nemas/200 g soil | | Reproduction factor (RF) RF = Pf/Pi | Av. no. of nemas/200 g soil | | Reproduction factor (RF) RF = Pf/Pi |
| | (Pi) | (Pf) | | (Pi) | (Pf) | |
| Beni-Sweif | 80 | 3 | 0.04 | 115 | 130 | 1.13 |
| Fayoum | 125 | 0 | 0 | 170 | 135 | 0.8 |
| Qalyobia | 270 | 0 | 0 | 100 | 107 | 1.07 |
| Ismailia | 75 | 0 | 0 | 90 | 136 | 1.51 |

Pi = Initial nematode population; Pf = Final nematode population.

Table 14.11. Measures of disease caused in strawberry plants by *M. hapla* after different soil fumigation with non-chemical alternatives.

| Treatment | Location: Cartaya[1] | | | Location: Moguer[1] | | |
|---|---|---|---|---|---|---|
| | Incidence[2] | Severity[3] | ♀/g[4] | Incidence[2] | Severity[3] | ♀/g[4] |
| Control | 54.2 a[5] | 2.01 a | 13.5 a | 15.0 a | 0.99 a | 13.1 a |
| Solarization | 36.7 a | 1.48 ab | 5.8 ab | 8.3 ab | 1.20 a | 10.3 a |
| Biofumigation | 32.5 ab | 1.10 bc | 5.0 ab | 0.8 b | 0.25 b | 0.2 a |
| Solarization + Metham sodium | 26.7 b | 0.63 bcd | 3.6 bc | 0.0 b | 0.00 b | 0.0 a |
| Solarization + MB/Telopic | 12.5 b | 0.72 cd | 1.7 bc | 0.0 b | 0.00 b | 0.0 a |
| Methyl Bromide | 0.8 c | 0.08 d | 0.0 c | 0.0 b | 0.00 b | 0.0 a |

1: Mean of 10 plants/plot, 3 blocks/year and 4 years; 2: Percentage of disease plants; 3: Severity (0=no symptoms; 4=more than 90% of roots affected); 4: Number of females/g of roots; 5: Means followed by the same letter are not different under a MDS test, ($P<0.05$).

### 14.16.3. Weeds

Weeds like Cyperus compressus, Digitaria and Eleusine indica are common in strawberry.

***(i) Mulching:*** Mulching with black polythene or hay is an important cultural operation not only to check weed growth but also to regulate soil temperature and improve soil structure (Porter, 1962). Opaque plastic mulch suppressed weeds, conserved soil moisture, increased soil temperature in cool weather and reduced them in warm weather and increased the yields in strawberry (Thompson, 1961; Srivastava and Agarwal, 1965).

Studies conducted at the Y.S. Parmar University of Horticulture and Forestry, Nauni, Solan (Himachal Pradesh) revealed that black polythene mulching gave 85-90% weed control in strawberry.

***(ii) Biofumigation:*** Different glucosinolate-containing *Brassica* sp., the varieties: 'ISCI-99' (received from ISCI-Bologna), 'Vitasso', both *B. juncea* and 'Petranova' (*B. napus*) were evaluated for biofumigation. The strongly weed-suppressing effect was remarkable. After incorporation, the germination of common chickweed (*Stellaria media),* shepherd's purse *(Capsella bursa-pastoris)* and deadnettle (*Lamium amplexicante)* was completely suppressed in the *Brassica* plots, while in the control plots chickweed was growing widely (Fig. 14.20). The interval of 14 days between incorporation and planting did not lead to phytotoxic reactions of the strawberries.

**Fig. 14.20.** Weed suppressing effect of biofumigation for 6 weeks after treatment and 4 weeks after planting of 'Elsanta'. The germination of weeds was completely suppressed in the biofumigated plots (left). In the control plots common clickweed was occurring area wide (right).

In the field, rotary incorporation of the biofumigant *Brassica* crop (8 tonnes/ha) did not produce detectable levels of ITCs in soil; yet, it suppressed the growth of emerging weeds by 40% and the growth of *P. cactorum* by 20%. The allelochemicals other than ITCs, such as nitriles, or other biological mechanisms might also play a role in biofumigation (Mattner *et al.*, 2008).

The effect of three types of seed meal, *Brassica napus* ('Athena'), *B. juncea* ('Pacific Gold'), and *Sinapis alba*, ('Ida Gold') was evaluated for herbicidal properties in glasshouse tests. Ida Gold seed meal has significantly more herbicidal activity on wild mustard and pigweed compared to the control or the other seed meals examined. Pacific Gold did have good herbicidal activity on wild oat seeds. Indeed, it was interesting to note that in Pacific Gold treatments the wild oats did not emerge, while in the Ida Gold treatment, wild oat emerged and later died after 21 days of growth. Similarly, weed biomass was significantly reduced by incorporation of either Ida Gold or Pacific Gold seed meals.

In strawberry field trials in 2002 and 2003, Ida Gold provided significantly better weed control than other treatments and was not significantly different from the weed-free plots. Weed populations were markedly reduced in the second year crop where the strawberry plants were fully established and competed better with weeds. Again, both showed high phytotoxicity on first year strawberry transplants. However, both treatments produced yields that were not significantly lower than the chemical treatment and that were always significant yield increases over the no-treatment control. Transplanting strawberry plants at the time of seed meal amendment caused significantly greater phytotoxicity compared to where transplanting was delayed until 14 days after meal incorporation. However, yield of strawberry from transplanting immediately after meal treatment produced significantly higher fruit yield. Overall, Ida Gold and Pacific Gold both showed high potential as an alternative bio-herbicide in strawberry production. Further field testing will be required to determine efficacy rates and timing of incorporation for maximum effect and productivity (Brown *et al.*, 2004).

Field studies were conducted to compare the effects of amending 2 MT/ha of canola (*Brassica napus* L.), Oriental mustard (*B. juncea* L.) and yellow mustard (*Sinapis alba* L.) seed meal into soils on strawberry crop health and weed control, compared to untreated and weed-free controls treatments. Yellow mustard meal-amended soils had significantly greater herbicidal activity with 95% and 88% reduction in weed emergence and biomass compared to the no-treatment control. Oriental mustard and canola treatments both reduced weed counts and biomass significantly compared to the no-treatment control. Strawberry plants showed most visible phytotoxicity when transplanted into Oriental mustard amended soils with 61% more damage than the weed-free control. Fruit yield from Oriental mustard side dress treatment was significantly higher than the weed-free control, and the other meal treatments were not significantly different from the weed-free control. Results showed that seed meal treatments can be used to control weeds in strawberry production without significant yield loss due to crop phytotoxicity.

***(iii) Solarization, Organic Amendments and Chemicals:*** The effect of solarization, and its combinations with dazomet (400 kg/ha), chicken manure (10 tonnes/ha) or straw (500 kg/ha), on weeds and crop were demonstrated at various sites. In Silifke, even solarization alone gave acceptable yields but *C. rotundus* control was not satisfactory. On the contrary, in Yaltir there was very good weed control but MB gave the best yield. However, solarization with dazomet seemed to be an alternative to MB in strawberry (Table 14.12).

Table 14.12. Effect of treatments on weeds and total yield in strawberry demonstration sites.

| Treatment | SILIFKE Weed intensity/m² | | | | YALTIR Weed intensity/m² | | | |
|---|---|---|---|---|---|---|---|---|
| | Early | Late | Pot | Yield (MT/ha) | Early | Late | Pot | Yield (MT/ha) |
| Solarization | 50 | 105 | 67 | 23 | 18 | 6 | 3 | 30 |
| Solarization + Dazomet | 40 | 35 | 46 | 40 | 24 | 49 | 4 | 44 |
| Solarization + Manure | 50 | 57 | 42 | 25 | 6 | 9 | 9 | 32 |
| Methyl Bromide | 1 | 0 | 8 | 22 | NA | NA | NA | 50 |
| Untreated check | NA | NA | NA | NA | NA | 33 | 69 | NA |

NA: Not applicable

## 14.17. POMOGRANATE, *PUNICA GRANATUM*

### 14.17.1. Weeds

***(i) Mulching:*** Mulching with paddy husk was found to be effective in controlling weeds in pomegranate.

***(ii) Intercropping:*** Intercropping can be done with fodders like berseem and lucerne, pulse crops like cowpea and green gram, and vegetables like cucurbits, cabbage, cauliflower, beans, pea, tomato, carrot, onion, radish, potato and brinjal for first 5-6 years to keep weeds under check and to get additional income.

## 14.18. BER, *ZIZIPHUS MAURITIANA*

### 14.18.1. Weeds

***(i) Mulching:*** Black polythene film and rice straw mulches have been found effective during the first year of planting to conserve soil moisture and to suppress weeds (Byun *et al.*, 1991). Mulching with black polythene gave effective control of weeds like *Aerva persica* and *A. tomentosa*. Straw mulch was found superior to black polythene in keeping the soil temperature low.

***(ii) Intercropping:*** Intercrops like green gram, black gram, cluster bean and cowpea were proved to be best in western Rajasthan. Under irrigated conditions, grain cumin and chillies are recommended as intercrops.

***(iii) Cover cropping:*** Cover cropping with short duration vegetable crops or forage crops is done in between trees to keep down the weed population and to conserve soil moisture.

## 14.19. AMLA, *EMBLICA OFFICINALIS*

### 14.19.1. Weeds

***(i) Mulching:*** During summer, the crop should be mulched with paddy or wheat straw at the tree base up to 15-20 cm, which not only conserves moisture, but also helps in keeping the weeds under check. Mulching with mustard straw (8 cm layer below tree canopy) also helps in reducing the weed population.

Studies conducted in Orissa University Agriculture & Technology, Bhubaneshwar (Orissa) revealed that mulching with black polythene gave 49% increase in yield as compared to control and 32% weed control.

***(ii) Cover Cropping*:** Cowpea and cluster bean can be used as cover crops to reduce the weed population.

## 14.20. CUSTARD APPLE, *ANNONA SQUAMOSA*

### 14.20.1. Weeds

***(i) Intercropping:*** In young orchards, vegetables and legumes can be grown as intercrops for the first 5 years. Besides economic benefit, intercrops also check soil erosion and weed growth.

***(ii) Mulching:*** Mulching the tree basins with dry grasses helps to avoid soil erosion due to wind and suppress weed growth.

*Chapter 15*

# VEGETABLE CROPS

China is the largest producer of vegetables contributing 47.0% of the total world production (Table 15.1). Vegetables occupy an area of 23.936 million ha in China with a production of 448.983 million MT. Productivity of vegetable crops is highest (33.0 MT/ha) in Spain.

Table 15.1. Major vegetable producing countries of the world (2007-08) (Bijay Kumar, 2009)

| Country | Area (in '000 ha) | Production (in '000 MT) | Productivity (in MT/ha) | % share of world production |
|---|---|---|---|---|
| China | 23936 | 448983 | 19 | 47 |
| India | 7803 | 125887 | 16 | 13 |
| USA | 1333 | 38075 | 29 | 4 |
| Turkey | 996 | 24454 | 25 | 3 |
| Russia | 970 | 16516 | 17 | 2 |
| Egypt | 598 | 16041 | 27 | 2 |
| Iran | 641 | 15993 | 25 | 2 |
| Italy | 528 | 13587 | 26 | 1 |
| Spain | 379 | 12676 | 33 | 1 |
| Japan | 433 | 11938 | 28 | 1 |
| Others | 16957 | 222625 | 13 | 24 |
| Total | 54573 | 946774 | | |

Source: FAO (except Indian data)
(Indian data source: Indian Horticulture Database 2008)

India is the second largest producer of vegetables next only to China contributing 13.0% of the total world production. Vegetables occupy an area of 7.803 million ha with a production of 125.887 million MT (Bijay Kumar, 2009) (Tables 15.2 and 15.3). Vegetable crops contribute 58.93% share to the total production of horticultural crops. India produces 18.0% of the world onions. In vegetables, India occupies the first position in peas; second in cabbage, cauliflower, onion, brinjal; third position in potato and tomato. The per capita availability of vegetables has increased from 76 g in 1951 to 256 g in 2001 as against 280 g/day recommended by ICMR for a balanced diet. Increase in production is mainly due to increase in productivity. Productivity is still lower in India than the world average productivity. There is scope for increasing the productivity of vegetables in India further. The vegetable production target for 2020 has been fixed at 150 million MT.

Table 15.2. Crop-wise area, production and productivity of vegetable crops in India (2007-08) (Bijay Kumar, 2009)

| Vegetable crops | Area ('000 ha) | Production ('000 MT) | Productivity (MT/ha) | % share of India production |
|---|---|---|---|---|
| Potato | 1786 | 34463 | 19.3 | 27.4 |
| Onion | 805 | 12157 | 15.1 | 9.7 |
| Tomato | 572 | 10261 | 17.9 | 8.2 |
| Brinjal | 566 | 9596 | 17.0 | 7.6 |
| Tapioca | 270 | 9054 | 33.5 | 7.2 |
| Cabbage | 265 | 5888 | 22.2 | 4.7 |
| Cauliflower | 321 | 5797 | 18.1 | 4.6 |
| Okra | 409 | 4193 | 10.3 | 3.3 |
| Peas | 314 | 2560 | 8.2 | 2.0 |
| Sweet Potato | 126 | 1146 | 9.1 | 0.9 |
| Others | 2370 | 30774 | 13.0 | 24.4 |
| Total | 7803 | 125887 | | |

(Source: Indian Horticulture Database 2008)

Table 15.3. State-wise area, production and productivity of vegetable crops in India (2007-08) (Bijay Kumar, 2009)

| State | Area ('000 ha) | Production ('000 MT) | Productivity (MT/ha) | % share of India production |
|---|---|---|---|---|
| West Bengal | 1313.2 | 22456.8 | 17.1 | 17 |
| Uttar Pradesh | 960.9 | 19790.3 | 20.6 | 15 |
| Bihar | 823.7 | 14067.7 | 17.1 | 11 |
| Orissa | 660.8 | 1214.6 | 12.4 | 7 |
| Tamil Nadu | 262.7 | 7975.7 | 30.4 | 6 |
| Gujarat | 411.7 | 7402.9 | 18.0 | 6 |
| Maharashtra | 455.3 | 6454.8 | 14.2 | 5 |

| | | | | |
|---|---|---|---|---|
| Karnataka | 406.0 | 5030.9 | 12.4 | 4 |
| Andhra Pradesh | 284.1 | 4769.6 | 16.8 | 4 |
| Assam | 328.9 | 4474.2 | 13.6 | 4 |
| Jharkhand | 238.9 | 3639.7 | 15.2 | 3 |
| Kerala | 166.9 | 3479.7 | 20.8 | 3 |
| Haryana | 274.7 | 3277.1 | 11.9 | 3 |
| Others | 915.2 | 2185.3 | 2.3 | 12 |
| Total | 7803 | 125887 | 16.1 | |

Potato, tomato, brinjal, chilli, cabbage, cauliflower, onion, peas, gourds, okra and melons are now grown on commercial scale with increasing demand, better output, and change in production system and development of suitable cvs. These crops have high potential for foreign exchange earning. Today the emphasis on vegetable breeding is on value addition, disease resistance and better quality.

The important vegetables exported include onions-7.1%, cabbage-4%, cauliflower-4%, okra-3%, peas-3%. Traditional vegetables like onion, potato, okra, bottle gourd and chilli are exported to Malaysia, Singapore, Gulf, Sri Lanka, Bangladesh, Pakistan and Nepal. Non-traditional vegetables like asparagus, cabbage, capsicum, sweet corn, baby corn, green peas, French bean, cucumber, gherkin, cherry tomato, Brussels sprout, broccoli, processed and frozen French bean and lima bean are exported to European countries, Australia, Gulf and S.E. Asian countries. French bean, lima bean, sweet corn and capsicum are exported to Gulf. Some of the new products introduced in export market are tomato paste in bulk aseptic packs, processed button mushrooms, freeze dried and instant frozen vegetables. Dehydrated onion and tomato products have great potential as export commodities.

## 15.1. POTATO, *SOLANUM TUBEROSUM*

### 15.1.1. Diseases

#### (i) Bacterial Wilt, *Ralstonia solanacearum*

Bacterial wilt of potato continues to be a major disease since $19^{th}$ century. In India, the disease was first recorded on potatoes. This was the first bacterial disease recorded in India. The disease is more severe in tropical, subtropical and warm climates. In Ranchi district of Jharkhand (India), cultivation of rain fed potato was abandoned due to this disease and similar instances are recorded in Indonesia and in

Peru. In India, the losses reported are up to 55% in Kumaon hills, 33-40% in Maharashtra, 20-25% in Hyderabad and over 75% in some localities of Karnataka.

**(a) Symptoms:** The disease affects both above and under ground parts of the plant. Infected plants show sudden wilting and complete collapse of plants (Fig 15.1). Vascular browning has been reported to be the characteristic symptom of the disease and the bacterial mass ooze out.

**Fig. 15.1.** Bacterial wilt disease in potato.

The infested leaves wilt during the (sunny) day and sometimes recover during cool hours. The wilting is similar to the result of lack of water. During the rapid development of the disease, the entire plant wilts quickly without yellowing. Other symptoms could be wilting of only a part of the stem, or one side of the leaf/stem. The stem wilts or dries up completely and the remainder of the plant remains healthy. When the diseased tuber is cut, it shows a browning of the vascular ring and the immediate surrounding tissues on the cut surface, a creamy fluid usually appears on the vascular ring.

## (b) Management

**Biofumigation:** On-station trials utilizing chopped (radish, broccoli, and cabbage) and shredded (broccoli and cabbage) crucifers applied at a rate of 50 t/ha or 5 kg/m$^2$ showed an initial *R. solanacearum* (Rs) population count of 81,050 cfu/g. The mean population densities after harvest were significantly lower in all treatments. Lowest mean population density after harvest of

45,341.95 cfu/g was noted in chopped broccoli followed by chopped cabbage (46,962.20 cfu/g). However, there was a dramatic increase in Rs population count of 190,641.55 cfu/g in untreated plots.

BW infection (6.59–9.95 %), number of infected tubers (18–25), and yield (7.81–8.75 t/ha) were comparable in all chopped and shredded treatments. However, untreated plots had the highest percentage of BW infection (32.48%), most number of infected tubers (31), and lowest yield (4.07 t/ha) (Abragan *et al.*, 2008).

On-farm trials utilizing chopped radish, cabbage, broccoli, cauliflower, and mixed crucifers showed that soil incorporated with crucifers had significantly lower BW incidence (21.13–44.27%) compared with untreated plots (57.79%). The lowest percentage of BW incidence was obtained with mixed crucifers at 21.13%. The lowest number of infected tubers was obtained with chopped radish at 2.28 t/ha.

The mixed treatment also gave the highest yield of 11.44 t/ha, but this was comparable to the yield in the cauliflower treatment (11.04 t/ha). The untreated plot had the lowest yield of 6.67 t/ha.

Economic analysis based on the on-farm trials for BW utilizing chopped radish, cabbage, broccoli, cauliflower and mixed crucifers showed that white potato grown in soil incorporated with crucifers gave higher yield (7.92–11.44 t/ha), and return on investment (ROI) (133–192.84%). Mixed crucifers obtained the highest marketable yield, and ROI. Untreated plots had the lowest yield (6.67 t/ha), and ROI (113.97%) (Abragan *et al.*, 2008).

In the US, farmers are using mustard green manure crops to replace synthetic fumigation for potatoes with huge savings on costs. Growing of radish, mustard and broccoli have reduced bacterial wilt significantly (50-60%) in most of the experiments, though researchers believe the treatment is more effective on sandy soil than on heavy clay soil.

Application of brassicas can effectively reduce incidence of bacterial wilt, with cauliflower being the most effective. Other biofumigants tested included broccoli, cabbage, Chinese cabbage, sunflower and untreated soil. In the study, brassica tissues were shredded and mixed with the soil and left to biofumigate for two weeks. The use of cauliflower showed the lowest disease incidence of 14.34% and was followed by broccoli, sunflower, cabbage and Chinese cabbage. The study also showed that biofumigation can be used as an option in an integrated and long-term management of bacterial wilt disease.

Potato yields can be increased by using mustard as a biofumigant green manure in intensive rotation. The mustard is grown for about 100 days, until early flowering. It is then mulched using a 4.5 m wide 4-row profiled pulverizer normally used to destroy potato stems in preparation for harvest. This implement is set up with short and long blades to follow the contours of the hilled potato beds. The pulverizer then works uniformly across the flat surface in which the mustard is grown. This maximizes the utility of the implement and avoids the need for a separate mulcher (Fig. 15.2).

**Fig 15.2.** Pulverizing mustard green manure.

Immediately after pulverizing, the mustard is incorporated into the soil using a mouldboard plough. It is then left for about one month before being disc ploughed at least 2-3 times at approximately two-week intervals.

A deeper cultivation is then carried out with a chisel plough. In preparation for planting potatoes, a four-row bed former is then used to form the hills, and the annual cycle resumes.

**Solarization:** Raising the soil temperature by covering the soil with polythene film has been found effective in reducing the bacterial wilt.

**Intercropping:** In some developing countries intercropping has been used as a means of reducing pathogen populations in the soil and root-to-root transmission. In Burundi, growing potato with bean showed a lower incidence of bacterial wilt, indicating that a bean intercrop, which has a dense root system and grows quickly, was better than a crop such as maize, which develops slower and has a more dispersed root system.

Intercropping of potato with maize and cowpea were the most effective in reducing the wilt incidence.

**Crop Rotation:** Adopting crop rotation for 2-3 years with non-host crops (maize, sorghum, cereals, garlic, onion, lupin, cabbage, cauliflower, knol-khol, carrot, wheat, barley, oat, finger millet, green manuring crops like sunnhemp, etc.) reduced the wilt incidence to the extent of 94%.

Studies undertaken in North Western hills using 6 cropping sequences revealed that the reduction in wilt incidence ranged between 25 to 81% by following 2 years rotations. Potato-finger millet-finger millet-potato gave maximum reduction in wilt incidence (81%) (Table 15.4). Roots of finger millet were free from *R. solanacearum* infection, whereas paddy, maize, wheat and beans carried the infection. However, rhizosphere population of the pathogen was found to be reduced.

Table 15.4. Effect of crop rotation on bacterial wilt incidence and yield of potato in North Western Hills ( Bhowali).

| Crop rotation | Wilt incidence (%) | Yield loss | Popn. of *R. solanacearum* in cfu/root system of different crops |
|---|---|---|---|
| Potato- Potato- Potato- Potato | 51.7 (+ 92) | 56.9 | 9.0 x $10^6$ (Potato) |
| Potato-Bean-Bean- Potato | 19.7 (- 25) | 18.3 | 9.2 x $10^4$ (Bean) |
| Potato-Paddy-Paddy- Potato | 12.0 (- 53) | 29.7 | 1.0 x $10^5$ (Paddy) |
| Potato-Wheat-Wheat- Potato | 9.2 (- 63) | 18.8 | 4.2 x $10^4$ (Wheat) |
| Potato-Maize-Maize- Potato | 8.7 (- 66) | 13.1 | 1.0 x $10^6$ (Maize) |
| Potato-Finger millet-Finger millet- Potato | 4.7 (- 81) | 7.1 | 0.0 (Finger millet) |

### (ii) Common Scab, *Streptomyces scabies*

**(a) Symptoms**: Scab spots are more or less circular brown, roughened areas, with irregular margins. Sometimes the ridged portions are in broken concentric rings. The spots may be raised or warty in appearance, level with the surface, or somewhat sunken. Superficial russetting to deep pitting may occur (Fig. 15.3). White grubs, wire worms, and millipedes are attracted to this affected tissue so that scab spots may be further enlarged or deepened. Scab begins when tubers start forming. Initially the spots may be so small that they are not noticed. As the tuber continues to grow, the areas of these reddish-brown spots also enlarge. An older tuber has too thick a protective layer on its surface to be invaded readily.

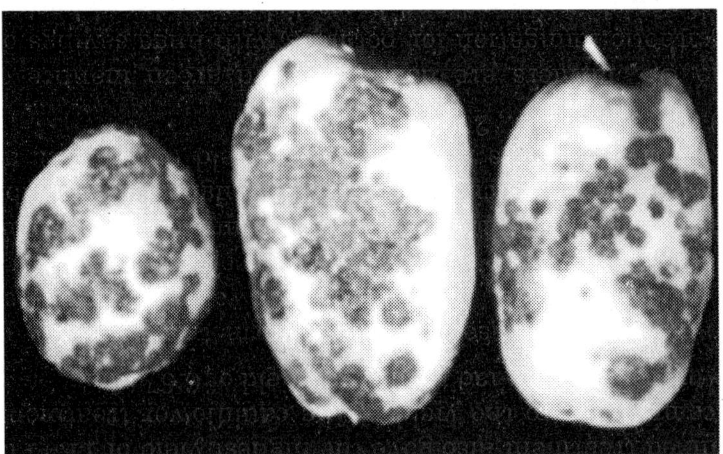

Fig. 15.3. **Common scab symptoms on potato tubers.**

## (b) Management

**Biofumigation:** In the greenhouse, soil incorporation of fresh and dry cabbage residues at rates of 0.25 and 0.1% (m/v), respectively, reduced common scab (*Streptomyces scabies*) significantly. Amendment of naturally infested soil in the field with any residues of cabbage, cauliflower, broccoli and Brussels sprouts, at 0.33% (m/v) similarly resulted in a significant reduction. These results were also verified in subsequent field trials where incorporation of dried cabbage residue significantly reduced common scab incidence in infested soil. All results obtained from the trials indicated reduction in scab incidence in soil with biofumigation practices. In summary, brassica residue incorporation as a disease control practice is an innovative means of managing common scab of potato that is effective, economically feasible and non-detrimental to the environment (Gouws, 2004).

To determine if volatile compounds released from residues of brassicaceous crops have a reducing effect on the incidence of common scab, field trials were conducted over three seasons with various brassica amendments. The treatments included green manuring, amendment with fresh residue, amendment with dry residue and no amendment (control). Amendment of naturally infested soil in the field with dry residues of cabbage at a rate of 1.56 % (m/v) resulted in a significant reduction in the disease. Green manuring with cabbage and amendment with fresh cabbage residue did not result in significant reduction of common scab on potato. No adverse effect on growth or yield of potatoes planted in soil amended with dry brassicaceous material was observed (Gouws and Wehner, 2006).

**Crop Rotation:** In field experiments, four different Brassica crops, including canola, rapeseed, yellow mustard, and Chinese (white) mustard, as well as standard rotation crops (ryegrass and/or oats), were evaluated for their effect on common scab disease in a subsequent potato crop when grown as a short-term rotation (green manure) crop.

Chinese mustard reduced common scab by 27% relative to a ryegrass control. These results indicate that Brassica crops have potential for use in the control of common scab disease (Tables 15.5 and 15.6).

Table 15.5. Effect of Brassica rotation crops on potato tuber diseases at a northern Aroostook farm

| Rotation | Common Scab | |
|---|---|---|
| | Incidence | Severity |
| Oats | 2.5 ab | 0.38 |
| Ryegrass - 'Lemtal' | 2.5 ab | 0.11 |
| Yellow Mustard-'Idagold' | 5.8 a | 0.19 |
| Canola | 0.0 b | 0.00 |
| Rapeseed-'Dwarf Essex' | 1.7 ab | 0.10 |
| Chinese Mustard (White) | 5.8 a | 0.43 |
| LSD (P=0.05) | 4.5 | 0.45 |

Table 15.6. Effect of Brassica rotation crops on potato tuber diseases at a central Aroostook farm.

| Rotation | Common Scab | |
|---|---|---|
| | Incidence | Severity |
| Ryegrass - 'Bahrenburg' | 1.7 | 0.03 |
| Ryegrass - 'Lemtal' | 2.5 | 0.03 |
| Yellow Mustard-'Ida Gold' | 3.3 | 0.03 |
| Rapeseed-'Dwarf Essex' | 2.5 | 0.03 |
| Chinese Mustard (White) | 1.7 | 0.02 |
| LSD (P=0.05) | 4.1 | 0.05 |

Crop rotation with wheat, peas, oat, barley, lupin, soybean, sorghum, bajra is effective in keeping the common scab disease under check.

**Green Manuring:** Significant reductions in common scab (*S. scabies*) were observed following brassica green manures, especially an

Indian mustard (*B. juncea*) green manure (Larkin and Griffin, 2007; Larkin *et al.*, 2006).

### (iii) Black Scurf, *Rhizoctonia solani*

**(a) Symptoms:** Perhaps the most readily apparent phase of *Rhizoctonia* disease is the black scurf or sclerotia present on tuber surfaces. These sclerotia can vary in size from very small, flat, superficial black specks to large, raised, irregularly shaped masses that can cover a major portion of the tuber. While black scurf can, under extreme conditions, affect the marketability of table stock/fresh pack potatoes, it is an extremely important seed-borne phase of the pathogen that deserves attention by all potato producers.

Lesions characteristic of *Rhizoctonia* on stems and stolons are brown to black and sunken. These cankers can continue to expand and are capable of girdling stems and stolons of young developing plants (Fig. 15.4). *Rhizoctonia* infection of older plants very seldom leads to girdled stems that die. However, the health of these plants can be severely compromised and they can frequently become more susceptible to other diseases, particularly early blight.

**Fig. 15.4.** Black scurf on collar region of potato.

### (b) Management

**Biofumigation:** The effect of biofumigation with canola and *Brassica napus* as green manure (20 days old plants) and cover crop (60 days old root tissues in soil) was evaluated against soil-borne *Rhizoctonia solani*, to distinguish the efficacy of allelochemicals produced by glucosinolates hydrolysis following soil incorporation of tender and old root tissues of *B. napus* in black scurf management of potato.

Green manuring with 20 days old canola crop gave significant protection against soil-borne inoculum of the fungus and significantly reduced inhibition of eyes germination, sprouts killing, stem canker index, stolon canker index, black scurf disease index and reduction in yield, when compared with incorporation of 60 days old canola root tissues and inoculated control (Table 15.7). Soil incorporation with 20 days old canola crop gave significant protection against the fungus when compared with the plants following incorporation of 60 days old canola root tissues and inoculated control.

Table 15.7. Response of green manuring and crop rotation (incorporation) on soil-borne inoculum of *Rhizoctonia solani* AG 3 Isolate CL-58 on black scurf disease of potato

| Treatment* | Eyes germination inhibition (%) | Sprouts killing (%) | Stem canker index (%) | Stolon canker index (%) | Black scurf disease index (%) | Yield reduction (%) |
|---|---|---|---|---|---|---|
| T1 | 73.89 a** | 53.33 a** | 70.00 a** | 53.33 a** | 74.44 a** | 64.99 a** |
| T2 | 35.55 b | 38.84 b | 40.87 b | 37.78 b | 47.83 b | 41.24 b |
| T3 | 64.45 a | 50.86 a | 67.22 a | 46.67 ab | 60.78ab | 60.78 a |
| CD (P = 0.05) | 9.99 | 11.05 | 19.44 | 14.93 | 20.62 | 10.27 |

Means within column followed by the same letters are not significantly different according CD test (P = 0.05).

* **T1** – Control (20 g mass inoculum + sowing of sprouted tuber), **T2** - 20 g mass inoculum + green manuring of 20 days old canola seedlings + sowing of sprouted tuber, **T3** - 20 g mass inoculum + incorporation of 60 days old canola plants root tissue in soil + sowing of sprouted tuber.

Soil application of mustard/groundnut cake at 2.5 t/ha significantly reduced the number of black scurf diseased tubers and intensity of the disease.

Early investigations in comparing "Fumus" (Indian mustard) and "BQ Mulch" (fodder rape/turnip) against a bare ground fallow indicated a substantial reduction in black scurf in the following potato crop. *Rhizoctonia* lesions were reduced from 75% in the bare ground fallow to 0-2% in the biofumigant treated areas. The use of "Fumus" and "BQ Mulch" has many other benefits in addition to biofumigation. Soil health benefits from an extensive fibrous and deep tap root system, and nutrients released from their rapid breakdown after incorporation into soil has helped to increase potato yields. When sown in autumn and winter under cooler conditions, excellent germination, high growth rates, low disease and insect pressure make them an easy crop to manage and ideal for a break crop between other horticultural crops in spring and summer in Tasmania (Aird and Long, 2008).

In field experiments, four different Brassica crops, including canola, rapeseed, yellow mustard, and Chinese (white) mustard, as well as standard rotation crops (ryegrass and/or oats), were evaluated for their effect on black scurf disease in a subsequent potato crop when grown as a short-term rotation (green manure) crop.

Canola and rapeseed rotations reduced the incidence and severity of black scurf by 54-73% relative to the oats rotation. Chinese mustard reduced black scurf by 50-85% relative to a ryegrass control. Greenhouse tests with these and additional Brassica crops indicated significant reductions in *Rhizoctonia* inoculum and potato seedling disease incidence and severity following growth of several Brassica crops. These results indicate that Brassica crops have potential for use in the control of black scurf and other soil-borne diseases (Tables 15.8 to 15.10).

Table 15.8. Effect of Brassica rotation crops on potato black scurf disease at a northern Aroostook farm

| Rotation | Black Scurf | |
|---|---|---|
| | Incidence | Severity |
| Oats | 32.5 b | 0.99 ab |
| Ryegrass - 'Lemtal' | 25.8 bc | 0.73 bc |
| Yellow Mustard-'Idagold' | 21.7 cd | 0.52 cd |
| Canola | 10.0 e | 0.30 d |
| Rapeseed-'Dwarf Essex' | 15.0 de | 0.26 d |
| Chinese Mustard (White) | 44.6 a | 1.16 a |
| LSD (P=0.05) | 8.40 | 0.34 |

Table 15.9. Effect of Brassica rotation crops on potato black scurf disease at a central Aroostook farm.

| Rotation | Black Scurf | |
|---|---|---|
| | Incidence | Severity |
| Ryegrass - 'Barenburg' | 11.7 a | 0.26 a |
| Ryegrass - 'Lemtal' | 3.3 bc | 0.03 c |
| Yellow Mustard-'Idagold' | 4.2 bc | 0.07 bc |

| | | |
|---|---|---|
| Rapeseed-'Dwarf Essex' | 1.7 c | 0.04 c |
| Chinese Mustard (White) | 5.8 b | 0.16 b |
| LSD (P=0.05) | 4.0 | 0.10 |

Table 15.10. Effects of different rotation crops on Rhizoctonia pathogen populations and disease levels in a subsequent potato crop in the greenhouse.

| Rotation | Propagule reduction | | Seedling disease | | | Plant Size |
|---|---|---|---|---|---|---|
| | Before | After | Incidence | Inc.>2% | Severity | |
| Turnip 'Purple Top' | 63.5 a | 48.9 b | 81.3 ab | 37.5 a | 1.25 a | 2.75 a |
| Barley | 54.2 ab | 64.1 a | 75.0 ab | 31.3 ab | 1.19 a | 2.63 a |
| Chinese mustard | 47.6 bc | 43.4 bc | 62.5 b | 12.5 ab | 0.75 b | 2.81 a |
| Rapeseed-'Dwarf Essex' | 22.0 e | 35.5 c | 50.0 b | 6.3 b | 0.63 b | 2.56 a |
| Canola | 39.4 cd | 24.4 d | 81.3 ab | 37.5 a | 1.19 a | 2.69 a |
| Oilseed radish | 31.9 de | 14.1 e | 56.3 b | 12.5 ab | 0.63 b | 2.75 a |
| Yellow mustard 'Idagold' | 22.9 e | 2.5 f | 68.8 ab | 18.8 ab | 0.88 ab | 2.63 a |
| Potato | 0.0 f | 0.0 f | 87.5 a | 37.5 a | 1.25 a | 2.13 b |

Researchers have documented consistent reductions in *Rhizoctonia* (canker and black scurf) on potato following either rapeseed green manure or canola grown for grain (Larkin and Griffin, 2007; Larkin *et al*., 2006).

**Crop Rotation:** For the management of *Rhizoctonia* canker, first and foremost is crop rotation. In this regard it is important to consider both the number of years between potato crops and the rotational crops used in the system. Most research has shown that one potato crop in three or four is adequate to reduce soil inoculum to economic levels. However, soil type can potentially impact the rate of pathogen decline. Observations suggest that the rate of *Rhizoctonia* decline in sandy soils exceed that of loamy soils with more organic matter. Cereal crops such as oat and barley are generally regarded as being good rotations with potato for the reduction of soil-borne inoculum.

**Solarization, Chemicals and Bioagents:** Physical, biological and chemical control was integrated to control soil-borne diseases of potato induced by *R. solani* (Elad *et al*., 1980). Although solar heating or methyl bromide fumigation significantly reduced the disease, the combination of either solar heating or methyl bromide soil fumigation followed by *T. harzianum* application improved control over solarization or fumigation alone and additionally controlled *Verticillium dahliae* and *Sclerotium rolfsii*.

Combination of soil solarization of black scurf infested fields and seed treatment with *T. viride* further improved the disease control.

### *(iv) Rhizoctonia* and Brown Spots

#### (a) Management

**Biofumigation and Solarization:** The effect of biofumigating the soil with four types of green manures (mustard, cabbage, corn, and millet) on a potato crop was compared with an untreated control. The method includes incorporating a biofumigant such as chopped green manure, into the soil, followed by coating the soil by more than one successive spraying of its surface with an appropriate polymer formulation that forms continuous plastic coats.

Application of mustard chopped green manure in combination with plastic coat gave maximum reduction in *Rhizoctonia* and brown spots followed by millet chopped green manure in combination with plastic coat. The maximum yield was obtained in corn chopped green manure in combination with plastic coat followed by mustard chopped green manure in combination with plastic coat (Table 15.11).

Table 15.11. Effect of biofumigation and solarization on *Rhizoctonia* and brown spots and yield of potato

| Treatment | Soil solarization | Yield | *Rhizoctonia* | Brown spots |
|---|---|---|---|---|
| Millet | Ecotex | 2.208 | 4.4% | 18.3% |
| | None | 1.943 | 21.5% | 44.3% |
| Cabbage | Ecotex | 2.187 | 15.1% | 43.8% |
| | None | 2.101 | 15.9% | 26.3% |
| Mustard | Ecotex | 2.279 | 2.0% | 6.8% |
| | None | 2.090 | 22.0% | 43.5% |
| Corn | Ecotex | 2.401 | 13.1% | 22.6% |
| | None | 2.067 | 17.2% | 35.2% |
| Control | Ecotex | 2.135 | 14.3% | 18.3% |
| | None | 1.996 | 35.1% | 44.3% |

**Green Manuring:** A study of brassica green manures in potato cropping systems compared winter rape (*Brassica napus*) and white mustard (*Sinapis alba*) to no green manure, with and without herbicides and fungicides. The winter rape system had a greater proportion of Rhizoctonia-free tubers (64%) than the white mustard (27%) and no green manure (28%) treatments in the non-fumigated plots.

### (v) Powdery Scab, *Spongospora subterranea*

Powdery scab disease has emerged as an important problem for potato growers in Maine and other potato-growing regions in recent years. This disease causes lesions on tubers, seriously affecting tuber quality and marketability, and is of special concern because the causal pathogen of powdery scab is also the sole vector and means of infection of potato mop-top virus (PMTV). Once the pathogen becomes established in an area, it will persist indefinitely.

#### (a) Management

**Biofumigation:** Early investigations in comparing "Fumus" (Indian mustard) and "BQ Mulch" (fodder rape/turnip) against a bare ground fallow indicated a substantial reduction in powdery scab in the following potato crop. Powdery scab lesions were reduced from 15% in the bare ground fallow to 2-5% in the biofumigants treated areas. The use of "Fumus" and "BQ Mulch" has many other benefits in addition to biofumigation. Soil health benefits from an extensive fibrous and deep tap root system, and nutrients released from their rapid breakdown after incorporation into soil has helped to increase potato yields. When sown in autumn and winter under cooler conditions, excellent germination, high growth rates, low disease and insect pressure make them an easy crop to manage and ideal for a break crop between other horticultural crops in spring and summer in Tasmania (Aird and Long, 2008).

**Crop Rotation:** In field experiments, four different Brassica crops, including canola, rapeseed, yellow mustard, and Chinese (white) mustard, as well as standard rotation crops (ryegrass and/or oats), were evaluated for their effect on powdery scab disease in a subsequent potato crop when grown as a short-term rotation (green manure) crop.

A soil bioassay developed to assess inoculum levels of the powdery scab pathogen indicated that inoculum reductions of 5-27% occurred after growth of the rotation crops and prior to planting potato.

Under high powdery scab disease pressure, all rotation crops reduced severity of disease, and rapeseed, ryegrass, and Chinese mustard reduced disease incidence by 19-39% relative to the oats rotation control. These results indicate that Brassica crops have potential for use in the control of powdery scab and other soil-borne diseases (Tables 15.12 to 15.14).

Table 15.12. Soil bioassay of powdery scab inoculum potential before and after planting various Brassica rotation crops at the central Aroostook site.

| Green Manure Crop | Root infection - tomato bioassay | | |
|---|---|---|---|
| | Pre-planting | Post-harvest 2001 Percent change | Pre-plant 2002 Percent change |
| Lemtal ryegrass | 65.4 | + 3.1 | - 6.8 |
| Barenbrug ryegrass | 63.3 | + 1.4 | - 8.4 |
| Dwarf Essex rapeseed | 64.8 | - 3.3 | - 14.2 |
| Idagold yellow mustard | 68.3 | - 3.8 | - 17.8 |
| Chinese white mustard | 64.6 | - 5.9 | - 26.8 |
| LSD (P=0.05) | NS | 6.7 | 15.9 |

Table 15.13. Effect of Brassica rotation crops on potato powdery scab disease at a northern Aroostook farm

| Rotation | Powdery Scab | |
|---|---|---|
| | Incidence | Severity |
| Oats | 58.8 a | 2.70 a |
| Ryegrass - 'Lemtal' | 45.0 bc | 1.62 b |
| Yellow Mustard-'Idagold' | 54.2 ab | 1.95 b |
| Canola | 49.9 ab | 1.88 b |
| Rapeseed-'Dwarf Essex' | 47.5 b | 1.99 b |
| Chinese Mustard (White) | 35.7 c | 1.66 b |
| LSD (P=0.05) | 11.1 | 0.40 |

Table 15.14. Effect of Brassica rotation crops on potato powdery scab disease at a central Aroostook farm.

| Rotation | Powdery Scab | |
|---|---|---|
| | Incidence | Severity |
| Ryegrass - 'Barenburg' | 67.5 bc | 1.97 b |
| Ryegrass - 'Lemtal' | 80.8 a | 2.53 a |
| Yellow Mustard-'Idagold' | 75.0 ab | 2.33 ab |
| Rapeseed-'Dwarf Essex' | 80.0 a | 2.43 ab |

| Chinese Mustard (White) | 60.8 c | 2.23 b |
|---|---|---|
| LSD (P=0.05) | 11.7 | 0.27 |

Significant reductions in powdery scab was observed following brassica green manures, especially an Indian mustard (*B. juncea*) green manure (Larkin and Griffin, 2007; Larkin *et al.*, 2006).

### (vi) Spraing/Stubby Root Nematodes, *Trichodorus* spp., *Paratrichodorus* spp. and Tobacco Rattle Virus/Corky Ringspot Disease Complex

Spraing is the term given to a range of internal and external symptoms in potato tuber flesh consisting of brown, necrotic rings, arcs and flecks. Tobacco rattle virus (TRV) cause these symptoms. In some cases external concentric rings may be visible on the surface of the tuber. The severity of symptoms in the tuber can range from mild (a few flecks or faint brown lines) to severe (heavy, dark chocolate coloured arcs with a corky texture). A few symptomatic tubers may be found or the stock could be heavily affected (>50%). Spraing symptoms may not be evident at harvest. Fluctuations in temperature during storage can exacerbate development of the internal tuber symptoms.

**Fig.15.5.** Left – Yellow ring spots induced by TRV. Right – Leaf twisting with necrosis induced by TRV.

#### (a) Symptoms

**Foliar Symptoms:** Generally TRV spraing infected tubers produce healthy plants. However, in some cases several healthy and symptomatic, stunted stems will be produced, which is termed as stem mottle. Symptoms may present as pinching and "scorching" of the leaf tip, mottles, distorted/wavy leaf margins and occasionally finely etched lines and rings or chevron type markings reminiscent of PMTV (Fig. 15.5).

Corky ringspot symptoms vary depending on virus strain, potato cultivar, and time of infection. Symptoms often include brown necrotic rings, arcs, and diffuse spots which are considered quality defects and may result in after-harvest devaluation or rejection of either table or processed potatoes.

Corky ringspot symptoms in potato include necrotic rings and pits on the tuber surface and range from diffuse brown spots to concentric rings or arcs of brown, necrotic tissue to dark-brown necrotic tissue which extends through tuber flesh. The virus is usually detectable when symptoms are seen but may also be present in asymptomatic tissue (Charlton, 2006).

Tubers from soil with a history of tobacco rattle virus serve as a reservoir. The virus may spread to daughter tubers when infected tubers are used as seed (Crosslin *et al.*, 1999). Newly formed potato tubers are quite vulnerable to tobacco rattle virus infection and tubers as small as 3-cm in diameter had corky ringspot blemishes in tobacco-rattle infested fields in Florida (Fig. 15.6) (Weingartner *et al.*, 1975).

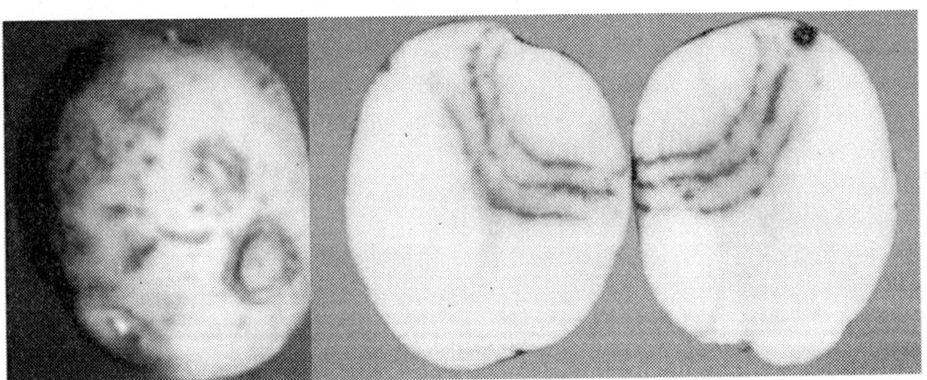

**Fig.15.6.** External and internal symptoms of corky ringspot virus on potato tubers.

Several species of *Trichodorus* and *Paratrichodorus* transmit M-type isolates of tobacco rattle virus (Harrison and Robinson, 1986). Stubby-root nematodes are migratory ectoparasites that are mobile during each stage of their life cycle (Stark and Love, 2003) and feed primarily on meristematic cells or root tips which hinders root elongation (Crow, 2005). Damaged root tips may swell and lateral roots may emerge behind them, resulting in root proliferation.

### (b) Management

**Cover Cropping:** Field studies were carried out during 2006-2007 to appraise brassicaceous amendments against trichodorid nematodes, the vectors of tobacco rattle virus (TRV) that causes spraing disease. The trials were conducted in a small grain-potato rotation in Central Sweden. Three cover crops, Caliente mustard (*Sinapis alba* and *Brassica juncea* blend, cv. Brandt 119), oil radish (*Raphanus sativus*, cv. Colonel) and arugula (*Eruca sativa* cv. Nemat) were tested and compared to a bare soil. The cover crops reduced nematode populations at all different locations although the suppression level differed among locations and cover crop species used. Substantial yield improvements were also recorded (Manduric, 2008).

## (vii) Early Dying/Verticillium Wilt, *Verticillium dahliae* and Lesion Nematode, *Pratylenchus thornei* Disease Complex

Severity of *Verticillium* wilt of potatoes is increased by the nematode, *Pratylenchus thornei*.

**(a) Symptoms:** A typical visual symptom of potato early dying is loss of plant vigour during mid- to late summer followed by senescence and death of the crop before normal maturity. Initially the disease causes uneven chlorosis of the lower leaves on a few plants. Areas between the leaf veins turn yellow and later brown. Later wilting of the lower leaflets of the plants may occur, and infection proceeds up the infected stems, which often remain erect (Fig. 15.7). The disease causes vascular discolouration of stems and their tuber ends and plants may die without reaching maturity. It is not uncommon for one or two stems of the plant to die and the other portions of the plant to remain healthy (Rangahau, 2003). The symptoms often occur on one side of the plant or on individual leaves (Powelson and Rowe, 1993). A light brown vascular discolouration in basal stem tissues is usually present in symptomatic plants. In severe cases, plants across an entire field will die over a period of several weeks (Rowe and Powelson, 2002).

A light brown vascular discolouration is often visible at the stem base when sliced (Fig. 15.8). Advanced symptoms usually do not occur until after flowering and may consist of a decline of isolated plants or in severe cases, the entire crop may mature early (Davis 1981; Krikun and Orion 1979; Mace *et al.*, 1981; Rowe 1983). Infected potato tubers with Verticillium show symptoms of light brown discolouration of the vascular ring (Fig. 15.9).

**Fig. 15.7.** Potato plants showing wilting symptoms due to early dying disease complex.

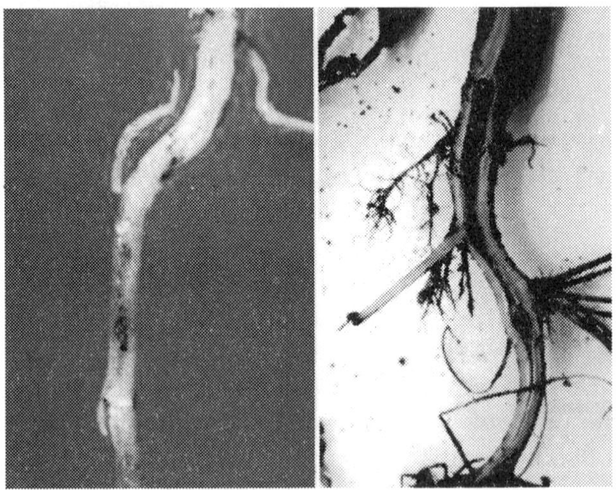

Fig. 15.8. Left - Root lesions, Right - Sliced potato root showing vascular discolouration.

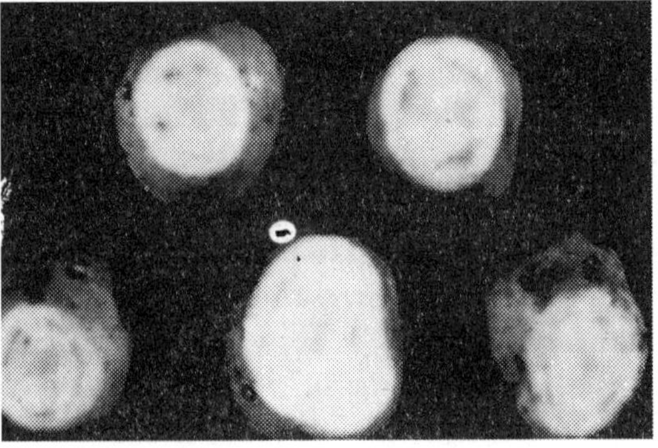

Fig. 15.9. Potato tubers infected with Verticillium showing symptoms of light brown discolouration of the vascular ring.

## (b) Management

**Solarization:** The efficiency of solarization in controlling *V. dahliae* and *P. thornei* on potatoes was tested in a naturally infested field and compared to a nematicide mixture of ethylene dibromide and chloropicrin. Solarization of soil killed microsclertia of *V. dahliae*, reduced diseased plants by 96-99 percent and the *P. thornei* population by 80-100 percent, controlled weeds, and increased the yields of potatoes by 35 percent.

**Crop Rotation:** In Ohio, potato-wheat rotation has been found to be useful as part of an overall control strategy (Joaquim *et al.*, 1988). Potato yield increases were observed with barley-corn rotation (Davis and McDole, 1979).

A single season of a Polynema marigold rotation crop significantly reduced *P. penetrans* recovery from roots of rye grown after the potato crop and increased marketable tuber yields. Saia oat, Rudbeckia and canola had greater numbers of saprophytic nematodes than potato, and tuber yield was correlated with increased recovery of saprophytes. In the second potato crop after a single year of rotation tuber yields were higher after Polynema marigold. Canola, while maintaining high lesion nematode numbers, resulted in high tuber yields. Rotation with Rudbeckia and canola resulted in the highest tuber yields.

In fields infested with root-knot nematodes and *V. dahliae,* Scholte (1990) found that a five-year rotation of maize-sugar beet-barley-barley-potato had significantly higher potato yields than continuous potato.

**Green Manuring**: The mustard green manure by itself can suppress *V. dahliae* sufficiently to maintain potato yields, and that metham sodium after a mustard green manure does not further increase crop yield. It is reasonable to assume that yields would have been below the county average if the *Verticillium* had not been suppressed by either the mustard green manure or the metham sodium.

In potato field infested with PED pathogens (*V. dahliae, Colletotrichum coccodes*) incorporation of broccoli, *Brassica juncea* ISCI20 and red giant mustard resulted in higher yields (113, 111, and 107% over the control) (Tsror *et al.*, 2006).

There was less *Verticillium* wilt incidence with winter rape incorporation (7%) than with white mustard (21%) or no green manure incorporation (22%) in non-fumigated plots.

Davis *et al.* (1996) reported that in Idaho, the disease was controlled after green manure treatments of either Sudangrass or corn and an increase in yield. Spent mushroom compost may be a means of reducing the effects of the disease on superior varities of potato and increasing the yield when one or both pathogens are present (LaMondia *et al.*, 1999). A study conducted in Washington showed that mustard green manure has the potential to replace the fumigant metham sodium for the production of potatoes in some cropping systems. The practice can also improve water infiltration rates and provide substantial savings to farmers (McGuire, 2003).

Rangahau (2003) reported that incorporating organic material such as soymeal into soil a few months before planting potatoes to encourage naturally occurring antagonists in the individual paddock will reduce potato early dying disease.

**Solarization, Chemicals and Bioagents:** Physical, biological and chemical control was integrated to control soil-borne diseases of potato induced by *V. dahliae* (Elad *et al.*, 1980). Although solar heating or methyl bromide fumigation significantly reduced the disease, the combination of either solar heating or methyl bromide soil fumigation followed by *T. harzianum* application improved control over solarization or fumigation alone and additionally controlled *R. solani* and *Sclerotium rolfsii*.

### 15.1.2. Nematodes

### (i) Columbia Root-knot Nematode, *Meloidogyne chitwoodi*

#### (a) Management

**Crop Rotation:** In potato field trials in Idaho and Washington, cropping sequences including the cultivation and incorporation of Sudan grass, a crop also known to release nematicidal compounds, dramatically reduced *M. chitwoodi* infestation levels in potato (Riga *et al.*, 2004).

In Oregon, mustard (cv Martigena) and radish (cv Colonel) or rapeseed (cv Jupiter) and Sudan grass (cv Trudan 8) had similar non-host effects on Columbia root-knot nematode (CRKN) suppression while these crops were growing. However, mustard and rapeseed had a stronger green manure effect on suppression after incorporation. Non-host effects were stronger with Sudan grass (cv Sordan 79) or canola than with radish (cv Rimbo) or mustard (Calinete 119), but green manure effects were stronger for mustard and radish. Without any nematicides, Sudan grass provided best suppression of *M.*

*chitwoodi*, eliminated any external symptoms in tubers, and reduced infection the most of any green manure crop tested (Ingham *et al.*, 2006).

The best rotation to control the Columbia root-knot nematode in potatoes involves planting a summer non-host crop, followed by a winter cover crop (rapeseed) incorporated as a green manure. Non-host crops include supersweet corn (Crisp and Sweet 710/711), pepper, lima bean, turnip, cowpea, muskmelon, watermelon, squash, rapeseed, canola, mustard, and Sudan grass (Trudan 8, Sordan 79).

**Biofumigation:** Mojtahedi *et al.* (1991, 1993) reported that incorporation of *B. napus* shoots into *M. chitwoodi* infested soil reduced nematodes to very low levels and that amending with *B. napus* was more effective than amending with wheat or corn. The effect of the amendment, however, did not extend into the soil layers below the zone of incorporation, and the amendment was more effective against second-stage juveniles than against egg masses, suggesting that biofumigation would be most effective when the nematodes are active. The amendment protected host plant roots growing in the zone of *B. napus* incorporation from nematode infestation for up to six weeks. Soil incorporation rates of 4% (w/w) killed nearly all second-stage juveniles, whereas rates of 6% were required to prevent hatching of juveniles from egg masses.

Mojtahedi *et al.* (1993) reported that *M. chitwoodi* was controlled by *Brassica* amendments at an average soil temperature of 19°C, and speculated that the slower decomposition of the plant tissues under cool conditions resulted in an extended period of nematode control.

The *Brassica* crop arugula (*Eruca sativa* var. Nemat) also received interest as a potentially useful biofumigant crop to manage root-knot nematodes, as it appears to act as a trap crop for *M. chitwoodi* (Riga *et al.*, 2006). A positive effect on yields of nematode-susceptible crops grown after arugula cultivation and biofumigation was reported for potato.

**Biofumigation and Chemicals:** Few studies have analyzed the cost of biofumigation as a nematode management strategy. Riga *et al.* (2006) reported a strategy in which a *Brassica* crop arugula (*Eruca sativa* var. Nemat) cover crop was combined with lower rates of a synthetic nematicide to manage in potato, reducing pest management costs by 50%.

Rapeseed, arugula and mustard were studied as alternatives to fumigation. The brassica cover crops are usually planted in late summer (August) or early fall and incorporated in spring before planting mustard. Results are promising, with nematodes reduced up to 80%, but because of the very low damage threshold, green manures alone cannot be recommended for adequate control of *M. chitwoodi* in potatoes. The current recommended alternative to fumigation is the use of rapeseed or mustard cover crop plus the application of nematicide (mocap). This regimen costs about the same as fumigation (2006 prices).

Arugula in combination with half the recommended rate of Telone reduced root knot nematode populations, *M. chitwoodi*. Arugula in combination with Temik and Mocap also reduced *M. chitwoodi*. In addition, Arugula did not reduce the beneficial free-living nematode populations and the non-pathogenic *Pseudomonas*. The cost of growing and incorporating Arugula and combining it with ½ rate of Telone and one fungicide was approximately half of the present commercial cost of Telone and fungicide applications.

Management of *M. chitwoodi* on potato was achieved using oil radish varieties Colonel, Commodore, and mustard variety Metex which were commercially planted in grower's field. In recent field studies it was found that two new oil radish varieties Defender and Comet significantly reduced the population of *M. chitwoodi* (99%) (Hafez and Sundararaj, 2006).

### (ii) Root-knot Nematode, *Meloidogyne incognita*

**(a) Symptoms:** Plants affected by root-knot nematodes are generally stunted, slightly yellow and may wilt during hot weather. The diagnostic symptoms of a *Meloidogyne* attack on potatoes are the galls found on both roots and tubers. Tuber surface becomes uneven and warty because of numerous blister-like galls (Fig. 15.10). When an infested tuber is transversely cut, the white glistening swollen females of the size of pin head can be easily seen embedded in the potato tissue. Such tubers are invariably small, unfit for marketing, rot quickly and cause considerable yield losses. Galling on tubers may render them unsalable.

### (b) Management

**Biofumigation:** On-station trials using shredded crucifers applied at a rate of 50 t/ha or 5 kg/m$^2$ as biofumigants against root knot nematodes (RKN) showed that at varying initial nematode counts ranging from 621 to 683 during pre-treatment, RKN population count was significantly reduced in all shredded treatments 1½ months after incorporation. Shredded mixed crucifers showed the highest percentage reduction in RKN population of 83.6% followed by radish (79.2%), broccoli (74.2%), and cabbage (68.4%). In untreated plots, however, a 24.1% increase in RKN population was noted.

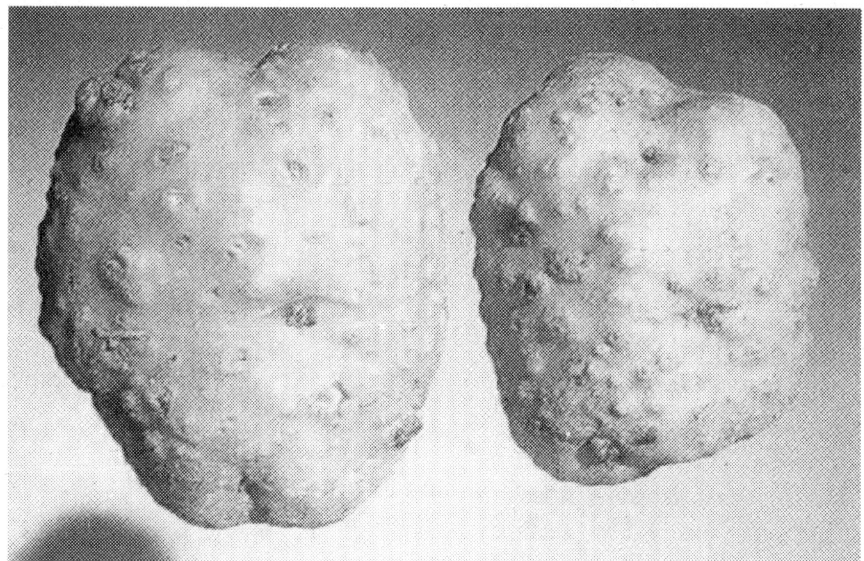

**Fig. 15.10.** Potato tubers with blisters incited by *Meloidogyne incognita.*

On-farm trials utilizing shredded crucifers as biofumigants against RKN in Lantapan, Bukidnon showed that at varying initial nematode counts ranging from 684 to 712 during pretreatment, nematode population decreased significantly with mixed crucifers by 86.7% followed by radish (85.4%), broccoli (82.2%), cauliflower (71.3%), and cabbage (70.9%). However, a significant increase in nematode population of 17.2% was observed in untreated plots. The use of mixed crucifers was recommended because it had the highest percentage reduction in nematode counts both on-station (83.57%) and on-farm (86.7%) and highest marketable yield (11.44 t/ha), and ROI of 192%.

**Intercropping:** Potatoes grown with white mustard in a pot of infested soil were less heavily attacked by root-knot nematodes than potatoes growing alone. Potato root diffusate was ineffective in the presence of leachings from the roots of mustard seedlings. Mustard oil increased the yield of potatoes by reducing the severity of nematode attack. The active principle involved in mustard is allyl isothiocyanate which is toxic to the nematodes. The interplanting of marigold with potato has resulted in a 7% increase in tuber yield.

Intercropping of onion and maize reduce galling by *M. incognita* on potato roots. Potatoes intercropped with maize have been found to be extremely vigorous with well developed root systems.

The nematode population on potato was checked significantly when mustard was grown intermixed with potato in alternate rows in combination with soil amendment of neem cake/leaves.

Growing of trap crop like French marigold, *Tagetes patula* in alternate rows with potato reduced the root-knot nematode larval population in soil, root galling and tuber infestation while the yields increased up to 123% over control. Growing of trap crops like *Tagetes patula* and *T. erecta* (African marigold) in between 2 to 3 rows of potato reduces the root-knot nematode population.

**Crop Rotation:** In North Western Hills, rotation of potato with French beans, leafy vegetables, maize and wheat helped in minimizing the nematode damage and increasing the productivity of the crop (Raj and Nirula, 1969). Crop rotation with non-host crops like maize, wheat, barley, beans, etc. reduces nematode population. Similarly, growing *Cassia* sp. also helped in reducing tuber infestations. potato crop can be followed by peanuts without any damage to peanuts and nematode populations will be reduced.

Following two years crop rotational sequence of maize-wheat-potato-wheat reduces root-knot damage significantly. Crop rotation with non-host crops like maize, wheat, barley, etc. reduces nematode population.

Rapeseed and Sudan grass green manures grown prior to potatoes at Prosser, Washington, provided between 72 and 86 percent control of the root-knot nematode in that crop (Stark, 1995).

Longer crop rotations that include mustards and non-host crops are also effective for nematode management. For example, a three-year rotation of potatoes > corn > wheat provides nearly complete control of the northern root-knot nematode (*M. hapla*) compared to methyl bromide and other broad-spectrum nematicides.

**Organic Amendments:** The oil cakes (neem, groundnut, mustard and castor) were effective in reducing parasitic nematode population (*Hoplolaimus*, *Tylenchorhynchus*, *Meloidogyne* and *Helicotylenchus* species) and in increasing yields of potato. Application of eucalyptus leaf waste from oil distillation plants at 2.5 t/ha helps in reducing the nematode population. Incorporation of saw dust at 2.6 t/ha also helped in reducing tuber infestations.

**Integrated Methods:** In an experiment, aldicarb at 2.5 kg a.i./ha and aldicarb + *Paecilomyces lilacinus* combined with crop rotation with maize reduced root-knot nematodes and increased tuber quality. Yield of maize, a rotation crop, in *P. lilacinus* and *P. lilacinus* +

aldicarb-treated plots was double than that in control plots (Jatala, 1985).

Several chemical and non-chemical alternatives to standardized cold or hot diffusion of MB (98-2), 50 g/m$^2$ are able to maintain adequate sanitary (against *M. incognita*) and productivity levels on potato cultivation (Table 15.15).

Table 15.15. Commercial yield of potato tubers (kg/m$^2$)

| Treatment | Potato yield (kg/m$^2$) | Relative yield |
|---|---|---|
| Soil solarization + MS | 2.80 a | 125 |
| Soil solarization + Telopic | 2.53 ab | 113 |
| Soil solarization + fresh chicken manure | 2.45 ab | 109 |
| Soil solarization alone | 2.40 ab | 107 |
| MB VIF | 2.24 bc | 100 |
| Control (no disinfestation) | 1.81 c | 81 |

**Fig. 15.11.** Cyst nematode on potato. Left-Infected, Right-Healthy.

## (iii) Cyst Nematode, *Globodera rostochiensis*

**(a) Symptoms:** The disease usually occurs in patches. The infested plants exhibit typical symptoms of patchy growth of weak and stunted plants (Fig. 15.11). The patches increase in time with continuous potato cropping. Under conditions of severe infestation, the plant growth is stunted and wilting occurs during hot part of the day. Plants show tufting of leaves at the top as the outer leaves turn yellow and die. Root system is infected with adults and cysts (Fig. 15.12).

Root system is smothered, secondary roots are induced at the collar region and the plants can be easily pulled out. Tubers formed are less in number and reduced in size. In extreme cases, tuber formation is arrested. The total photosynthesis per plant is also significantly reduced as a result of reduced leaf area and this is reflected in reduced potato yield.

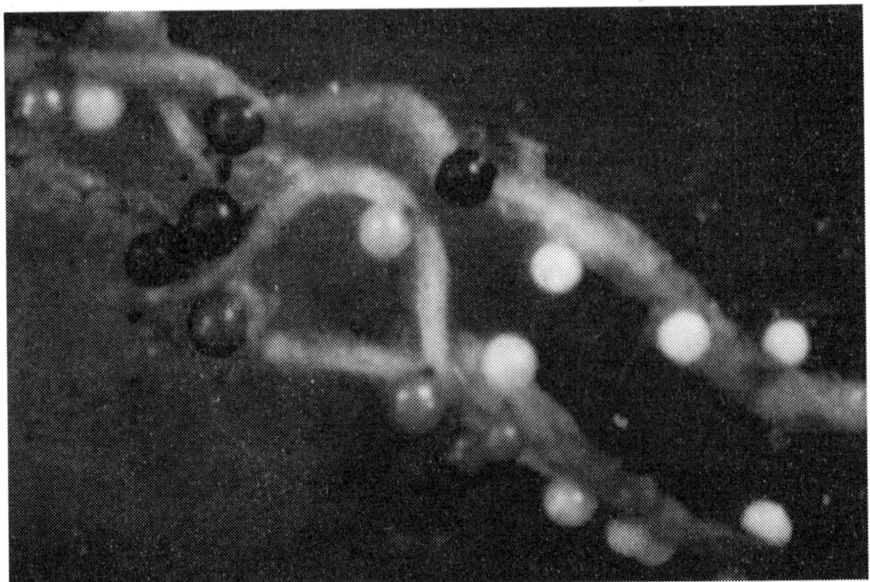

**Fig. 15.12.** Potato roots infected with *Globodera rostochiensis.*

### (b) Management

**Crop Rotation:** Continuous cropping with potatoes increases the cyst population of *G. rostochiensis.* More than three year rotations with wheat, strawberry, cabbage, cauliflower, peas, maize and beans reduces the nematode population to a safe level. Growing non-host crops like radish, garlic, beet root, French bean, cruciferous vegetables, turnip or green manuring crops bring down the cyst nematode population by more than 50%. A crop rotation pattern (using potato-French bean-peas-potato), wherein potato is grown once after 3-4 other crops, decreased the nematode population by 98 to 99% and increased yields of potato by 90%. In Nilgiris, crop rotation with cabbage and carrot gave effective control of cyst nematodes. Growing of vegetables and resistant potato in a long term rotation trial brought down the cyst propagule population up to 92% and was effectively used for nematode management at Nilgiris.

**Intercropping:** Potatoes grown with white mustard in a pot of infested soil were less heavily attacked by nematodes (*Globodera* *rostochiensis)* than potatoes growing alone. Potato root diffusate was ineffective in the presence of leachings from the roots of mustard seedlings. In field trials, mustard oils increased the yield of potatoes by reducing the severity of nematode attack.

**Organic Amendments:** Incorporation of eucalyptus litter at 10 t/ha at the time of preparation of land reduces the population of the potato cyst nematodes. Application of eucalyptus distillation waste at 2.5 t/ha helps to reduce the cyst nematode population in soil.

**Biofumigation:** Black Mustard, Caliente Mustard, Marigolds and Sudan Grass have the potential to reduce PCN by biofumigation. Chopped and incorporated into soil, the green manure releases chemicals that 'fumigate' the soil killing the cysts as well as fungal spores and weed seeds.

The effect of two concentrations (0.2 and 0.05 µmoles.100 g-1 dry weight) of total glucosinolates from 6 plants extracts (broccoli, cauliflower, collards of *Brassica rapa*, collards of kale, Portuguese cabbage and watercress) were evaluated on the potato cyst nematodes (*G. rostochiensis*). The results showed a highly significant effect ($P<0.001$) of the type of plant extract and concentration on the suppression of cyst nematodes. On average, the minor reproductive factor (Rf) was observed for water cress, followed by cauliflower and *Brassica rapa*, based on the levels of 2-phenthyl and 2- propenyl glucosinolates. The results suggest also that the incorporation of plant extracts, simultaneously to the potato cultivation could be an effective strategy to suppress the potato cyst nematode population (Aires *et al.*, 2006).

**Solarization, Organic Amendments and Bioagents:** Better control of *G. rostochiensis* by *Pasteuria penetrans* was possible with soil solarization combined with application of FYM or karanj cake which resulted in greater kill of nematodes (Sitaramaiah and Naidu, 2003).

### (iv) Root-lesion Nematode, *Pratylenchus* spp.

**(a) Symptoms:** High populations of lesion nematodes cause areas of poor growth; plants are less vigorous, turn yellow and cease to grow. Damage is often caused by direct feeding and, usually, only cortical tissues are affected. Large nematode populations cause extensive lesion formation and cortex destruction of unsuberized feeder roots (Mai *et al.*, 1981). Tubers are often attacked and small lesions are formed on the surface.

## (b) Management

**Crop Rotation:** For root lesion nematode control on potatoes, researchers found that forage pearl millet (Canadian Hybrid 101) and marigold (Crakerjack) as rotation crops with potatoes resulted in fewer root lesion nematodes and increased potato yields than rotation with rye (Ball-Coelho *et al.*, 2003).

**Green Manuring:** Oil radish as a green manure has dramatically reduced stubby root nematode (*Trichodorus*). and root lesion nematode (*Pratylenchus*) in Idaho potato fields (Anon, 2001).

On-farm research in western Idaho showed that rapeseed green manures decreased soil populations of root lesion nematodes to a greater extent than did Sudan grass green manures. Fall Sudan grass should be ploughed down after it is stressed (i.e., the first frost, stopping irrigation). Winter rapeseed and canola should be incorporated in very early spring (Cardwell and Ingham, 1996).

## (v) Potato Rot Nematode, *Ditylenchus destructor*

*D. destructor* is the most important pest of potato tubers and is responsible for dry rot of tubers. High yield losses occur in the areas where climatological conditions favour establishment of the potato rot nematodes. The effect of nematodes will manifest itself at harvest or storage when infected tubers will rot.

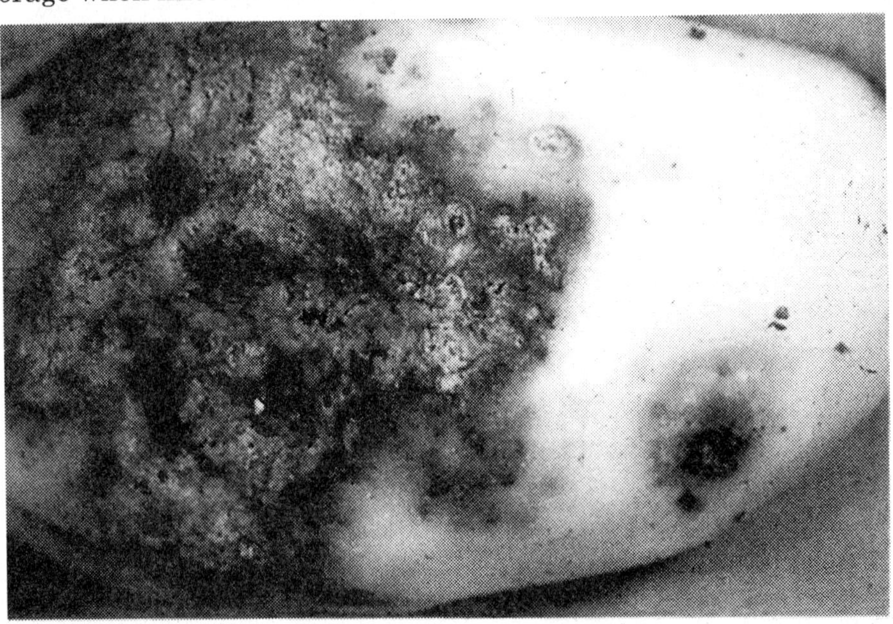

**Fig. 15.13.** Dry rot of potato tuber incited by *Ditylenchus destructor*.

**(a) Symptoms:** *D. destructor* enters potato tubers through lenticels and initially causes small white mealy spots just below the surface that are only visible if the skin is removed. Infested areas enlarge and coalesce and light brown lesions, consisting of dry granular tissue, may be visible beneath the skin. As the infestation progresses, the tissues dry and shrink and the skin becomes cracked and papery (Fig. 15.13). Internal tissues gradually darken and there are often secondary invasions of fungi, bacteria, mites, etc. If stored in moist conditions, a general rot may ensue and spread to neighbouring tubers.

### (b) Management

**Crop Rotation:** Two to three years rotation of potato with beans, corn and fallow kept free of weeds gave a very marked reduction in nematode populations. Satisfactory control of *D. destructor* can also be obtained by crop rotation with small grains, vetch and lupine provided potatoes are grown once in 3-4 years. Rotation of potatoes with sugar beet and other non-host crops can reduce nematode populations.

## 15.1.3. Weeds

***(i) Soil Solarization:*** Soil solarization helps in reducing the weed density. In solarized soil, number of weeds and fresh weight of weeds per plot of 1.5 m$^2$ were, 5g and 4.29g, respectively. Whereas, corresponding figures for control were 265g and 416g, respectively (Table 15.16). Reduction in weed population in solarized plots could be due to direct thermal killing of weed seeds which might have occurred because of increase in temperature below the tarp.

Table 15.16. Effect of soil solarization and organic mulch on seed potato tuber yield and Weeds.

| Treatment | Tuber | | Weed ( 45 DAP) | |
|---|---|---|---|---|
| | Number/1.5 m$^2$ | Yield ( t/ha) | Number/1.5 m$^2$ | Weight ( t/ha) |
| Soil solarization | 370 | 33.83 | 5 | 0.43 |
| Organic mulch | 304 | 22.20 | 259 | 41.07 |
| Control | 263 | 20.33 | 265 | 41.61 |
| CD at 5% | 44.6 | 3.37 | 31.6 | 5.08 |

***(ii) Crop Rotation:*** In potatoes, with fall-planted and spring-incorporated "Juniper" winter rape (*Brassica napas*), weed biomass was reduced by 96% when common lambsquarter (*Chenopodium* *album*) and 50% when redroot pigweed (*Amaranthus retroflexus*) were the major weed species.

***(iii) Biofumigation:*** In greenhouse trials, *Brassica hirta*, *B. juncea*, and *B. napus* chopped residues added to soil inhibited emergence of several small-seeded weed species from 20 to 95% and reduced growth of weed seedlings from 8 to 90%. Redroot pigweed and Barnyardgrass, *Echinochloa crusgalli,* germination and growth were inhibited to various degrees by eight individual isothiocyanates, with methyl- and allyl isothiocyanates being the most inhibitory. Meadow foam, *Limnanthes alba* seed meal added to soil in greenhouse studies at 1% or more by weight reduced fresh weight of redroot pigweed by greater than 90%.

In subsequent field trials, 4,500 to 18,000 kg/ha meadow foam incorporated in the top 10 cm of soil reduced weed emergence in one of two years in potato, but total weed biomass at potato harvest was not reduced (Boydston *et al.*, 2004).

## 15.2. TOMATO, *LYCOPERSICON ESCULENTUM*

### 15.2.1. Diseases

#### (i) Bacterial Wilt, *Ralstonia solanacearum*

This is the most serious bacterial disease of solanaceous crops. The disease is prevalent in warmer climate having high humidity under tropical and subtropical conditions.

**Fig. 15.14.** Bacterial wilt affected tomato plants and bacterial ooze from cut surface of stem.

**(a) Symptoms:** Bacterial wilt has often been reported as the most serious handicap for tomato in the tropics. The disease causes rapid wilting and death of the entire plant without any yellowing or spotting of leaves. Total collapse of the plant usually occurs when temperature reach 32°C and above. Plant wilts while still green. The lower leaves may droop before wilting occurs. The vascular system becomes brown, root and rootlets rots leaving behind only corky portion of main root. When the stem of a wilting plant is cut across, the pith has a darkened water-soaked appearance, and on squeezing the cut stem, a white, yellow or grayish, slimy exudate may appear (Fig. 15.14). In later stages of the disease, decay of the pith may cause extensive hollowing of the stem.

### (b) Management

**Biofumigation:** The control efficacy of *Brassica* variety TB 675 as a green manure was evaluated in pot and field trials. Results of preliminary pot trials showed that amendment using chopped tissues with a 28-day incubation period or amending macerated tissues with a three-day incubation resulted in a population density of *R. solanacearum* of 10 or 100 cell per g soil compared with $10^6$ cell per g soil in the control. This suggests both short-term biofumigation and longer-term organic matter effects can suppress the pathogen population. Preliminary field trial results showed a TB 574 green manure could delay disease development. Growing Indian mustard as a green manure to control tomato bacterial wilt can be a component of an integrated management program. However, this can only be achieved by producing sufficient biomass of the green manure and understanding of other factors that could affect the control efficacy, such as soil type and moisture content (Wang *et al.*, 2008).

The suppression of bacterial wilt in soil resulted from two separate mechanisms. The first mechanism was related to the "biofumigation" as usually defined - the short-term release of isothio-cyanates. It was most obvious on sandy soils with low organic matter content, and could be enhanced using strategies to increase ITC release and persistence in the soil. The second mechanism occurred over 3-4 weeks and was unrelated to GSL content of the added tissue but was related to the added organic matter. Further experiments utilizing additions of pure C and N sources confirmed that the N components of the amendments were most active in suppression and N sources alone could reproduce the suppressive effects. Investi-gations suggested this suppression was not related to a general shift in the soil biological community but to more specific changes in the microbial community.

Freeze-dried powder of a higher GSL plant that was selected in this programme reduced tomato bacterial wilt by *R. solanacearum* more than lower GSL lines (Maeda) (Hashizume *et al.*, 2008).

The reduction and long-term elimination of bacterial diseases from the soil by incorporating especially mustard or radish plants in large amounts into the soil immediately before planting tomatoes has been developed. This practice is reported to reduce incidence of bacterial wilt by 50-70% in the Philippines.

Indian mustard (*Brassica juncea*) and Fall Raab (*B. rapa*) were evaluated as winter cover crops/spring biofumigation treatments in plasticulture tomato production. Biofumigation treatments significantly affected both number and total weight of marketable fruit ($P < 0.01$) (Fig 15.15, Table 15.17). The highest tomato yields were produced on plots treated with the 'Fall Raab' biofumigation treatment. Plots with 'Fall Raab' incorporated into the soil had 40% more marketable tomato fruit than plots with the rye (control) incorporated into the soil. The 'Fall Raab' treated plots also produced 19% more marketable fruit than plots with soil incorporated with the Indian mustard cover crop. Plots with Indian mustard incorporated into the soil produced 18% more marketable fruit yield than plots with the rye (control) incorporated into the soil.

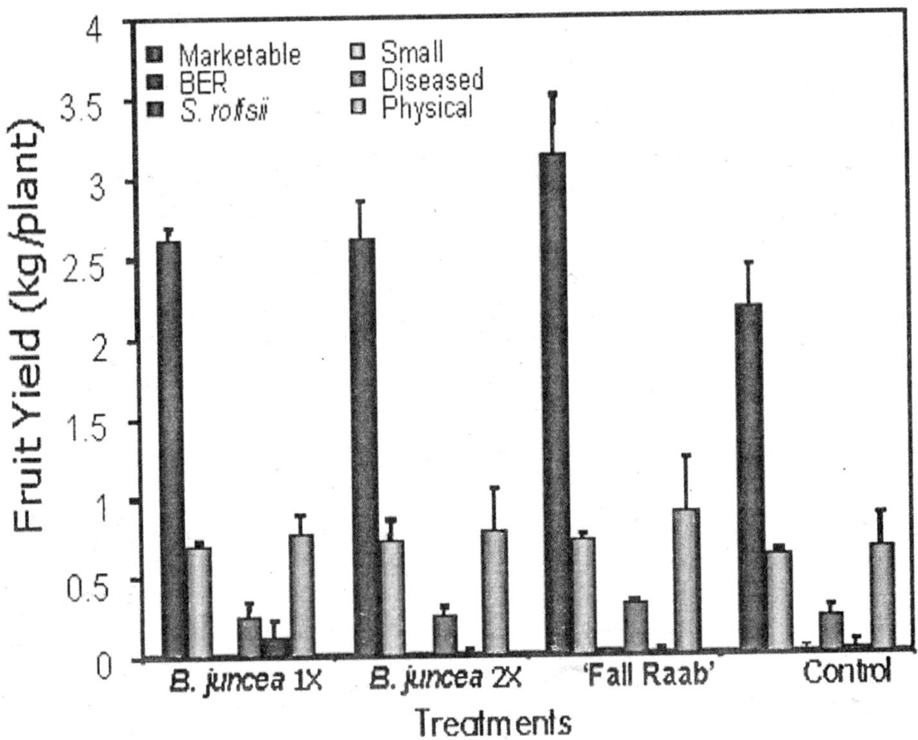


**Fig. 15.15.** Distribution of fruit yield per plant among treatments

Table 15.17. Marketable tomato fruit yield per plant by treatment.

| Treatment | Number of fruits/plant* | Weight of fruits (kg/plant)* |
|---|---|---|
| Indian mustard- 1x | 14.6 ± 3.4 b | 2.6 ± 0.7 b |
| Indian mustard- 2x | 14.7 ± 2.7 b | 2.6 ± 0.5 b |
| Fall raab – 1x | 17.3 ± 2.2 a | 3.1 ± 0.5 a |
| Rye (control) | 11.9 ± 2.8 c | 2.2 ± 0.5 c |
| | P < 0.01 | P < 0.01 |

* Letters represent mean separation (CD P =0.05).

**Cover Cropping:** Studies have suggested that cover crops, especially brassicas, can provide some of the benefits of a fumigant. This test looked at > Green Sprouting = broccoli raab > Florida Broadleaf = mustard and pasture wheat as cover crops compared with standard fumigation using methyl bromide.

Tomatoes grown on methyl bromide plots tended to have the highest yields; production on the wheat plots had the lowest yields. Tomato yields on methyl bromide and mustard plots were higher than yields with broccoli cover crop (Table 15.18).

Table 15.18. Effect of cover crops on marketable yield of tomato

| **Treatment** | **Total Marketable Yield (lb/acre)** | **Significance** |
|---|---|---|
| Mustard | 1,430ab | *** |
| Broccoli | 1,210b | |
| Wheat | 450c | |
| Methyl Bromide | 1,770a | |
| Methyl Bromide | 13,000a | |

**Crop Rotation:** Following 3 years rotation with cereals/crucifers are effective for the management of wilt. Crop rotation with cowpea-maize-cabbage, okra-cowpea-maize, maize-cowpea-maize and finger millet-brinjal (Pusa Purple Cluster)-French bean are reported effective in reducing bacterial wilt.

**Intercropping:** Intercropping of tomato with sorghum, maize, onion, garlic and marigold resulted in better crop stand.

Green Manuring and Crop Rotation: Soil amendment with green manuring crops, transplanting disease-free seedlings and following 3 years rotation with cereals/crucifers are effective for the management of wilt.

**Green Manuring and Bioagents:** The combined treatment of *Pseudomonas fluorescens* at 1 x $10^{10}$ cfu/ml (seed treatment) + green manuring with sunhemp was highly effective in reducing the bacterial wilt incidence to 6.75% and significantly increasing the fruit yield to 25.5 MT/ha as compared to high wilt incidence of 55.6% and low yield of 7.8 MT/ha recorded in control (Gopalakrishnan and Ajit Kumar, 2006).

### Integrated Methods

- Crop rotation with French beans reduces bacterial wilt.
- Soil solarization of nursery beds for 2-3 weeks using transparent polythene sheets (100-160 gauge) kills soil-borne pathogens.
- Deep summer ploughing helps in exposing pathogens and their resting stages to sun's heat.
- Green manuring with sunnhemp in July – August.

### (ii) Bacterial Canker, *Clavibacter michiganensis* sub sp. *michiganensis*

**(a) Symptoms:** Temporary and later permanent wilting of leaflets of tomato plants are the first symptoms of the disease. Light streaks appear at the juncture of petiole and stem extending down the internodes and up to the petiole. As these streaks become darker, epidermal tissues may rift and longitudinal canker like opening may appear in stem, petiole and midrib. When the stem of diseased plants cut longitudinally, a creamy white, yellow or brown line follows the phloem. Discolouration of vascular system can be traced up to the fruit, placenta and seeds. Secondary infection appears on green fruits as water-soaked spots with a white halo. The spots remain less than 3 mm in diameter exclusive of halo. Halo is the distinguishing character of bacterial canker of tomato (Fig. 15.16). The pith is easily separated from the wood along this line.

The vascular bundles within the pith are destroyed and cavities are formed in the pith. Young affected fruits are small, deformed and seeds are aborted. Spots may also develop on older fruits. These spots are circular, up to three mm wide, with slightly raised brown centre surrounded by a pronounced white halo resembling a birds' eye. The halo is flat. The bacterium is seed-borne and can survive up to three years in soil in crop debris. Spread within a crop is by rain splash, sprinklers and pruning knives.

**Fig. 15.16.** Bacterial canker on leaves, stem and fruits of tomato.

**(b) Management**

**Solarization:** The bacterial canker of tomato (*C. michiganensis* sub sp. *michiganensis*), could be successfully controlled by the application of soil solarization. The temperatures developed during soil tarping are sufficient to reduce significantly or eliminate bacterial inocula.

**(iii) Bacterial Speck, *Pseudomonas syringae* pv. *tomato***

**(a) Symptoms:** The disease starts as small, round, water-soaked spots, which gradually become yellowish brown and finally blackish brown with sunken ashy centre and pale yellowish green halo. Later on, spots become irregular to circular, measuring 1 mm in diameter, distributed over the entire leaf lamina and in severe cases forming large blotches. Spots on petioles are more or less oval.

**(b) Management**

**Solarization:** In non solarized soil, the average number of bacterial speck lesions per tomato plant were 1.05, in contrast to only 0.24 in solarized soil. Also, there were 104 infected transplants in non solarized soil, whereas solarized soils harboured only 39 infected transplants. These results suggested a reduction in bacterial speck disease incidence of 63% by soil solarization in tomato.

**(iv) Damping-off, *Pythium aphanidermatum***

**(a) Symptoms:** In the pre-emergence phase of the disease, the young seedlings are killed even before they emerge out of the soil surface. However, this disease mostly occurs at post-emergence stage which is characterized by the toppling over of the infected seedlings anytime after they emerge from the soil until the stem has hardened sufficiently to resist invasion by the pathogen. Infestation usually occurs at ground level or through roots. The infected tissue appears soft and water-soaked. As the disease advances, the stem becomes constricted at the base of the plants that collapse later. Seedlings that are healthy looking one day may have collapsed the next morning.

**(b) Management**

**Organic Amendments:** Cultural practices like application of well decomposed FYM should be adopted.

**Solarization:** Soil solarization using polythene sheet for 30 days during peak summer was most effective and resulted in least occurrence of damping off (3.9%) when compared to control (30.64%).

**Solarization and Bioagents:** Combination of the seed/root application of *T. harzianum* or *P. fluorescens* with soil solarization was very effective in management of damping-off of tomato in nursery at farmers' field.

**(v) Southern Blight, *Sclerotium rolfsii***

It has wide host range and causes severe problem on tomato and other vegetable crops.

**(a) Symptoms:** The first symptom of the disease is observed as soft tissue necrosis of bark of the stem near soil line. White growth of cottony mycelium is clearly visible on the affected portion just below the soil surface. Later, dense silvery fungus growth along with white to light brown mustard-like sclerotia is observed on the same portion. Progressive drooping, yellowing or wilting of the entire plant is observed. Sometimes the plant is collapsed soon after infection. The disease is soil-borne where the pathogen survives in the form of sclerotia.

**(b) Management**

**Solarization and Bioagents:** Control by solar heating has been successful in Israel. Studies by Elad and Katan (1980) have demonstrated that 2-4 week soil solarization applications during the summer were able to reduce sclerotial numbers and limit disease. Additionally, these researchers demonstrated that combining solar heating with applications of *Trichoderma harzianum* was more effective than either treatment alone. Punja (1985) contended that by altering the composition or activities of soil microflora, including the addition of biocontrol agents such as antagonistic microorganisms, that there is a potential for disease control. *T. harzianum*, which has been described as a destructive mycoparasite that competes aggressively against other soil fungi, is acknowledged to suppress Southern Blight when introduced to field soils (Jenkins and Averre, 1986). The growth of the biological control agent *T. harzianum* is enhanced with solar heating (Jenkins and Averre, 1986).

Studies in North Carolina showed that soil solarization combined with *T. virens* reduced the disease incidence by 49% during the first season after solarization and 60% during the second season (Ristaino *et al.*, 1991).

**Biofumigation, Solarization and Chemicals:** Sixty percent of control plants developed symptoms of Southern Blight (Fig.15.17). This percentage was significantly reduced in all of the experimental treatments. The chemical fumigant, Basamid, reduced disease to 13% of all plants. Biofumigation, utilizing mustard meal or a Brassica cover crop, reduced disease to 24-29% of all plants. Interestingly, the rate of meal used did not impact disease suppression. The 1000, 2000, and 4000 lb/acre rates had similar results. In contrast to fruit yield, the integrated treatments were not more effective at controlling disease than each treatment alone.

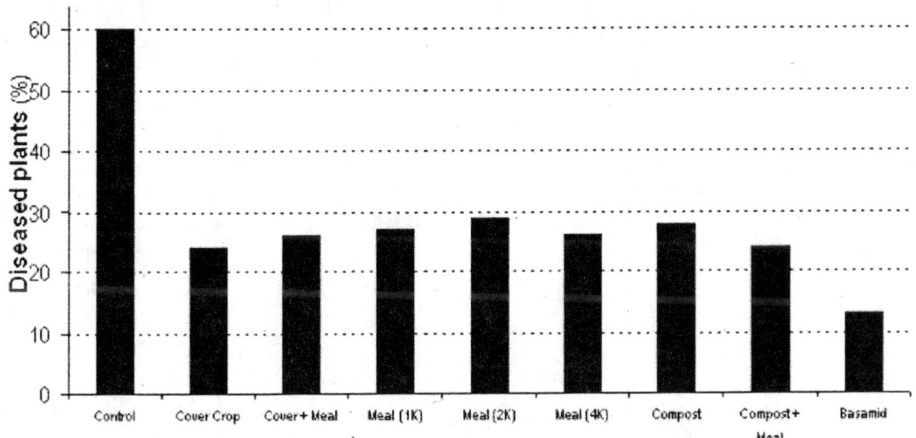


**Fig. 15.17.** Effect of different treatments on southern blight disease incidence in tomato.

Tomato yields of the plants grown in the treated plots were higher than the control. Fruit yields for the controls averaged 9.6 lbs per plant. Plots amended with an incorporated Brassica cover crop averaged 11.6 lbs of fruit per plant, a 21% increase above the control plants. When meal was also added to the incorporated cover crop, yields increased to 13.1 lb/plant, 37% above controls. Mustard meal used as an amendment alone also resulted in yields increases (Fig. 15.18). The 4000 lb/acre rate had the greatest impact. This treatment elevated yields to 13.7 lb/plant, a 43% increase above controls. The largest increases in tomato yields came from the plants grown in soil that had been amended with spent mushroom compost as well as mustard meal (2000 lb/acre). These plants produced 54% more fruit than the control plants. The data supports the concept that adding organic matter to production soils will enhance yields. Integrating two or more organic treatments, such as Brassica (mustard) plant tissue and compost is favourable. The superior yields can be attributed to an increase in available nutrients in combination with improved physical and chemical soil properties.

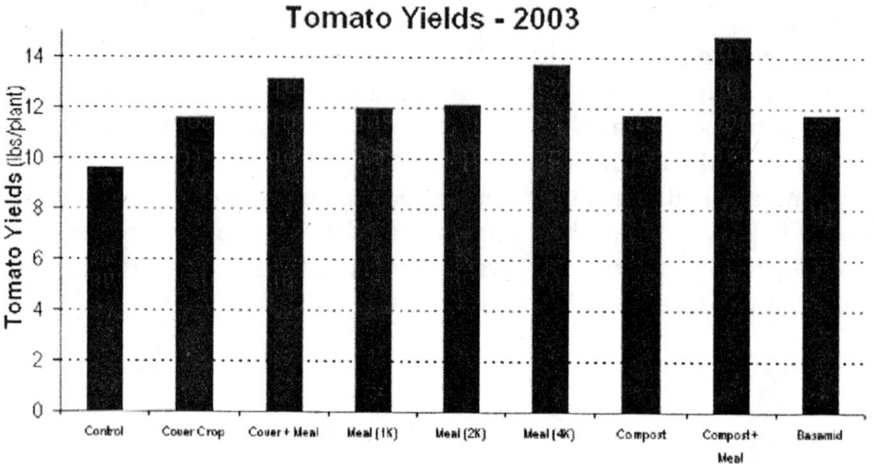


**Fig. 15.18.** Effect of different treatments on tomato fruit yield.

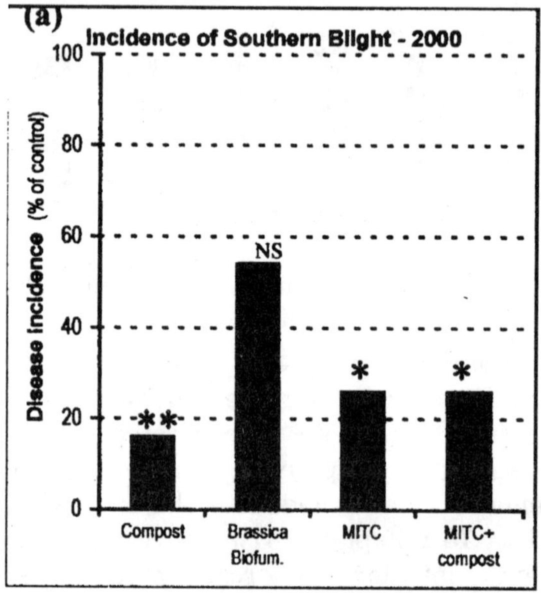


**Fig. 15.19.** Incidence of southern blight in different treatments

Pre-plant soil treatments utilizing composted amendments, biofumigation, solarization, and low dosage dazomet (Basamid) produced higher yields and resulted in a lower incidence of the soil-borne disease, southern blight, than did controls (Fig. 15.19). The compost amended plots produced yields 59% higher than controls (Fig. 15.20). MITC and biofumigation plots had 39 and 40% higher yields than controls. However, the combined MITC + compost plots produced the highest yields - 82% above control. These results suggest that the most effective treatments for prevention of tomato diseases and enhanced yields may be those that integrate two or more control mechanisms while also favouring the growth of beneficial soil organisms.

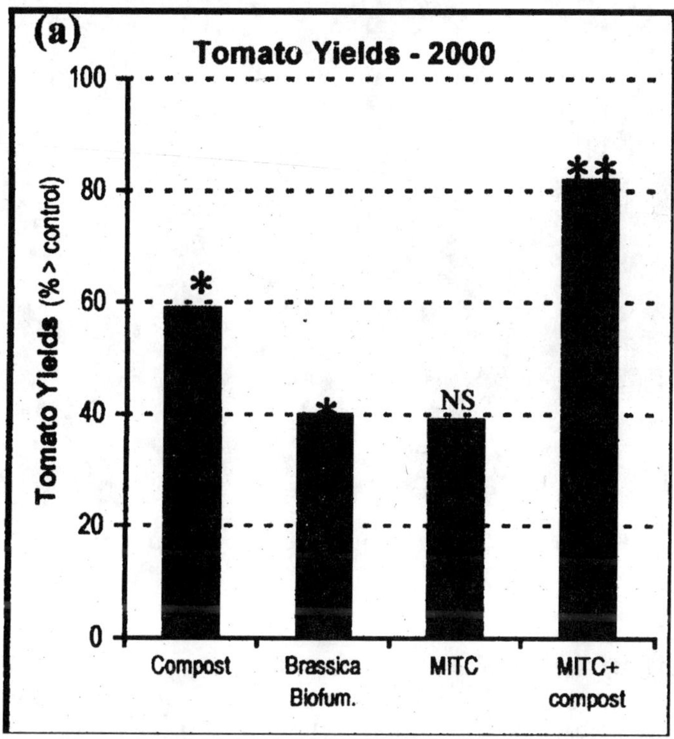


**Fig. 15.20.** Tomato yields in different treatments

The composts were applied at 30 MT/acre

The MITC (Basamid) was applied at 350 lbs/acre.

**Crop Rotation:** Long rotations with cereals viz., corn, sorghum and millets will reduce the disease incidence.

**Solarization:** Summer ploughing for 3 times and irrigating the field destroys the germinating sclerotia. Maintaing proper aeration near collar region and avoiding dense planting helps to keep the disease incidence low.

### (vi) Wilt, *Fusarium oxysporum f. sp. lycopersici*

**(a) Symptoms:** The wilt disease is becoming a major disease of concern in many parts of the country. The disease is characterized by yellowing and wilting of leaves and finally the entire plant wilts and dies prematurely (Fig. 15.21). Stem tissue often is discoloured throughout the plant. Vascular browning takes place in the root. The fungus can persist in the soil for many years and is virulent at moderate temperature (26-28°C).

**Fig. 15.21.** Tomato plant infected by *Fusarium* wilt.

### (b) Management

**Solarization**: In Morocco, solar heating of the soil proved to be efficient in controlling *F. oxysporum* f. sp. *lycopersici* on tomato (Besri and Drame, 1982). Typical maximal temperature in the open field of non-solarized soil, recorded at depth of 20 cm, was 35°C at Ismailia and Giza, while temperatures of the solarized soil were 8-15°C higher at Giza and Ismailia, respectively.

### (vii) Wilt, *Verticillium dahliae*

### (a) Management

**Solarization and Fertilization:** In Morocco, solar heating of the soil proved to be efficient in controlling *V. dahliae* on tomato (Besri and Drame, 1982). The mean maximum temperature at 15 cm depth in the solarized plot was 42°C (9°C higher than non-covered control soil), and the mean minimum was 34°C (7°C higher than the control). Maximum temperatures at 15 cm depth in the fine sandy loam plot were 44°C and 36°C in solarized and control plots, respectively.

Factorial ANOVA revealed significant activity of fertilizers against *V. dahliae* in the loam soil, but not in the fine sandy loam (Table 15.19). Fungicidal activity of solarization was highly significant in both soils. A highly significant fertilizer-solarization interaction was found against *V. dahliae* in the loam soil.

Table 15.19. Populations of *Verticillium dahliae* in soil treated with ammonia-based fertilizers and/or solarization[1]

| Treatment | Loam soil | Fine sandy loam soil |
|---|---|---|
| Aqua ammonia | 5.5 | 17.4 |
| Aqua ammonia + Solarization | 0.1 | 5.2 |
| Urea | 1.6 | 23.0 |
| Urea + Solarization | 0 | 5.7 |
| Ammonium sulphate | 3.0 | 0.5 |
| Ammonium sulphate + Solarization | 0 | 0 |
| Ammonium phosphate | 8.9 | 14.0 |
| Ammonium phosphate + Solarization | 0 | 0.4 |
| Non-treated Control | 9.5 | 19.2 |
| Solarization | 0.1 | 0 |
| LSD | 3.6 | 13.1 |
| Factorial analysis for variance | | |
| Fertilizers | P<0.01 | NS |
| Solarization | P<0.01 | P<0.01 |
| Fertilizers[3] x Solarization[4] | P<0.01 | NS |

1 Fertilizers were applied at 305 kg N/ha. 2 Populations are colony-forming units (cfu)/g of oven-dried soil at 0.15 cm soil depth 3 P<0.05 or 0.01; NS = no significant differences. 4 Interaction of treatments.

There were no differences in number of surviving plants in either soil, and no symptoms of *Verticillium* wilt were observed. Factorial ANOVA showed both fertilizers and solarization to be highly significant in increasing all growth parameters in the loam soil; however, no fertilizer-solarization interactions were found (Table 15.20). No significant differences in tomato growth due to fertilizers were found in the sandy loam plot. Solarization, however, was highly significant in increasing numbers of fruit and fruit fresh weight.

Table 15.20. Fresh market tomato ("Early Pak 7") growth in soil treated with ammonia-based fertilizers and/or solarization

| Soil Treatment[1] | Loam soil | | | Fine sandy loam soil | | |
|---|---|---|---|---|---|---|
| | Number of fruit[2] | Fruit fresh wt.(g) | Vegetative fresh wt. (g) | Number of fruit | Fruit fresh wt. (g) | Vegetative fresh wt.(g) |
| Aqua ammonia | 5.9 | 90.8 | 604 | 22.7 | 398.3 | 1386 |
| Aqua ammonia + Solarization | 14.0 | 210.4 | 1102 | 30.5 | 493.6 | 1384 |
| Urea | 8.7 | 151.8 | 538 | 26.0 | 490.6 | 1130 |
| Urea + Solarization | 14.5 | 206.4 | 1 080 | 30.7 | 605.0 | 1246 |
| Ammonium Sulphate | 12.3 | 175.1 | 690 | 22.6 | 404.8 | 1160 |
| Ammonium Sulphate + Solarization | 21.4 | 382.6 | 1 582 | 32.3 | 552.0 | 1660 |
| Ammonium Phosphate | 18.5 | 321.5 | 1 162 | 27.9 | 359.7 | 1754 |
| Ammonium Phosphate + Solarization | 26.1 | 502.7 | 1 668 | 28.3 | 533.0 | 1766 |
| Non-treated Control | 11.7 | 132.1 | 638 | 22.4 | 360.3 | 1180 |
| Solarization | 15.0 | 215.7 | 1 078 | 33.5 | 574.4 | 1448 |
| LSD | 8.0 | 171.9 | 437 | 11.9 | 247.6 | 670 |
| Factorial analysis of variance Fertilizers | P<0.01 | P<0.01 | P<0.01 | NS | NS | NS |
| Solarization | P<0.01 | P<0.01 | P<0.01 | P<0.01 | P<0.01 | NS |
| Fertilizers[3] x Solarization[4] | NS | NS | NS | NS | NS | NS |

1 All fertilizers were applied at 305 kg N/ha. 2 All values are given on a per-plant basis. 3 P = 0.05 or 0.01; NS = no significant differences. 4 Interaction of treatments

### *(viii) Pythium ultimum*

#### (a) Management

**Solarization and Fertilizers:** Solar heating of the soil proved to be efficient in controlling *P. ultimum* on tomato (Besri and Drame, 1982).

Composted chicken manure, but not ammonium phosphate reduced numbers of *P. ultimum* (Tables 15.21 and 15.22). Fungal numbers and galling index generally decreased with increasing fertilizer dosage. Soil solarization reduced cfu of *P. ultimum* in all cases. No interaction between ammonium phosphate fertilization

Table 15.21. *In vitro* effects of ammonium phosphate fertilization and soil heating on numbers of *Pythium ultimum*

| Treatment and dosage | *P. ultimum* (cfu/g soil) |
|---|---|
| Ammonium phosphate-100 mg/1 | 8.1 |
| 100 mg/1 + Solarization | 4.5 |
| Ammonium phosphate-200 mg/1 | 8.3 |
| Ammonium phosphate-200 mg/1 + Solarization | 3.6 |
| Ammonium phosphate-300 mg/1 | 9.5 |
| Ammonium phosphate-300 mg/1 + Solarization | 5.1 |
| Ammonium phosphate-400 mg/1 | 4.8 |
| Ammonium phosphate-400 mg/1 + Solarization | 6.0 |
| Control | 17.0 |
| Heat only | 5.7 |
| LSD | 8.1 |
| Factorial ANOVA | NS |
| Fertilizer | |
| Solarization | 0.05 |
| Fertilizer x Solarization | NS |

Table 15.22. In vitro effect of amending soil with composted chicken manure' and/or soil heating[2] on numbers (cfu) of *Pythium* sp.

| Treatment manure dosage | *Pythium* sp. (cfu/g soil) |
|---|---|
| Control | 14.4a[3] |
| Solarization only | 12.0b |
| Chicken manure at 2690 kg/ha | 9.6b |
| Chicken manure at 5381 kg/ha | 1.1c |
| Chicken manure at 2690 kg/ha + Solarization | 1.8c |
| Chicken manure at 5381 kg/ha + Solarization | 0 c |

1 Containing 3280 mg $NH_4$-N/kg. 2 Three day incubation. Diurnal healing regime: 42°C high; 18°C low. 3 Values within columns followed by different letters are different at P<0.05 according to Duncan's Multiple Range Test.

The mean maximum temperature at 15 cm depth in the solarized plot was 42°C (9°C higher than non-covered control soil), and the mean minimum was 34°C (7°C higher than the control). Maximum temperatures at 15 cm depth in the fine sandy loam plot were 44°C and 36°C in solarized and control plots, respectively.

Factorial ANOVA revealed significant activity of fertilizers against *P. ultimum* in the loam soil, but not in the fine sandy loam (Table 15.23). Fungicidal activity of solarization was highly significant in both soils.

Table 15.23. Populations of *Pythium ultimum* in soil treated with ammonia-based fertilizers and/or solarization[1]

| Treatment | Loam soil | Fine sandy loam soil |
|---|---|---|
| Aqua ammonia | 131.2 | 38.3 |
| Aqua ammonia + Solarization | 34.8 | 14.8 |
| Urea | 101.7 | 39.2 |
| Urea + Solarization | 18.8 | 5.2 |
| Ammonium sulphate | 117.8 | 41.8 |
| Ammonium sulphate + Solarization | 46.9 | 7.0 |
| Ammonium phosphate | 150.0 | 48.8 |
| Ammonium phosphate + Solarization | 40.2 | 12.2 |
| Non-treated Control | 150.0 | 66.1 |
| Solarization | 29.5 | 7.0 |
| LSD | 32.5 | 28.3 |
| Factorial analysis for variance | | |
| Fertilizers | $P<0.05$ | NS |
| Solarization | $P<0.01$ | $P<0.01$ |
| Fertilizers[3] x Solarization[4] | NS | NS |

1 Fertilizers were applied at 305 kg N/ha. 2 Populations are colony-forming units (cfu)/g of oven-dried soil at 0.15 cm soil depth 3 $P<0.05$ or 0.01; NS = no significant differences. 4 Interaction of treatments.

There were no differences in number of surviving plants in either soil. Factorial ANOVA showed both fertilizers and solarization to be highly significant in increasing all growth parameters in the loam soil; however, no fertilizer-solarization interactions were found (Table 15.24). No significant differences in tomato growth due to fertilizers were found in the sandy loam plot. Solarization, however, was highly significant in increasing numbers of fruit and fruit fresh weight.

Table 15.24. Fresh market tomato (*Lycopersicon esculentum* "Early Pak 7") yield in soil treated with ammonia-based fertilizers and/or solarization

| Soil Treatment[1] | Loam soil | | | Fine sandy loam soil | | |
|---|---|---|---|---|---|---|
| | Number of fruit[2] | Fruit fresh wt.(g) | Vegetative fresh wt. (g) | Number of fruits | Fruit fresh wt.(g) | Vegetative fresh wt. (g) |
| Aqua ammonia | 5.9 | 90.8 | 604 | 22.7 | 398.3 | 1386 |
| Aqua ammonia + Solarization | 14.0 | 210.4 | 1102 | 30.5 | 493.6 | 1384 |
| Urea | 8.7 | 151.8 | 538 | 26.0 | 490.6 | 1130 |
| Urea+ Solarization | 14.5 | 206.4 | 1 080 | 30.7 | 605.0 | 1246 |
| Ammonium Sulphate | 12.3 | 175.1 | 690 | 22.6 | 404.8 | 1160 |
| Ammonium Sulphate + Solarization | 21.4 | 382.6 | 1 582 | 32.3 | 552.0 | 1660 |
| Ammonium Phosphate | 18.5 | 321.5 | 1 162 | 27.9 | 359.7 | 1754 |
| Ammonium Phosphate + Solarization | 26.1 | 502.7 | 1 668 | 28.3 | 533.0 | 1766 |
| Non-treated Control | 11.7 | 132.1 | 638 | 22.4 | 360.3 | 1180 |
| Solarization | 15.0 | 215.7 | 1 078 | 33.5 | 574.4 | 1448 |
| LSD | 8.0 | 171.9 | 437 | 11.9 | 247.6 | 670 |
| Factorial analysis of variance Fertilizers | P<0.01 | P<0.01 | P<0.01 | NS | NS | NS |
| Solarization | P<0.01 | P<0.01 | P<0.01 | P<0.01 | P<0.01 | NS |
| Fertilizers[3] x Solarization[4] | NS | NS | NS | NS | NS | NS |

1 All fertilizers were applied at 305 kg N/ha. 2 All values are given on a per-plant basis. 3 P = 0.05 or 0.01; NS = no significant differences. 4 Interaction of treatments

### (ix) Stem Canker, *Didymella lycopersici*

It has been shown that the pathogen (*D. lycopersici*) survives mainly in the reed canes (*Arundo donax*) and in the *Eucalyptus* sp. stakes used to trellis the tomato plants. Besri (1982 a, b) reported that the incidence of the pathogen is higher in the plots where the plants are staked with supports regularly used in the tomato farms than in the plots where the tomato plants are supported with new stakes. With the stakes which have been frequently used, it was also observed that 90 percent of the *Didymella* stem cankers are located at the contact point between the tomato stem and the stakes. The pathogen was isolated from the stakes and its pathogenicity confirmed by tomato plant inoculation (Besri, 1982 a, b).

### (a) Management

**Solarization of the Tomato Supports**: Infested tomato supports are exposed in June on wet soil and then covered with a white transparent polyethylene sheet (180 gauge) which has been in use for three years as a cover to a plastic house. The maximum temperatures recorded under the polyethylene sheet were always higher than 5° to 20°C, the maximum air temperature varying between 26°C and 34°C (Besri, 1983). The tomato stakes were solarized for three months (June-August), and then used to support the tomato plants. It was observed that solarization of the tomato stakes considerably decreased the incidence of *D. lycopersici* and delayed the disease occurrence (Besri, 1983). The high temperatures obtained under the polyethylene mulch killed the fungus and suppressed the primary inoculum.

In Morocco, tomatoes for export are grown both in plastic houses (November - May) and in the open air (September - January). From June to October, the plastic houses are empty and the temperatures in these houses are higher than 50°C. Therefore, these houses could be used to solarize the stakes (from June to August) before their use as tomato supports. Besri (1983) reported that the incidence of *Didymella* stem canker is considerably reduced when the tomato stakes are stored in the plastic houses instead of their storage in the open air. This new procedure of exposing the tomato stakes to high temperatures during summer months is simple and cheap. It presents excellent prospects for controlling *D. lycopersici* and maybe other pathogens which survive in the stakes, in Morocco and in other countries with similar climates.

### (x) Corky Root, *Pyrenochaeta lycopersici*

### (a) Management

**Biofumigation and Solarization:** Soil solarization alone or in combination with *Brassica* green manures resulted in a significant ($P < 0.05$) decrease of corky root (*P. lycopersici*) severity of tomato in all three years of trial (1999 through 2002). *Brassica oleracea* ("Senshi") green manure treatment reduced significantly ($P < 0.05$) the disease in two years out of three. *B. rapa* local ecotype caused a significant (P <

0.05) reduction of corky root, and tomato yield increased in one year of testing out of three. Soil manuring with *B. oleracea* ("Senshi") resulted in a significant ($P < 0.05$) increase of tomato yield in two years out of three.

The effects of soil-solarization and green manures on soil population density of fluorescent *Pseudomonas* sp., *Bacillus* sp., *Streptomyces* sp., total bacteria, and total fungi were also investigated. Immediately after its application, soil-solarization alone or in combination with green manures significantly ($P < 0.05$) reduced fluorescent *Pseudomonas* sp. population. However, *Pseudomonas* population increased at the end of tomato cropping in most cases. At the end of tomato cultivation, *Bacillus* sp. population was higher in solarized plots than in not solarized ones. Solarization and green manuring had no effect on *Streptomyces* sp., total bacteria, and total fungi. Corky root severity was negatively correlated with *Bacillus* sp. population density in all trials (Amenduni *et al.*, 2004).

### *(xi) Root Rot, Rhizoctonia solani*

#### (a) Management

**Biofumigation:** Biofumigation with amendments of broccoli, roquette and mustard (10% v/v) against *R. solani* was demonstrated in tomato greenhouse experiments. Yields were increased and disease was suppressed. Controlling root-knot nematodes by solarization alone was observed, but the combination of solarization with green manure residues resulted in even better disease suppression (Tsror *et al.*, 2006).

### (xii) Fruit Rot, *Alternaria alternata*

**(a) Symptoms:** Lesions sunken or slightly flattened, water soaked appearance with or without definite margin on fruits. Affected areas soon become brown to black in colour. Surface of lesions become covered with dense dark grey to olive green to black conidia and conidiophores.

#### (b) Management

**Biofumigation:** The effect of benzyl-isothiocyante (BITC) on *A. alternata* infecting tomato fruit packaged in low density polyethylene film (LDPF) was tested, at concentrations of 0.28 and 0.56 mg/ml. It was observed that 85% reduction in the tomato fruit infection by *A.*

*alternata* was observed without any negative effects on fruit quality and ripening physiology (Troncoso-Rojas *et al.*, 2008).

### 15.2.2. Nematodes

#### (i) Root-knot Nematodes, *Meloidogyne incognita, M. javanica*

**(a) Economic Importance:** Lal and Ansari (1960) reported *M. arenaria* for the first time on tomato from Bihar, India (Fig. 15.22). *M. incognita* was responsible for 30.57 to 46.92% loss in fruit yield of tomato (Bhatti and Jain, 1977; Reddy, 1985; Darekar and Mahse, 1988), while *M. javanica* caused 77.5% loss in yield (Anon, 1993).

**Fig. 15.22.** Heavy galling of tomato roots with *Meloidogyne incognita.*

#### (b) Management

**Solarization:** Soil solarization of nursery beds with 100 gauge LLDPE clear plastic film during summer (in between April 15$^{th}$ and

June $15^{th}$) for 15 days is recommended. It increased the number of transplantable seedlings by 615% giving net profit of Rs.7,661 from 1000 sq.m. (ICBR 1: 5.65) and decreased the root-knot disease and weeds by 66% and 93%, respectively (Table 15.25).

Table 15.25. Effect of soil solarization of nursery beds on root galling and number of transplantable seedlings of tomato in 1000 sq. m.

| Treatment | No. of seedlings | Weight of 100 seedlings | Root-knot index | Net profit | Cost: Benefit ratio |
|---|---|---|---|---|---|
| With plastic | 2,60,400 | 36 g | 1.37 | Rs. 7,661 | 1: 5.65 |
| Without plastic | 36,400 | 26 g | 3.97 | --- | --- |
| % Increase (+)/ decrease (-) over control | (+) 615 | (+) 38 | (-) 66 | --- | --- |

Red plastic mulch suppresses root-knot nematode damage in tomatoes by diverting resources away from the roots (and nematodes) and into foliage and fruits (Adams, 1997).

Tomato roots were heavily galled and rotting by the end of the growing season in control plots, while they were nearly free of nematodes in plots treated with methyl bromide. In solarized plots root infestation was intermediate because of the late nematode attack, but no rotting of the root was observed (Cartia *et al.*, 1988). Moreover, alternating methyl bromide with solarization treatments inhibited root infestation as did soil treated with methyl bromide for two consecutive years. In the same solarized plots, yield increases averaged 60 percent over the control and were similar to those obtained with methyl bromide treatments.

The reproduction factor in the solarized treatment was 0 on tomato plants for *Meloidogyne* spp., *Tylenchorhynchus* spp., and *Trichodorus* spp. The corresponding RF value in non-solarized soil was 0.89 on the same crop and the same nematodes (Table 15.26).

Table 15.26. Effect of solarization on the control of nematodes on tomato

| Treatment | Average number of nematodes per 200 g soil[1] | | Reproduction factor (RF) RF = Pf/Pi |
|---|---|---|---|
| | (Pi) | (Pf) | |
| Tomato - solarized | 80 | 0 | 0 |
| Tomato - non - solarized | 156 | 140 | 0.89 |

Pi = Initial nematode population; Pf = Final nematode population.

In Japan and other Asian countries, several farmers growing successive crops, such as tomato and melon susceptible to root-knot nematodes use solarization in plastic tunnels for 30 days in summer as an alternative to methyl bromide fumigation (Sano, 2002).

Ploughing during summer not only leads to disturbance and instability in nematode community but also causes their mortality by exposing them to solar heat and desiccation. Three summer ploughings each at 10 days interval during June (average atmospheric temperature ranging between 40-46°C) at Hisar (Haryana) led to 96.5 % reduction in *M. javanica* population (Jain and Bhatti, 1987). Additional use of plastic sheets for covering soil either in nursery beds or in field further enhanced nematode reduction. Such an approach also helps in reducing the intensity of weeds, fungi and bacteria in the soil.

**Crop Rotation:** To control *M. incognita* and *M. javanica*, a tomato crop can be followed by peanut without risk of damage to the peanuts. While the peanuts are growing, the nematodes can not reproduce. Instead many of the larvae in the soil die or become non-infective because of starvation and the attacks of predators, fungi and diseases. If the population is reduced sufficiently, tomatoes can be grown again after peanuts without serious injury (Taylor and Sasser, 1978).

Tomato rotated with cotton-wheat-cotton resulted in low gall index due to *M. javanica* (root-knot index 1-2) and gave maximum yield (38% increase over continuous tomato) (Hashmi and Hashmi, 1990).

Sesamum-tomato sequence produced the least reproduction rate (0.47), least root-knot index (1.33) and highest yield (86 q/ha) of tomato which was at par with niger-tomato sequence that had reproduction rate of 0.56, root-knot index of 1.66 and highest yield of 75.5 q/ha of tomato (Sahoo *et al.*, 2004).

In cropping sequence studies for the management of *M. javanica* on tomato, nematode population decreased to low levels when carrot, capsicum and onion were grown (Kanwar and Bhatti, 1992).

The population of *M. incognita* was reduced considerably under groundnut-mustard-tomato cropping sequence (Sharma *et al.*, 1980).

Tomato nursery beds previously planted with trap crop (marigold) gave maximum reduction in root-knot nematode population in soil which also increased the germination of tomato seeds and production of more healthy (nematode-free) seedlings (Rangaswamy *et al.*, 1999).

Inclusion of rice, maize, groundnut and *Stylosanthes* in rotation sequences controlled root-knot nematodes and increased yields of tomato. In crop rotation trials, it was found that a crop of groundnut or strawberry grown prior to tomato in soil heavily infested with root-knot nematodes greatly increased yields of tomato. Rotation of tomato with leek (*Allium porrus*), onion, groundnut and sunnhemp reduced the population of root-knot nematodes.

Decrease in *M. incognita* population on tomato occurs following crop rotation with marigold, spinach and bottle gourd (Khan *et al.*, 1975). Crop rotation with non-hosts (maize, wheat, sorghum) or antagonistic crops (marigold, mustard, sesame) is also effective.

**Intercropping:** Intercropping with marigold or mustard with tomato reduced the damage of root-knot and reniform nematodes. The root-knot development on tomato was low when interplanted with *Tagetes erecta*. The population of *Tylenchorhynchus, Helicotylenchus, Hoplolaimus, Rotylenchulus* and *Pratylenchus* was also markedly reduced (Alam *et al.,* 1977; Khan *et al.,* 1971). α-Terthienyl is the active principle in *Tagetes* spp. which is toxic to these nematodes.

Marigold intercropped with tomato at 1: 4 and 1: 6 ratios and mustard at 1: 2 ratio was found to be effective in reducing root galls, egg masses and nematode population at harvest and gave cost: benefit ratio of 1: 8.36, 1: 7.88 and 1: 3.31, respectively (Rangaswamy *et al.*, 1999).

Interculture of onion with tomato reduced gall index due to *M. javanica* significantly and improved tomato fruit yield. Intercropping with 2 rows of onion with one row of tomato was the most effective treatment in reducing the nematode infection and population and enhancing tomato yield (Ram and Gupta, 2001).

In pots having *Zinnia elegans* with tomato, the population of *M. incognita* and *R. reniformis* declined. The root-knot index on tomato was reduced to 1.08 compared to 3.75 in control. The nematode reproduction factor of *M. incognita* and *R. reniformis* was reduced to 0.88 and 0.80, respectively compared to 3.86 and 2.43, respectively in control (Tiyagi *et al.*, 1986).

Cultivation of tomato with castor results in considerable reduction in number of root-knot galls on tomato (Hackney and Dickerson, 1975).

Intercropping with marigold, mustard, sesame, maize, wheat or sorghum with tomato reduced the damage of root-knot and reniform nematodes.

**Green manuring:** Green manuring with naffatia/besharmi or neem leaves (ICBR 1: 2.87) or water hyacinth (ICBR 1: 2.52) or calotropis (ICBR 1:2.26) or congress grass (ICBR 1: 2.09) each at 3 kg/m$^2$ 15 days prior to seeding is recommended for economic and effective management of root-knot disease in tomato nursery. Incorporation of green materials of naffatia and neem leaves each at 2 kg/1.44 m$^2$ (ICBR 1: 15.7) or congress grass and neem leaves each at 2 kg/1.44 m$^2$ (ICBR 1: 15.6) or calotropis, naffatia, congress grass and neem leaves each at 1 kg/1.44 m$^2$ (ICBR 1: 14.7) 15 to 20 days prior to tomato seeding gave effective management of root-knot nematodes and production of more healthy seedlings in tomato nursery during kharif season.

**Biofumigation:** In order to verify the effectiveness of biofumigation strategy in managing nematode *M. incognita*, field trials on tomato were undertaken during the two years in an unheated plastic-house in the southern part of Sicily (Italy). In the first trial, two accessions of mustard species of the *Brassicaceae* family were selected for their glucosinolate content (*Brassica juncea* sel. ISCI 99 and *Eruca sativa* cv Nemat). These were rotated to ascertain their effectiveness against the root-knot nematode in comparison to the addition of pellets derived from the *Brassicaceae* species. In the second trial the tomato crop was treated with oil extract of *Brassicaeae* species, distributed by drip irrigation when the maximum development of nematode population was registered.

All techniques succeeded in reducing the juvenile nematode population in the soil and gall index in the roots at the end of the tomato's growth cycle. All the techniques produced a tomato yield significantly higher than with untreated soil. In particular, *Brassica juncea* sel. ISCI 99 acted mainly as a biofumigant while *Eruca sativa* cv Nemat revealed an interesting trap crop effect. Also the use of pellets derived from *Brassicaceae* species, incorporated into the soil before the transplanting has shown a lower gall index in the roots. The oil extract of *Brassicaceae* species reduced the juvenile nematode population in the soil during the whole tomato cultivation cycle (Colombo *et al.*, 2008).

In a 3-year field study on the effect of biofumigation with 5 different winter-grown crops on *M. incognita* infestation of a summer grown tomato crop in Southern California, Lopez-Perez *et al.* (2008) reported that broccoli reduced tomato root galling and increased yields compared to the other non-*Brassica* treatments, but generally did not lower *M. incognita* soil population levels. They suggested that biofumigation with broccoli tissue prevented an immediate attack of the tomato transplants, allowing plants to become established and rendering them more tolerant to the nematodes.

The *Brassica* crop arugula (*Eruca sativa* var. Nemat) also received interest as a potentially useful biofumigant crop to manage root-knot nematodes, as it appears to act as a trap crop for *M. incognita* (Curto *et al.*, 2006). Positive effects on yields of nematode-susceptible crops grown after arugula cultivation and biofumigation were reported for tomato.

The effects of soil incorporation of dry leaves of *Brassica macrocarpa* four months before transplanting and repeated after three weeks from transplanting was compared with soil incorporation of dry leaves of *B. macrocarpa* only three weeks after transplanting for the management of root-knot nematodes (*Meloidogyne* spp.) and on tomato plant growth and development. The results of both trials showed a significant reduction of root-knot as consequence of nematodes soil infection and of *B. macrocarpa* insertion into the soil. The use of *B. macrocarpa* permitted to control the pest and to increase fruit yield, above-all in terms of fruit number and weight (Branca and Furtado, 2004).

Three doses of canola *Brassica napus* seed meal at 4, 6 and 8 MT/feddans used as organic amendment incorporating into the soil were evaluated against root-knot nematode *M. incognita* inhabiting field planting with tomato. Super strain B at all three doses significantly affected root-knot population in soil, galls formation and yield increase (39% over control for the highest dose) (Ameen and Mohamed. 2006).

Four accessions of mustard species (*Brassica juncea* sel. ISCI 99, *Eruca sativa* cv. Nemat, *Crambe abyssinica* cv. Mario, and *Raphanus sativus* cv. Boss), selected for their glucosinolate content, were rotated with host crops of the southern root-knot nematode, *M. incognita*, to ascertain their effectiveness in managing the nematode, in comparison with a rotation using a susceptible crop *Vicia villosa* cv. Haymaker Plus. All the species used significantly suppressed numbers of the nematode juveniles in the soil and in the roots in comparison with a susceptible intercrop. *B. juncea* and *C. abyssinica* acted mainly as biofumigants while *R. sativus* and *E. sativa* revealed a clear catch crop effect. The results showed a significant tomato yield increase after mustard crops, except for the rotation with *B. juncea*. The beneficial effect of the rotation with mustard species was greater in biocidal crops sown in late spring rather than in late summer (Curto *et al.*, 2006).

In greenhouse trial, three natural plant products based on *Quillaja saponaria* extract, *Tagetes erecta* oil and *Brassica* defatted meal were applied and compared with a chemical control (Cadusafos) and an untreated one. The results showed a statistically significant effectiveness among treatments, in nematode (*M. incognita*) reproduction factor (0.9<R<11), infestation root index (2.4-1.9 in the untreated control and in *Q. saponaria*, 1.3-0.5 in *T. erecta*, 1 in *Brassica* defatted meal and 0.4-0.2 in cadusafos) and fruit weight (from 3.8-4.2 kg per plant in the control, *Brassica* defatted meals and cadusafos, to 5.7-4.9 kg per plant in *Q. saponaria* and in *T. erecta*), while the factorial analysis registered a statistical relevance of nematicidal green manure in decreasing the initial nematode charge in the soil till safe levels for the following tomato crop (Curto *et al.*, 2006).

Singh and Sitaramaiah (1967) reported that application of 5-10% (w/w) of leaves of neem, *Melia azedirach, Cassia fistula, Sesbania aculeata* and *Crotalaria juncea* to the infested soil reduced the incidence of root-knot nematode on tomato.

Soil amendment with chopped shoot parts of jack, *Ficus carica, F. elastica, F. glomerata, Ipomea fistulosa, Nerium indicum, Taberna-emontana coronaria* gave good control of *M. incognita* and *R. reniformis* on tomato (Siddiqi *et al.,* 1987).

Soil amendment with pure latex and chopped shoot parts of *Calotropis procera* significantly reduced the root-knot development caused by *M. incognita* and the population of *Hoplolaimus indicus, T. brassicae, T. filiformis* and improved the plant growth of tomato (Alam, 1986; Maqbool *et al.*, 1982).

Soil incorporation of leaves of mustard, *Pedilanthus tithymaloides,* castor, neem and calotropis at 80 g/kg/pot were as effective as carbofuran at 30 mg/kg/pot for the management of *M. incognita* and in increasing the plant growth parameters of tomato (Rao and Parvatha Reddy, 1992).

Technical feasibility and effectiveness of *Brassica* green manure biofumigation for the control of root-knot nematodes were investigated in a greenhouse experiments on a tomato - melon succession. Green manures of *Brassica juncea* and *Eruca sativa* were compared with a *Brassica* seed meal commercial pellet, alone or combined with *E. sativa* manure, two commercial formulations of quillay or tagetes, chemical nematicide cadusaphos and an untreated soil. In the tomato crop, green manure of *E. sativa* resulted in the lowest *M. incognita* population densities and gall formation, as its results were not significantly different from chemical treatment but were lower than *B. juncea* green manure, the pellet and quillay formulation. This last treatment was statistically similar to the untreated control. *E. sativa* green manure, either alone or combined with the pellet, provided a tomato yield two-three times higher than control, whereas *B. juncea* and cadusaphos almost doubled the yield of untreated soil.

In the following melon crop, the lowest nematode densities were recorded in soil treated with cadusaphos or the commercial pellet. Green manure of *E. sativa*, alone or combined with the pellet, resulted in a higher suppressivity than *B. juncea*, whereas a limited nematicidal activity emerged for the quillay formulation. The highest melon yield was provided by the seed meal pellet, followed by the combination of *E. sativa* green manure and pellet and by both single green manures (De Mastro *et al*., 2008b).

The technical applicability and effectiveness for the control of the root-knot nematode *M. incognita* of soil biofumigation treatments with *Brassica juncea* and *Eruca sativa* green manures, *B. juncea* hay, and a *Brassica* seed meal commercial pellet were investigated over two years in field experiments on a carrot - tomato rotation. Biofumigation treatments were compared with the standard nematicide fenamiphos, a quillay-derived nonchemical formulation and an untreated control. At the end of the first cycle, soil population density of *M. incognita* were significantly lower in the plots treated with green manure or hay of *B. juncea* or with green manure of *E. sativa*, either alone or combined with seed meal pellet, compared with untreated control. Nematicidal effects of biofumigation treatments did not differ from that of fenamiphos. Moreover, commercial pellet and *B. juncea* green manure and hay also resulted in carrot yield almost double the control, whereas lower increases were provided by *E. sativa* green manure, either alone or combined with the pellet.

In the following tomato crop the lowest *M. incognita* densities were recorded in soil treated with *B. juncea* hay or with *E. sativa* green manure plus pellet. Hay and green manure of *B. juncea* and integrated treatment with *E. sativa* green manure and pellet resulted in the highest tomato yields, always significantly higher than fenamiphos.

**Organic Amendments:** Singh and Sitaramaiah (1967) applied finely divided oil cakes to root-knot infested soil and noted a reduction in disease incidence in tomato.

Olive pomace (FOP) and chicken manure (CM) significantly increased tomato yield and reduced *M. incognita* population in the soil, compared to untreated control, at all the tested dosages (Table 15.27). Combinations of 100 MT/ha olive pomace with 2 and 4 MT/ha chicken manure resulted in a further yield increase and were more suppressive than fenamiphos on *M. incognita*.

Table 15.27. Effect of olive pomace (FOP) and chicken manure (CM) soil amendments on *Meloidogyne incognita* on tomato.

| Treatment | Crop yield (MT/ha) | Final nema popn. (eggs & J2/ml soil) |
|---|---|---|
| Untreated control | 44.8 a | 49.1 a |
| Fenamiphos 300 kg/ha | 65.0 def | 20.7 cdef |
| FOP 25 MT/ha | 57.1 bcd | 27.6 bc |
| FOP 50 MT/ha | 53.0 ab | 26.8 cd |
| FOP 100 MT/ha | 56.5 bcd | 25.5 cde |
| CM 1 MT/ha | 53.5 ab | 37.9 b |
| CM 2 MT/ha | 54.3 bc | 26.1 cd |
| CM 4 MT/ha | 53.8 bc | 20.4 cdef |
| FOP 25 MT/ha + CM 1 MT/ha | 56.5 bcd | 25.4 cde |
| FOP 25 MT/ha + CM 2 MT/ha | 56.1 bcd | 23.2 cde |
| FOP 25 MT/ha + CM 4 MT/ha | 60.5 bcd | 23.0 cde |
| FOP 50 MT/ha + CM 1 MT/ha | 60.8 bcd | 20.4 cdef |
| FOP 50 MT/ha + CM 2 MT/ha | 63.3 cde | 21.0 cdef |
| FOP 50 MT/ha + CM 4 MT/ha | 61.0 bcd | 20.6 cdef |
| FOP 100 MT/ha + CM 1 MT/ha | 64.1 def | 16.0 def |
| FOP 100 MT/ha + CM 2 MT/ha | 73.9 f | 14.5 ef |
| FOP 100 MT/ha + CM 4 MT/ha | 71.8 ef | 11.3 f |

* Means followed by the same letters in the same column are not significantly different according to Duncan's Multiple Range Test (P = 0.01).

## Integrated Methods

**Organic Amendments and Bioagents:** Application of *Aspergillus niger* and *P. lilacinus* along with mustard cake gave maximum reduction in nematode population both in root and soil with enhanced plant vigour. *A. niger* being toxic agent, kills the second stage juveniles present in the rhizosphere; while *P. lilacinus* being the egg parasite, invade the eggs of *M. incognita* which escaped from the toxicity of *A. niger*. As a result, there is an overall reduction in root and soil population. Further, addition of mustard cake also helps to maintain the general plant health in addition to possessing nematicidal properties (Goswami *et al.*, 1998) (Table 15.28).

Table 15.28. Effect of *Aspergillus niger, Paecilomyces lilacinus* and mustard cake on biomass and multiplication of *Meloidogyne incognita* infecting tomato

| Treatment | Biomass | No. of galls/ plant | No. of egg masses /plant | No. of eggs/egg mass | Nematode popn./500 g soil |
|---|---|---|---|---|---|
| *Aspergillus niger* | 9.8 | 34 | 18 | 272 | 470 |
| *Paecilomyces lilacinus* | 10.2 | 42 | 23 | 262 | 580 |
| *A. niger* + *P. lilacinus* | 12.2 | 26 | 16 | 264 | 270 |
| *A. niger* + *P. lilacinus* + Mustard cake | 12.6 | 22 | 10 | 255 | 290 |
| Control | 9.0 | 98 | 72 | 320 | 2880 |
| CD at 5% | 2.33 | 7.24 | 3.24 | 27.18 | 162.80 |

Combined application of *P. penetrans* + *T. viride* + neem/castor cake at 1/3 doses was significantly superior compared to their individual applications in terms of increased tomato plant growth and reduced root galling, egg mass production and final populations of *M. incognita*. Similarly, total parasitization by bioagents was higher in plants treated with these bioagents in terms of number of juveniles encumbered and adult females infected with *P. penetrans* and egg masses parasitized by *T. viride*, compared to their individual applications (Rangaswamy *et al.*, 1999).

The integration of castor leaves with *P. lilacinus* increased the efficiency of nematode control (*M. javanica*) in tomato. The residual effect on the second crop of tomato revealed the reduction in root-knot index (29.6 to 356.8%) and J2 population in soil (26.4 to 57.9%) (Zaki, 1998).

**Organic Amendments and AMF:** Application of calotropis leaves at 400 g/m$^2$ along with *G. fasciculatum* (250 g/m$^2$ containing 25-30 chlamydospores/g) in nursery beds gave significant reduction in root galling and fecundity of *M. incognita*. The above treatment also gave maximum increase in plant growth and root colonization with *G. fasciculatum* (Rao *et al.*, 1996). Organic amendments (oil cakes, calotropis leaves) in combination with AMF enhance the colonization of AMF on tomato roots which further increased plant growth and reduced gall index.

Inoculation of *G. mosseae* or *G. fasciculatum* in the nursery beds amended with neem cake/castor cake/neem leaf/calotropis leaf reduced the infestation of root-knot and reniform nematodes to the maximum extent on tomato. Amendment of botanicals in the nursery beds increased the multiplication of endomycorrhizae and colonization of tomato root which in turn could protect the crop from these nematodes in the main field resulting in increased yields (Rao *et al.*, 1996; 1997).

**Biofumigation, Intercropping, Grafting, Solarization and Bioagents:** The effects of biofumigation, intercropping, grafting, solarization or bioagents against *M. incognita* in organic tomato production were evaluated in unheated greenhouse conditions.

In both seasons root galling caused by *M. incognita* was lower (1.90 in autumn and 2.95 in spring) in the grafted tomato plants on the rootstock Beaufort than the other treatments. In terms of yield, among the tested treatments, grafting (0.922 kg/plant in autumn and 1.785 kg/plant in spring) and biofumigation with *R. communis* in autumn (0.924 kg/plant) and biofumigation with broccoli in the spring (2.315 kg/plant) gave higher yields. The overall results showed that biofumigation with broccoli and grafting could be used efficiently against *M. incognita* in organic agriculture production.

In both seasons, the treatment of grafting with Beaufort was placed statistically ($P<0.05$) in a different group compared to the untreated plots, reducing dramatically root galling. However, the highest root galling indices were found in the soil solarization treatment (4.63±0.45) in autumn and in the control plots (8.08±0.34) in the spring season (Table 15.29).

Table 15.29. Effect of different treatments on root galling of tomato plants infected with *Meloidogyne incognita*

| Treatment | Galling index (1-10 scale) | |
|---|---|---|
| | 2007 autumn | 2008 spring |
| Control | 3.50 ± 0.32 b* | 8.08 ± 0.34 c |
| Grafted plants | 1.90 ± 0.19 a | 2.95 ± 0.36 a |
| Biofumigation with Broccoli | 3.55 ± 0.43 b | 4.33 ± 0.36 b |
| *Tagetes erecta* | 2.28 ± 0.29 a | 6.70 ± 0.43 c |
| Soil solarization | 4.63 ± 0.45 c | --- |
| Biofumigation with castor | 2.83 ± 0.33 ab | --- |
| BioAct | --- | 7.48 ± 0.39 c |
| Nestin | --- | 7.25 ± 0.27 c |

* Standard error of mean is given after ±

In terms of the reproduction ratio, there was a difference between the seasons. In autumn, the lowest reproduction ratio of root-knot nematodes in the soil was assessed from the treatments of biofumigation with *R. communis* (0.42) and broccoli (0.67), followed by control (1.37) and grafting (1.94). However, with the root galling index evaluation, the highest reproduction ratio was obtained in the soil solarization plots (3.40)

(Table 15.30). In the spring, the reproduction ratio of root-knot nematodes in the soil was the lowest with the treatment of BioAct (0.36) and followed by grafting (1.32) and Netisin (1.36) whereas the highest ratio was in the control plots (6.62).

Table 15.30. The effect of different treatments on reproduction ratio of *Meloidogyne incognita* in 100 cm³ soil

| Treatment | 2007 autumn | | | 2008 spring | | |
|---|---|---|---|---|---|---|
| | Pi | Pf | Ro | Pi | Pf | Ro |
| Control | 465 b | 635 b | 1.37 | 65 a | 430 a | 6.62 |
| Grafted plants | 80 a | 155 a | 1.94 | 155 a | 255 a | 1.32 |
| Biofumigation with Broccoli | 455 b | 305 ab | 0.67 | 215 a | 455 a | 2.12 |
| *Tagetes erecta* | 80 a | 260 ab | 3.25 | 90 a | 205 a | 2.28 |
| Soil solarization | 100 a | 340 ab | 3.40 | --- | --- | --- |
| Biofumigation with castor | 455 b | 190 b | 0.42 | --- | --- | --- |
| BioAct | --- | --- | --- | 290 a | 105 a | 0.36 |
| Nestin | --- | --- | --- | 175 a | 175 a | 1.34 |

Reproduction ratio (Ro) = Final population (Pf)/ Initial population (Pi)

There was a significant difference in terms of yield between the seasons. In autumn, the tomato yield recorded from all treatments was statistically placed at the same significance level ($P<0.05$). The lowest tomato yield was obtained from the *T. erecta* plots (0.771±0.15 kg/plant), however, the highest yield was in the control treatment (0.989±0.81 kg/plant), followed by the grafting (0.922±0.76 kg/plant) and biofumigation with *R. communis* plots (0.924±0.90 kg/plant) ($P<0.05$). While the yield was the lowest in the control (1.118±0.21 kg/plant) in the spring placing statistically ($P<0.05$) in a different group compared to all other treatments. The highest yield was recorded only from the biofumigation with broccoli (2.315±0.11 kg/plant) followed by the grafting (1.785±0.14 kg/plant) at a statistically different significance level ($P<0.05$) (Table 15.31).

In conclusion, grafting of tomato on Beaufort rootstock or biofumigation with broccoli controlled *M. incognita* and increased tomato yield to some extent. Therefore, these treatments could be used for root-knot nematode control in an organic greenhouse tomato production.

Table 15.31. The effect of different treatments on yield assessment of tomato plants

| Treatment | Tomato fruit yield (kg/plant) | |
|---|---|---|
| | 2007 autumn | 2008 spring |
| Control | 0.989 ± 0.81 a | 1.118 ± 0.21 c |
| Grafted plants | 0.922 ± 0.76 a | 1.785 ± 0.14 ab |
| Biofumigation with Broccoli | 0.785 ± 0.15 a | 2.315 ± 0.11 a |
| *Tagetes erecta* | 0.771 ± 0.15 a | 1.487 ± 0.27 bc |
| Soil solarization | 0.780 ± 0.13 a | --- |
| Biofumigation with castor | 0.924 ± 0.90 a | --- |
| BioAct | --- | 1.437 ± 0.21 bc |
| Nestin | --- | 1.592 ± 0.14 bc |

* Standard error of mean is given after ±

**Biofumigation and Solarization:** Fewer root-knot nematode eggs were recovered on the tomatoes grown in solarized soil than in the nonsolarized soil. Fewer root-knot eggs (*M. javanica*) were found in the soil amended with Dwarf Essex (*Brassica napus*) than with Aspen (*B. napus*). The tomato plants grown in the rapeseed amended soil were larger than those grown in the unamended soil.

Root knot nematodes were generally suppressed by the incorporation of any organic material whether *Brassica* or not, and suppression was often unrelated to GSL content. *Brassica* tissues rarely outperformed other forms of green manure unless experimental conditions favoured the impacts of volatile compounds - such as in sealed containers. An experiment in which the solid, liquid and vapour phases were separated and applied individually or in combination showed all of these components had activity against nematodes. A significant outcome of the nematode work was the revelation that among brassicas only some radish lines (*Raphanus sativus* ) were poor hosts of the nematodes and during their growth could reduce populations to a similar extent as traditional sorghum green manures (Wang *et al.*, 2008).

Tsror *et al.* (2006) reported much improved control of *Meloidogyne* spp. on tomato by combining solarization with biofumigation compared to solarization alone.

**Biofumigation, Organic Amendments and Trap Cropping:** In commercial greenhouse trials in Spain an integrated management system was developed, including biofumigation with sheep manure and mushroom residue and the cultivation of short-cycle vegetables acting as trap crops. Using this strategy, initial very high levels of *M. incognita* were reduced to near zero in the main susceptible tomato crop (Bello, 1998).

**Chemicals, Crop Rotation and Solarization:** Chemical soil treatments, crop rotation and solarization are the most commonly used and most effective control practices that are employed to keep nematode populations at a minimum level. Therefore, the effect of combinations of a susceptible [sunflower (*Helianthus annus*)], moderately susceptible [watermelon (*Citrullus vulgaris*)] or resistant [maize (*Zea mays*)] hosts, and the successive tomato crop on the *Meloidogyne* population density and its control was studied. Data showed that the population density of *Meloidogyne* spp. was greatly influenced by the preceding summer crops and cropping system (Table 15.32). The greatest percentage of nematode control was obtained when a resistant crop preceded tomato. When tomato was preceded by a susceptible host, the nematode population density was at a maximum level. However, significant nematode control was observed when soil solarization was introduced to the system, ranging from 40-51 percent. The result was more or less the same, whether soil solarization was applied before or after the susceptible host, or both. Percent nematode control increased when the soil was chemically treated, and particularly when solarization was included in the system, it ranged between 75 percent and 88 percent.

Nematode control was between 62 percent and 83 percent when a moderately susceptible host preceded tomato. Suppression of nematode populations was more pronounced in this case than in the two above-mentioned cases, i.e. when a susceptible or a resistant host preceded tomato. Nevertheless, nematode population reductions after chemical treatment of soil were not as great, due to the fact that nematode percent control in this rotation was generally high (87 to 96 percent). It was reported that solarization consistently surpassed fumigation in reducing nematode population density only within the top 7.3 cm of soil (Heald and Robinson, 1987).

Table 15.32. The effect of short-term soil solarization in the integrated control of *Meloidogyne* spp. under laizemeter conditions

| Cropping system[1] | Egg mass/5g of roots (% of control) | Females/5g of roots (% of control) | Nematode control (%) |
|---|---|---|---|
| S-T | 100 | 100 | 0 |
| M-T | 41 | 35 | 62 |
| R-T | 0 | 0 | 100 |

| | | | |
|---|---|---|---|
| Z-S-T | 64 | 55 | 40 |
| Z-M-T | 29 | 26 | 73 |
| Z-R-T | 0 | 0 | 100 |
| S-Z-T | 60 | 55 | 43 |
| M-Z-T | 23 | 21 | 78 |
| R-Z-T | 0 | 0 | 100 |
| Z-S-Z-T | 50 | 48 | 51 |
| Z-M-Z-T | 18 | 17 | 83 |
| Z-R-Z-T | 0 | 0 | 100 |
| S-Tr | 20 | 30 | 75 |
| M-Tr | 9 | 17 | 87 |
| R-Tr | 0 | 0 | 100 |
| Z-S-Tr | 11 | 21 | 84 |
| Z-M-Tr | 4 | 6 | 95 |
| Z-R-Tr | 0 | 0 | 100 |
| S-Z-Tr | 15 | 25 | 80 |
| M-Z-Tr | 1 | 10 | 95 |
| R-Z-Tr | 0 | 0 | 100 |
| Z-S-Z-Tr | 10 | 15 | 88 |
| Z-M-Z-Tr | 3 | 6 | 96 |
| Z-R-Z-Tr | 0 | 0 | 100 |

S = Susceptible host; M = Moderately susceptible host; R = Resistant host; Z = Solarization; T = Tomato; and Tr = Tomato + chemical treatment.

**Organic Amendments and Chemicals:** Integration of chopped castor leaves (40 to 60 g/kg soil) 1 or 2 weeks before transplanting with application of aldicarb or carbofuran each at 2 kg a.i./ha at transplanting, significantly reduced number of galls due to *M. javanica* and enhanced the growth of tomato (Dutt and Bhatti, 1986).

Integration of soil application of castor leaves with inorganic fertilizer (75 kg N/ha) enhanced plant growth of tomato and reduced *M. javanica* infestation (95% and 78% reduction in root galls and egg mass production, respectively) (Zaki and Bhatti, 1989).

**Solarization and Fertilization:** Addition of ammonium phosphate fertilizer to soil significantly reduced galling of tomato roots by *M. incognita* (Table 15.33). Galling index generally decreased with increasing fertilizer dosage. Soil solarization reduced nematode galling in all cases.

Table 15.33. *In vitro* effects of ammonium phosphate fertilization and soil solarization on galling of tomato roots by *Meloidogyne incognita*

| Treatment and dosage | *M. incognita* (Gall rating[1]) |
|---|---|
| Ammonium phosphate-100 mg/1 | 2.0 |
| Ammonium phosphate-100 mg/1 + Soil solarization | 0.8 |
| Ammonium phosphate-200 mg/1 | 1.0 |
| Ammonium phosphate-200 mg/1 + Soil solarization | 0.3 |
| Ammonium phosphate-300 mg/1 | 0.9 |
| Ammonium phosphate-300 mg/1 + Soil solarization | 0.2 |
| Ammonium phosphate-400 mg/1 | 0.7 |
| Ammonium phosphate-400 mg/1 + Soil solarization | 0.0 |
| Control | 3.5 |
| Soil solarization only | 1.8 |
| LSD | 0.7 |
| Factorial ANOVA | 0.01[2] |
| Fertilizer | |
| Soil solarization | 0.01 |
| Fertilizer x Soil solarization | NS[3] |

Based upon 0-4 rating scale. 2 Significance level.
3 NS = no significant difference.

The mean maximum temperature at 15 cm depth in the solarized plot was 42°C (9°C higher than non-covered control soil), and the mean minimum was 34°C (7°C higher than the control). Maximum temperatures al 15 cm depth in the fine sandy loam plot were 44°C and 36°C in solarized and control plots, respectively.

There were no differences in number of surviving plants in either soil. Factorial ANOVA showed both fertilizers and solarization to be highly significant in increasing all growth parameters in the loam soil; however, no fertilizer-solarization interactions were found. No significant differences in tomato growth due to fertilizers were found in the sandy loam plot. Solarization, however, was highly significant in increasing numbers of fruit and fruit fresh weight (Tables 15.34).

Table 15.34. Fresh market tomato (*Lycopersicon esculentum* "Early Pak 7") yield in soil treated with ammonia-based fertilizers and/or solarization

| Soil Treatment[1] | Loam soil | | | Fine sandy loam soil | | |
|---|---|---|---|---|---|---|
| | Number of fruits[2] | Fruit fresh wt.(g) | Veget-ative fresh wt. (g) | Number of fruits | Fruit fresh wt.(g) | Veget-ative fresh wt. (g) |
| Aqua ammonia | 5.9 | 90.8 | 604 | 22.7 | 398.3 | 1386 |
| Aqua ammonia + Solarization | 14.0 | 210.4 | 1102 | 30.5 | 493.6 | 1384 |
| Urea | 8.7 | 151.8 | 538 | 26.0 | 490.6 | 1130 |
| Urea+ Solarization | 14.5 | 206.4 | 1 080 | 30.7 | 605.0 | 1246 |
| Ammonium Sulphate | 12.3 | 175.1 | 690 | 22.6 | 404.8 | 1160 |
| Ammonium Sulphate + Solarization | 21.4 | 382.6 | 1 582 | 32.3 | 552.0 | 1660 |
| Ammonium Phosphate | 18.5 | 321.5 | 1 162 | 27.9 | 359.7 | 1754 |
| Ammonium Phosphate + Solarization | 26.1 | 502.7 | 1 668 | 28.3 | 533.0 | 1766 |
| Non-treated Control | 11.7 | 132.1 | 638 | 22.4 | 360.3 | 1180 |
| Solarization | 15.0 | 215.7 | 1 078 | 33.5 | 574.4 | 1448 |
| LSD | 8.0 | 171.9 | 437 | 11.9 | 247.6 | 670 |
| Factorial analysis of variance | 0.01[3] | 0.01 | 0.01 | NS | NS | NS |
| Fertilizers | | | | | | |
| Solarization | 0.01 | 0.01 | 0.01 | 0.01 | 0.01 | NS |
| Fertilizers[3] x Solarization[4] | NS | NS | NS | NS | NS | NS |

1 All fertilizers were applied at 305 kg N/ha. 2 All values are given on a per-plant basis. 3 P = 0.05 or 0.01; NS = no significant differences. 4 Interaction of treatments

### (ii) Reniform Nematode, *Rotylenchulus reniformis*

*R. reniformis* was responsible for 42.25 to 49.02% loss in fruit yield of tomato (Subramanyam *et al.*, 1990)

### (a) Management

**(b) Organic Amendments:** Soil amendment with fresh glyricidia leaves, sawdust and chicken manure controlled the root-knot and reniform nematode population on tomato (Castillo, 1985). The yield increases and the nematode suppressive effect of fresh chicken manure surpassed even phenamiphos (Table 15.35).

Table 15.35. Effect of soil amendment with organic materials on *R. reniformis* population and tomato fruit yield.

| Treatment | No. of nematodes/300 ml soil & 1 g roots | | | Yield (t/ha) |
|---|---|---|---|---|
| | No. of weeks after treatment | | | |
| | 2 | 6 | 11 | |
| Phenamiphos at 10 kg a.i./ha | 96 | 366 | 465 | 31.1 |
| Rice straw at 10 t/ha | 215 | 500 | 1921 | 29.1 |
| Glyricidia leaves at 10 t/ha | 201 | 533 | 1885 | 27.3 |
| Saw dust at 10 t/ha | 342 | 859 | 958 | 25.3 |
| Chicken manure at 10 t/ha | 17 | 169 | 1078 | 32.6 |
| Control | 552 | 816 | 2515 | 22.2 |

### (iii) Stubby Root Nematode, *Paratrichodorus minor*

#### (a) Management

**Soil Solarization:** Soil solarization using clear plastic mulch or photo selective (black) polyethylene film reduce populations of *P. minor* on tomato as effectively as fumigation with a mixture of methyl bromide and chloropicrin in warm climates.

**Inter Cropping:** *Asparagus officinalis* would not support populations of *P. minor* for more than 40 to 50 days. Tomato, normally a good host of this nematode, supported only a low population when asparagus was growing in the same pots. A glycoside (asparagusic acid) is the active principle involved which is toxic to *P. minor* and several other nematode species.

### (iv) Wilt, *Fusarium oxysporum* f. sp. *lycopersici*/*F. solani* and Root-knot Nematode, *Meloidogyne incognita* Disease Complex

Jenkins and Coursen (1957) induced wilting in *Fusarium* wilt-resistant tomato variety 'Chesapeake' only when root-knot nematodes were present along with fungal inoculum. Furthermore, when *M. hapla* was combined with the fungus, only 60% of the plants wilted, whereas *M. incognita acrita* promoted wilt in 100% of the plants.

**(a) Management**

**Solarization and Bioagents:** Deep ploughing and exposing soil to hot sun in summer, removal and burning of crop debris, soil application of *T. viride* and *P. lilacinus* help in controlling the disease complex.

**Biofumigation and AMF:** Bhagawati *et al.* (2000) demonstrated that although the mustard cake and AM fungus, *Glomus etunicatum* were effective in reducing the damage caused by *M. incognita* and *F. o.* f. sp. *lycopersici* on tomato, the performance of concomitant application of both (bioagent and botanical) was much better than the individual application.

### 15.2.3. Weeds

***(i) Soil Solarization:*** Soil solarization by covering of the top of well irrigated soil by transparent polyethylene sheet for 6-7 weeks has eliminated *Orobanche* and majority of weeds from the treated plots of tomato. The tomato fruit yield has increased from 7.8 t/ha in untreated control plots to 21.08 t/ha on plots treated with solarization in the 1995/96 growing season.

Gutal *et al.* (1992) reported that the 25μ black LDPE film reduced weed growth by 90%, saved soil moisture by conserving to the extent of 28% and increased tomato fruit yield by 55% compared with the control.

***(ii) Mulching and Drip Irrigation:*** The treatment combining drip irrigation at 0.4 PE with black plastic mulch reduced weed infestation by 95%, increased yield by 53% and resulted in a 44% saving in irrigation water. Highest tomato crop yield (51 t/ha) and 44% saving in irrigation water was obtained using the combination of drip irrigation at 0.4 PE and a mulch of sugarcane trash (Srivastava *et al.*, 1995).

Plant basins covered with red coloured plastic mulch of 25 μ thickness along with drip irrigation helped in obtaining superior tomato fruit yield. The yield of tomato in red, white and black plastic mulch plots were 45.875, 36.562 and 31.218 t/ha, respectively, while it was 23.072 t/ha in unmulched control plot. The increased yields were 98, 58 and 35%, respectively over control. Mulching reduces the weed population and improves the microbial activity of the soil by improving the environment around the root zone.

## 15.3. BRINJAL, *SOLANUM MELONGENA*

### 15.3.1. Diseases

#### (i) Bacterial Wilt, *Ralstonia solanacearum*

**(a) Symptoms:** Characteristic symptoms of bacterial wilt are rapid and complete wilting of normal grown up plants (Fig. 15.23). Pathogen is mostly confined to vascular region. In advanced cases, it may invade the cortex and pith and cause yellowish-brown discolouration of tissues. Infected plant parts when cut and immersed in clear water, a white streak of bacterial ooze coming out from cut ends is visible.

**Fig. 15.23.** Bacterial wilt affected plants of brinjal.

This disease occurs with bad drainage, especially in the hot, wet season and is often combined with symptoms of root-knot nematodes. The pathogen is soil-borne with a wide host range. Root-knot nematode infestation aggravates the disease development.

The initial symptom is a wilting of the terminal leaves, which after 2 to 3 days becomes permanent when the whole plant wilts due to the active development of the disease. Then the whole plants wilt and die suddenly. Total collapse of the plant usually occurs when temperatures reach 32°C and above. Plant wilts while still green. In the case of a slow development of the disease, the plant stunts and produces large numbers of adventitious roots on the stem. Bacterial wilt diagnosis in the field can be done easily. Cut a piece of the stem 2 to 3 cm long from the base. Suspend the cut stem in clear water in a glass container. Hold the stem with an improvised tong to maintain a vertical position. Within a few seconds, milky bacterial threads are discharged from the cut stem.

**(b) Management**

**Crop Rotation:** Crop rotation with cowpea-maize-cabbage, okra-cowpea-maize, maize-cowpea-maize and finger millet-brinjal (Pusa Purple Cluster)-French bean are reported effective in reducing bacterial wilt.

**Biofumigation:** Studies have suggested that cover crops, especially brassicas, can provide some of the benefits of a fumigant. This test looked at > Green Sprouting = broccoli raab, > Florida Broadleaf = mustard and pasture wheat as cover crops compared with standard fumigation using methyl bromide.

Eggplants grown on Methyl Bromide plots tended to have the highest yields followed by mustard; production on the wheat plots had the lowest yields (Table 15.36).

Table 15.36. Effect of cover crops on marketable yield of brinjal

| Treatment | Total Marketable Yield (lb/acre) | Significance |
|---|---|---|
| Mustard | 15,400 | NS |
| Broccoli | 16,100 | |
| Wheat | 8,900 | |
| Methyl Bromide | 17,400 | |

**(ii) Damping-off, *Pythium aphanidermatum***

**(a) Symptoms:** The pathogen causes death of plants in the nursery beds.

**(b) Management**

**Soil Solarization:** Soil solarization using polythene sheet for 30 days during peak summer was most effective and resulted in least occurrence of damping off (1.5%) when compared to control (30.64%).

**Soil Solarization and Bioagents:** Combination of the seed/root application of *T. harzianum* or *P. fluorescens* with soil solarization was very effective for the management of damping-off in brinjal nursery at farmers' field.

**(iii) Verticillium Wilt, *Verticillium albo-atrum***

**(a) Management**

**Biofumigation:** Brassica cover crops can successfully be integrated into eggplant production systems. After five harvests, the highest yield was recorded in methyl bromide plots with 666 bushels of marketable fruit (Table 15.37). This was followed by treatments with oilseed radish, oriental mustard, and yellow mustard. Harvest index, the proportion of fruit to total plant biomass was greatest in yellow mustard plots, followed by methyl bromide, oriental mustard, oilseed radish, SoilBuilder, and finally the control.

Table 15.37. Eggplant 'Classic' yield in bushels (33 lb)/A, fruit number and harvest index (fruit weight divided by total plant weight) as affected by the previous brassica cover crops and soil

| Treatment | Marketable fruit yield | | Total fruit yield | | Weight (lb/plant) | Harvest index |
|---|---|---|---|---|---|---|
| | Yield (bushels/A) | Count (fruits/A) | Yield (bushels/A) | Count (fruits/A) | | |
| Methyl Bromide | 666 | 33189 | 703 | 36017 | 2.38 | 0.63 |
| Oilseed radish | 565 | 33189 | 611 | 35451 | 2.55 | 0.58 |
| Oriental mustard | 541 | 30171 | 570 | 31869 | 2.16 | 0.61 |
| Yellow mustard | 530 | 30831 | 572 | 33189 | 1.88 | 0.65 |
| SoilBuilder | 417 | 26023 | 439 | 27437 | 2.26 | 0.54 |
| Control | 427 | 25269 | 449 | 26966 | 2.58 | 0.50 |
| CD (P = 0.05) | 112 | 6365 | 110 | 6288 | NS | 0.09 |

Disease evaluation conducted throughout the season confirmed the presence of *Verticillium* wilt in the site. Data collected on individual plants and visual ratings of the plots indicated high efficacy of methyl bromide (Table 15.38). Symptoms were lowest in methyl bromide plots throughout the season. When compared with the control plot, both oriental mustard and oilseed radish treatments showed lower symptoms. Surprisingly, yellow mustard plots showed the highest symptoms at most evaluation dates.

Table 15.38. Verticillium wilt of 'Classic' eggplant as affected by the previous brassica cover crops and soil treatment (methyl bromide and SoilBuilder).

| Treatment | Plants showing disease symptoms Sep. 20 (#/plot) | Overall plot health* | | |
|---|---|---|---|---|
| | | Aug. 7% | Aug. 21% | Sep. 20% |
| Methyl Bromide | 3.00 | 2.50 | 3.25 | 9.50 |
| Oilseed radish | 4.75 | 18.75 | 9.25 | 22.00 |
| Oriental mustard | 2.75 | 18.75 | 11.00 | 13.75 |

| | | | | |
|---|---|---|---|---|
| Yellow mustard | 8.00 | 12.50 | 16.25 | 42.50 |
| SoilBuilder | 3.25 | 22.50 | 6.75 | 13.75 |
| Control | 4.25 | 16.25 | 15.50 | 27.50 |
| LSD (0.05) | 2.60 | 6.50 | 7.03 | 17.50 |

*Overall plot health was assessed visually using a scale of 0 to 100% (0% = no symptom and 100% = complete death due to yellowing and wilting).

Use of brassica cover crops as biofumigants improved eggplant yield. Despite the substantial suppression of *Verticillium* wilt in the oilseed radish and oriental mustard treatments, the level was still generally low compared with methyl bromide. Therefore it is difficult at this stage to determine if benefits observed were due to cover crop effect or to biofumigation effect or to both. These preliminary results are promising; however, growers interested in brassica cover crops for biofumigation purposes should keep in mind that (1) the amount of toxic compounds produced by the cover crops is generally low compared to commercial products like methyl bromide, (2) benefits may vary greatly depending on the cash crop and growing conditions, (3) the cover crops should be used as a component of an integrated strategy to replace methyl bromide, and (4) all rotation restrictions that are applicable to cash crops also apply to cover crops, including avoiding planting species of the same botanical family back to back.

### (iv) Fusarium Wilt/Root Rot, *Fusarium oxysporum, F. solani*

#### (a) Management

**Solarization, Organic Amendments and Bioagents:** The effect of solarization, and its combinations with dazomet (400 kg/ha), chicken manure (10 tonnes/ha) or *Trichoderma*, on soil borne diseases (*Fusarium* spp.) were detected in eggplant. Solarization with dazomet and solarization and manure gave better disease control compared to other treatments (Table 15.39).

Table 15.439. Average disease incidence for eggplant in the demonstration sites.

| Treatment | % Disease |
|---|---|
| Solarization + Dazomet | 20 |
| Solarization + Manure | 20 |
| Solarization + *Trichoderma* | 31.2 |
| Check | 57.5 |

### (v) Collar Rot, *Sclerotinia sclerotiorum*

**(a) Symptoms:** Infection takes place on leaves, twigs, flowers and fruits. Water-soaked lesions develop and the infected tissue is macerated by the pathogen. White mass of mycelium grow on the surface and later embedded sclerotia are formed. These sclerotia become black after drying. The pathogen is soil-borne.

#### (b) Management

**Solarization, Organic Amendments and Chemicals:** The effect of solarization, and its combinations with dazomet (400 kg/ha), chicken manure (10 tonnes/ha) or *Trichoderma*, on soil borne diseases. *Sclerotinia sclerotiorum* was observed in the field. Solarization + *Trichoderma* controlled diseases better than solarization + dazomet or solarization + manure (Table 15.40).

Table 15.40: Yield results in eggplant demonstration fields

| Treatment | Yield |
|---|---|
| Solarization + Dazomet | 105 |
| Solarization + Manure | 112 |
| Solarization + *Trichoderma* | 91 |

**Solarization and Bioagents:** Summer ploughing and seedling dip treatment with *T. viride* effectively controlled the disease incidence at all the growth stages up to the maturity of crop (185 to 205 DAT).

## 15.3.2. Nematodes

### (i) Root-knot Nematodes, *Meloidogyne* spp.

*M. incognita* was responsible for 27.30 to 48.55% loss in fruit yield of brinjal (Bhatti and Jain, 1977; Parvatha Reddy and Singh, 1981; Darekar and Mahse, 1988) (Fig. 15.24).

#### (a) Management

**Solarization:** Ploughing not only leads to disturbance and instability in nematode community but also causes their mortality by exposing them to solar heat and desiccation. Generally, 2 to 3 summer ploughings each at 10 days interval during April-May (40-46°C) have been recorded to reduce 96.5% *M. javanica* population. Additional use of plastic sheets for covering soil either in nursery beds or in field further enhanced nematode reduction. Such an approach also helps in reducing the intensity of weeds, fungi and bacteria in the soil.

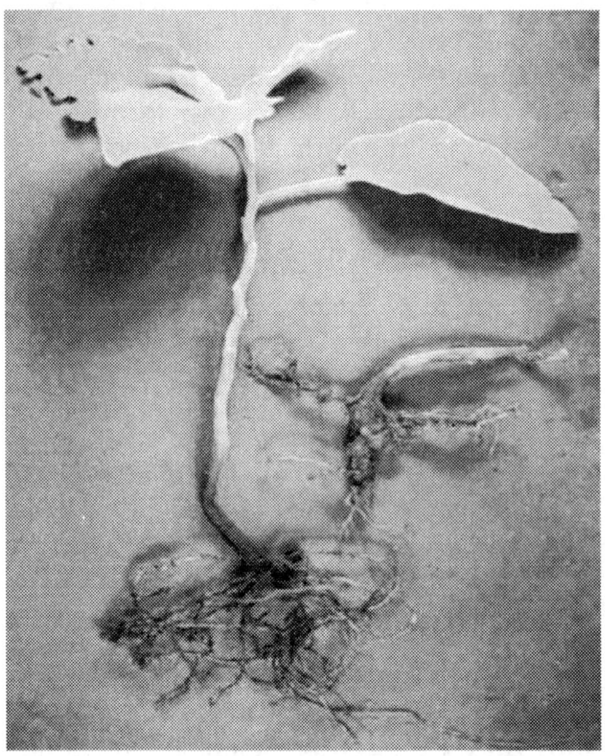

**Fig. 15.24.** Root-knot nematode on brinjal. Left – Healthy; Right – Infected.

The maximum reduction in *M. javanica* population and increase in fruit yield was recorded in brinjal by summer ploughing + covering with polythene sheet, followed by summer ploughing + exposure to sun (Anon, 1989).

**Crop Rotation:** Rotation of brinjal with sweet potato (cv. Sree Bhadra) reduced the root-knot nematode population by 47% and increased the fruit yield by 22%. Since the sweet potato cv. Sree Bhadra is a high yielding one, the farmer will get a good crop within 3 months. Thus it becomes an additional income generating practice (Sheela *et al.*, 2002). Crop rotation with sorghum, wheat and chilli reduced the root-knot nematode population. The population of *M. incognita* larvae decrease in combination of brinjal-wheat, brinjal-chilli-wheat and red gourd-brinjal-mustard. The root-knot larvae were completely eliminated in marigold-marigold-fallow and chilli-cauliflower-cauliflower. Garlic following brinjal also brought about considerable reduction in larvae of root-knot nematode. The population levels of *M. incognita* on highly infested land dropped to zero after 18 months of continuous cultivation of *Panicum maximum*. Growing of brinjal after *P. maximum* was completely free from root-knot nematodes (Netscher, 1983).

Decrease in *M. incognita* population on brinjal occurs by following crop rotation with marigold, spinach and bottle gourd (Khan *et al.*, 1975).

**Intercropping:** Intercropping brinjal with marigold, garlic or resistant tomato (SL-12) improved plant growth and reduced number of galls and final soil population of nematode (Jain *et al.*, 1990). Marigold (cv. Calcutta Yellow)-brinjal cropping sequence has been recommended to the farming community for management of *M. javanica* affecting brinjal (Singh, 1991). Intercropping with garlic is also recommended.

**Trap Cropping:** Growing of trap crop like knol-khol with brinjal but removed and destroyed before new generation of nematode could emerge helps in reducing the root-knot nematode population (Ayyar, 1926).

**Organic Amendments:** Spot application of mustard cake at 25g/spot and furrow application of saw dust at 210g/furrow of 3m length were found most effective in reducing the nematode infestation and increasing the yield of brinjal. Similarly, Hazarika (1990) found that spot application of castor/mustard/neem cake at 15g/spot and furrow application of the above cakes at 100g/3m furrow reduced the infestation of *M. incognita* and increased the yield of brinjal. The yield was maximum in spot application of neem cake along with maximum reduction of *M. incognita* population in soil.

Soil amendment with chopped shoot parts of jack, *Ficus carica, F. elastica, F. glomerata, Ipomea fistulosa, Nerium indicum, Tabernaemontana coronaria* gave good control of *M. incognita* on brinjal (Siddiqi *et al.,* 1987).

Soil amendment with pure latex and chopped shoot parts of *Calotropis procera* significantly reduced the root-knot development caused by *M. incognita* and the population of *Hoplolaimus indicus, T. brassicae, T. filiformis* and improved the plant growth of brinjal (Alam, 1986; Maqbool *et al.*, 1982).

## Integrated Methods

**Organic Amendments and Bioagents:** Borkakaty (1993) observed that inoculation of *P. lilacinus* at 4 g/kg of soil in combination with mustard oil cake at 0.5 and 1.0 MT/ha increased plant growth with corresponding decrease in number of galls, egg masses and eggs/egg mass of *M. incognita* on brinjal.

**Organic Amendments and AMF:** Inoculation of *G. mosseae*/*G. fasciculatum* in the brinjal nursery beds and subsequent application of 5% aqueous extracts of neem cake/castor cake/neem leaf/calotropis leaf in the nursery beds resulted in the effective management of root-knot and reniform nematodes and yielded healthy brinjal seedlings which would withstand the attack from nematodes after transplanting in the main field (Rao *et al.*, 1998).

**Organic Amendments, Bioagents and Chemicals:** Soil application of poultry manure at 10 MT/ha a week prior to transplanting + *P. lilacinus* ($10^9$/conidia/g) at 25 kg spore dust with carrier/ha at the time of transplanting; or neem cake at 2 MT/ha a week prior to transplanting + *P. lilacinus* ($10^9$ /conidia/g) at 25 kg spore dust with carrier/ha at the time of transplanting; or carbofuran at 2 kg/ha in 2 equal splits (one at the time of transplanting and the other after 75 days) + *P. lilacinus* ($10^9$/conidia/g) at 25 kg spore dust with carrier/ha at the time of transplanting considerably reduced root galling and also gave higher brinjal fruit yield over control (Vyas *et al.*, 2009) (Table 15.41).

Table 15.41. Effect of different organic amendments, bioagent and nematicide for the management of *Meloidogyne incognita* infecting brinjal

| Treatment | Gall index (0-5 scale) | Yield MT/ha | ICBR |
|---|---|---|---|
| *P. lilacinus* at 25 kg/ha + poultry manure at 10 MT/ha | 2.0 | 42.088 | 1: 18.5 |
| *P. lilacinus* at 25 kg/ha + Mustard cake at 2 MT/ha | 2.0 | 44.801 | 1: 7.7 |
| *P. lilacinus* at 25 kg/ha + Neem cake at 2 MT/ha | 1.8 | 41.745 | 1: 5.8 |
| *P. lilacinus* at 25 kg/ha + carbofuran at 2 kg a.i./ha | 1.9 | 41.498 | 1: 13.6 |
| *P. lilacinus* at 25 kg/ha | 2.5 | 39.941 | 1: 77.6 |
| Poultry manure at 10 MT/ha | 2.3 | 40.000 | 1: 18.5 |
| Mustard cake at 2 MT/ha | 1.9 | 41.810 | 1: 6.7 |
| Neem cake at 2 MT/ha | 1.9 | 39.954 | 1: 5.2 |
| Carbofuran at 2 kg a.i./ha | 2.0 | 40.530 | 1: 14.9 |
| Control | 3.9 | 28.107 | --- |

### (ii) Reniform Nematode, *Rotylenchulus reniformis*

#### (a) Management

**Organic Amendments:** Soil amendment with chopped shoot parts of jack, *Ficus carica, F. elastica, F. glomerata, Ipomea fistulosa, Nerium indicum, Tabernaemontana coronaria* gave good control of *R. reniformis* on brinjal (Siddiqi *et al.*, 1987).

**Organic Amendments, Bioagents and AMF:** Inoculation of *G. fasciculatum* in the castor cake amended nursery beds followed by the root-dip treatment of mycorrhizal seedlings of brinjal in spore suspension of *P. lilacinus* is effective for the management of *R. reniformis* (Rao *et al.*, 1993).

### (ii) Root-knot Nematode, *Meloidogyne incognita* and Bacterial Wilt, *Ralstonia solanacearum* Disease Complex

Egg plant is prone to many soil borne diseases among which the bacterial wilt (*R. solanacearum*) in combination with root-knot nematode (*M. incognita*) takes heavy toll every year all over the world (Naik, 2004).

**(a) Symptoms:** The root-knot nematode, *M. incognita*, present along with *R. solanacearum*, greatly increases the incidence of bacterial wilt of brinjal (Fig. 15.25). The root-knot nematode is responsible for breaking bacterial wilt resistance in "Pusa Purple Cluster" cultivar of brinjal (Parvatha Reddy *et al.*, 1979). *M. incognita* interacts with *R. solanacearum* in increasing the speed of development and severity of wilt disease of eggplant. Maximum disease occurred when the bacterium and nematode were inoculated simultaneously. Inoculation with *R. solanacearum* alone resulted in less severe disease. It was concluded that root-knot nematodes modify the plant tissues facilitating bacterial colonization (Nayar *et al.*, 1988).

#### (b) Management

**Solarization, Organic Amendments and Chemicals:** Integration of two summer ploughing during May-June, half recommended dose each of carbofuran (0.75 kg a.i./ha), neem cake (7.5 g/spot), streptocycline (250 ppm at 30 ml/spot) and full dose of *T. harzianum* (150 g/spot) improved plant growth parameters and fruit yield with corresponding decrease in the nematode reproduction rate (*M. incognita*), bacterial population in the soil and bacterial wilt incidence (*R. solanacearum*) in brinjal (Hussain and Bora, 2008) (Table 15.42).

**Fig. 15.25.** Root-knot nematode and bacterial wilt disease complex in brinjal.

Table 15.42. Effect of Integration of different practices for the management of disease complex in brinjal

| Treatment | No. of galls/root system | No. of egg masses/root system | % wilt incidence | Yield (kg/4m$^2$) |
|---|---|---|---|---|
| 2 Summer ploughings + full dose of carbofuran (1.5 kg a.i./ha) | 65 | 59 | 27.77 | 3.21 |
| 2 Summer ploughings + full dose of neem cake (15 g/spot) | 71 | 69 | 27.77 | 3.10 |
| 2 Summer ploughings + full dose of streptocycline (500 ppm at 30 ml/spot) | 128 | 112 | 38.88 | 2.43 |
| 2 Summer ploughings + full dose of *T. harzianum* (150 g/spot) | 67 | 63 | 27.77 | 3.10 |
| 2 Summer ploughings + ½ dose carbofuran + full dose *T. harzianum* | 35 | 26 | 5.53 | 3.90 |
| 2 Summer ploughings + ½ dose neem cake + full dose *T. harzianum* | 57 | 47 | 11.10 | 3.30 |
| 2 Summer ploughings + ½ dose strep-tocycline + full dose *T. harzianum* | 59 | 55 | 24.99 | 3.20 |
| 2 Summer ploughings + ½ dose carbofuran + ½ dose neem cake + ½ dose streptocycline | 28 | 20 | 5.53 | 4.20 |

| | | | | |
|---|---|---|---|---|
| 2 Summer ploughings + ½ dose carbofuran + ½ dose neem cake + ½ dose streptocycline + full dose *T. harzianum* | 24 | 17 | 5.53 | 4.90 |
| Control | 362 | 207 | 66.66 | 1.6 |
| CD (P = 0.05) | 6.69 | 6.75 | 11.15 | 0.83 |

## 15.4. CHILLI, *CAPSICUM ANNUUM*

### 15.4.1. Diseases

#### (i) Damping-off, *Pythium aphanidermatum*

**(a) Symptoms:** The pathogen is responsible for death of seedlings in nursery beds.

#### (b) Management

**Soil Solarization:** Soil solarization using polythene sheet for 30 days during peak summer was most effective and resulted in least occurrence of damping off (1.28%) when compared to control (30.64%).

**Soil Solarization, Organic Amendments and Bioagents:** Integration of soil solarization, application of neem cake and *T. viride* gives effective control of the disease.

### 15.4.2. Nematodes

#### (i) Root-knot Nematodes, *Meloidogyne spp.*

Root-knot nematodes induce much smaller galls on the roots of chillies. *M. incognita* was responsible for 24.54 to 28.00% loss in fruit yield of chillies (Singh *et al.*, 2003).

**(a) Symptoms:** Root galls on chilli are frequently small.

#### (b) Management

**Crop Rotation:** Decrease in *M. incognita* population on chillies occurs by following crop rotation with marigold, spinach and bottle gourd (Khan *et al.*, 1975).

**Intercropping:** Growing of chilli along with marigold gave highest degree of reduction in root-knot nematode population on chilli roots, followed by onion, garlic and asparagus (Trivedi and Tiagi, 1984).

**Organic Amendments:** Soil application of decaffeinated tea leaves effectively reduced root galling on chilli plant (Trivedi and Tiagi, 1984).

### 15.4.3. Weeds

The major weeds of chilli fields are *Cynodon dactylon, Digitaria marginata, Cyperus* spp., *Euphorbia hirta, Oldenlandia umbellate, Aegyratum conyzoides, Phyllanthus niruri* and *Amaranthus* spp.

***(i) Mulching:*** Mulching with maize and sorghum residues (10 t/ha) increased chilli yields. This is attributed to reduced weed growth due to mechanical hindrance to emerging weed seedlings and availability of more nutrients to crop plants. Among different mulch materials tested, black plastic film and straw mulch produced greater green fruit yields (10.7 t/ha). An increase of 66.40 to 67.19% in yield of chillies was recorded over unmulched control. The above mulches showed better weed control which eventually led to smother the weeds and encouraged the crop growth.

## 15.5. SWEET PEPPER, *CAPSICUM ANNUUM*

### 15.5.1. Diseases

#### (i) Damping-off, *Pythium aphanidermatum*

**(a) Symptoms:** The pathogen is responsible for death of seedlings in nursery beds.

#### (b) Management

**Soil Solarization and Bioagents:** Combination of the seed/root application of *T. harzianum* or *P. fluorescens* with soil solarization was very effective in management of damping-off of capsicum in nursery at farmers' field.

#### *(ii) Phytophthora* Root Rot, Fruit Rot and Leaf Blight, *Phytophthora capsici*

**(a) Symptoms:** This is a disease of rainy season, characterized by small, water-soaked spots appearing on fruits leading to complete rot of fruits. On leaves, water-soaked bleached spots appear, resulting in blighting. It causes damage to plants and affect fruit yield. Infection of the root and lower portion of the stem leads to plant wilting (Fig. 11.14), which is the most conspicuous symptom of the disease. In furrow-irrigated fields, there is a row-delimited pattern of wilted plants (Fig. 15.26). Over time, wilting is displayed over the entire field. *Phytophthora capsici* produces specialized swimming cells, known as zoospores, which enable spread of the pathogen from plant to plant.

**Fig. 15.26.** Wilting of chilli plant and a field severely affected by *Phytophthora* root rot.

This is a disease of rainy season, characterized by small, water-soaked spots appearing on fruits leading to complete rot of fruits. On leaves, water-soaked bleached spots appear, resulting in blighting. It causes damage to plants and affect fruit yield.

## (b) Management

**Biosolarization:** Biosolarization experiments were initiated using local organic amendments (sheep and chicken manure) to look for non chemical alternatives equally efficient as MB to manage, *Phytophthora* spp. infecting bell pepper. The biosolarization had a positive effect on the control of *Phytophthora* when carried out in the middle of September. The development and production of the plants were comparable with MB treatment in the second year. Non pathogenic fungi were not affected, physical and chemical properties of the soil were improved and lixiviates were not produced (Guerrero *et al.*, 2008).

**Biofumigation and Solarization:** Biofumigation with fresh sheep manure (if it is done from July to October) promise to be an efficient disinfectant although the parameters recorded (plant growth, yield and pathogen incidence) were not at the same level of control (Table 15.43). An improvement of the results can be obtained if biofumigation is repeated for the following seasons. This practice is a good solution for soil disinfection in organic agriculture (Bello *et al.*, 1997).

Table 15.43. Effects of biofumigation on 5 plastic greenhouses (Season 1998/99).

| Treatments | Plants height (cm) | Commercial yield (kg $m^{-2}$) | % plants with *Phytophthora* |
|---|---|---|---|
| MB 60 g/$m^2$ | 163.8 a | 9.7 a | 0.4 a |
| Biofumigation and solarization (4 weeks) | 135.3 b | 8.7 b | 9.4 b |
| Untreated | 122.4 c | 4.3 c | 37.9 c |

No significant differences between values with the same letter (P> 0.05)

The best results were obtained when biofumigation with solarization was applied at the end of August or the beginning of September. Sweet pepper yield using fresh sheep manure with soybean flour, or fresh sheep manure with chicken manure treatments, was similar to sweet pepper yield when MB was used. The reiteration of biofumigation with solarization over two or more years led to an improvement in pathogen and weed control, higher plants, an increase in yield and improvements to the soil physical properties, with a higher level of macro- and micronutrients as well as electrical conductivity.

### (iii) Southern Blight, *Sclerotium rolfsii*

#### (a) Management

**Solarization and Bioagents:** Solarization of fallow soil in raised beds for 6 weeks and application of a bran prill formulation of *G. virens* significantly reduced the southern blight disease incidence in bell pepper and survival of sclerotia of *S. rolfsii* to depths of 30 cm (Ristaino *et al.*, 1996).

### (iv) Post-Harvest Disease, *Alternaria alternata*

**(a) Symptoms:** Leaf spots with concentric rings are characteristic feature of the disease. On leaves, disease spots are dark and leathery and often coalesce to cover large areas of leaf blade. Affected leaves die and are shed prematurely. On affected fruits, characteristic black spore masses develop. Infected fruits drop off prematurely.

#### (b) Management

**Biofumigation and Solarization:** Methyl-isothiocyante (MITC) at 0.56 mg/ml in combination with low density polyethylene film (LDPF) reduced the fungal infection (*A. alternata*) in the bell pepper fruit better than a commercial fungicide (Captan) without any detrimental effects on fruit quality (Troncoso-Rojas *et al.*, 2008).

### 15.5.2. Nematodes

#### (i) Root-knot Nematode, *Meloidogyne incognita*

**(a) Symptoms:** Root galls caused by root-knot nematodes on sweet pepper are frequently small.

#### (b) Management

**Crop Rotation:** Crop rotation with non-hosts (maize, wheat, sorghum) or antagonistic crops (marigold, mustard, sesame) proved to be reasonably effective and economical to check multiplication of root-knot nematodes on capsicum.

**Solarization:** Ploughing not only leads to disturbance and instability in nematode community but also causes their mortality by exposing them to solar heat and desiccation. Generally, 2 to 3 summer ploughings each at 10 days interval during April-May (40-46°C) have been recorded to reduce 96% *M. javanica* population. Additional use of plastic sheets for covering soil either in nursery beds or in field further enhance nematode reduction. Such an approach also helps in reducing the intensity of weeds, fungi and bacteria in the soil.

Integrating summer ploughing with soil solarization with polythene mulching effectively reduced *M. javanica* population in brinjal (Jain and Gupta, 1991) (Table 15.44).

Table 15.44. Integrated management of root-knot nematodes by summer ploughing/ fallowing and nursery bed treatment in capsicum.

| Treatment | Reduction in nematode population (%) | | Yield in kg/m$^2$ | |
|---|---|---|---|---|
| | Hisar | Vellayani | Hisar | Vellayani |
| Ploughing + Fallowing for 15 days | 87.00 | 66.20 | 28.98 | --- |
| Ploughing + Covering with polythene sheet for 15 days | 94.90 | 77.00 | 18.84 | 78.00 |
| No ploughing + No covering | 77.60 | --- | --- | --- |

**Intercropping:** Intercropping with marigold, onion and garlic, proved to be reasonably effective and economical to check multiplication of root-knot nematodes on capsicum.

**Biofumigation:** Biofumigation experiments have been carried out in Murcia (Spain), in a greenhouse affected by a population of *M. incognita*, race 1. The fumigant effect of sheep manure plus 25% chicken manure was analyzed after one, two and three years of treatment. Doses were decreased every year from the initial dose of 100 t/ha, to 75 t/ha in the 2nd year and to 60 t/ha in the 3rd year. Results were compared to a methyl bromide (MB) (98: 2) treatment at a dose of 30 g/$m^2$ and to the control plots. Biofumigation is as effective in controlling *M. incognita* as MB, and saprophagous nematodes increased after biofumigation, while nematode fauna biodiversity in the MB treatments was null. The average indexes of *M. incognita* root galling at the end of the crop in the month of August are as follows, according to treatments: 3rd year of biofumigation (1.6), 2nd year (1.9), 1st year (4.5), MB treatment (1.2) and control group (6.8). There are no significant differences between biofumigation from the 2nd (11.5 kg $m^2$) and 3rd (11.7 kg $m^2$) years and MB (11.5 kg $m^2$) treatments in sweet pepper production.

Biofumigation with fresh sheep manure (if it is done from July to October) is promising to be an efficient disinfectant although the parameters recorded (plant growth, yield and pathogen incidence) were not at the same level of control (Table 15.45). An improvement of the results can be obtained if biofumigation is repeated for the following seasons. This practice is a good solution for soil disinfection in organic agriculture (Bello *et al.*, 1997).

Table 15.45. Effects of biofumigation on 5 plastic greenhouses (Season 1998/99).

| Treatments | Plants height (cm) | Commercial yield (kg/$m^2$) | % plants with *Meloidogyne* |
|---|---|---|---|
| MB 60 g/$m^2$ | 163.8 a | 9.7 a | 8.2 a |
| Biofumigation and solarization (4 weeks) | 135.3 b | 8.7 b | 39.0 b |
| Untreated | 122.4 c | 4.3 c | 64.8 c |

No significant differences between values with the same letter ($P > 0.05$)

Biofumigation experiments were initiated using local organic amendments (sheep and chicken manure) to look for non chemical alternatives equally efficient as MB to manage *M. incognita* infecting bell pepper. Biofumigation with manure was not effective during first year. When applied for a second year the control of *Meloidogyne* was random but was improved when repeated applications were used. The development and production of the plants were comparable with MB treatment in the second year. Physical and chemical properties of the soil were improved and lixiviates were not produced (Guerrero *et al.*, 2008).

**Biofumigation and Solarization:** Biofumigation combined with soil solarization in a greenhouse in Spain provided levels of *M. incognita* control in bell pepper, those were similar in levels achieved using methyl bromide (Bello *et al.*, 2004).

The best results were obtained when biofumigation with solarization was applied at the end of August or the beginning of September. Sweet pepper yield using fresh sheep manure with soybean flour, or fresh sheep manure with chicken manure treatments, was similar to sweet pepper yield when MB was used. The reiteration of biofumigation with solarization over two or more years led to an improvement in pathogen and weed control, higher plants, an increase in yield and improvements to the soil physical properties, with a higher level of macro- and micronutrients as well as electrical conductivity.

**Solarization and Chemicals:** Experiments in Sicily (Cartia and Greco, 1987; Cartia *et al.*, 1988, 1989), aimed at the control of soil-borne pathogens of pepper in greenhouse culture, yield and fruit size from solarized soil averaged 12.2 and 2.3 times, respectively, over those in controls. Higher yield was obtained with soil solarization, alone or in combination with reduced dosages of methyl bromide, and with methyl bromide alone (6 times over control) than with DD (5.4 times the control).

Table 15.46. Effect of different organic amendments, bioagent and nematicide for the management of *Meloidogyne incognita* infecting capsicum

| Treatment | Gall index (0-5 scale) | Yield (MT/ha) | ICBR |
|---|---|---|---|
| *P. lilacinus* at 25 kg/ha + Poultry manure at 10 MT/ha | 2.0 | 42.088 | 1: 18.5 |
| *P. lilacinus* at 25 kg/ha + Mustard cake at 2 MT/ha | 2.0 | 44.801 | 1: 7.7 |
| *P. lilacinus* at 25 kg/ha + Neem cake at 2 MT/ha | 1.8 | 41.745 | 1: 5.8 |
| *P. lilacinus* at 25 kg/ha + Carbofuran at 2 kg a.i./ha | 1.9 | 41.498 | 1: 13.6 |
| *P. lilacinus* at 25 kg/ha | 2.5 | 39.941 | 1: 77.6 |
| Poultry manure at 10 MT/ha | 2.3 | 40.000 | 1: 18.5 |
| Mustard cake at 2 MT/ha | 1.9 | 41.810 | 1: 6.7 |
| Neem cake at 2 MT/ha | 1.9 | 39.954 | 1: 5.2 |
| Carbofuran at 2 kg a.i./ha | 2.0 | 40.530 | 1: 14.9 |
| Control | 3.9 | 28.107 | --- |

**Organic Amendments, Bioagents and Chemicals:** Soil application of poultry manure at 10 MT/ha a week prior to transplanting + *P. lilacinus* ($10^9$/conidia/g) at 25 kg/ha spore dust with carrier at the time of transplanting; or neem cake at 2 MT/ha a week prior to transplanting + *P. lilacinus* ($10^9$ conidia/g) at 25 kg/ha spore dust with carrier at the time of transplanting; or carbofuran at 2 kg/ha in 2 equal splits (one at the time of transplanting and the other after 75 days) + *P. lilacinus* ($10^9$ conidia/g) at 25 kg/ha spore dust with carrier at the time of transplanting considerably reduced root galling and also gave higher capsicum fruit yield over control (Vyas *et al.*, 2009) (Table 15.46).

**Solarization, Organic Amendments, Bioagents and Chemicals:** The effect of solarization, and its combinations with dazomet (400 kg/ha), chicken manure (10 tonnes/ha) or straw (500 kg/ha), on soil borne nematodes and crop were demonstrated at various sites. Pepper fields were heavily infested with *Meloidogyne* spp. Combinations of solarization with basamid, manure or *Trichoderma* were the most effective applications for the management of root-knot nematodes infecting bell pepper. MB and solarization + straw were partially effective. However, it can be said that all treatments effectively controlled nematodes (Table 15.47). Galling index overall averages of six pepper plastic houses were 0.10, 0.89, 0.98, 1.92, 2.04, and 5.95 for solarization + *Trichoderma*, solarization + manure, solarization + dazomet, MB, solarization + straw, and check, respectively.

Table 15.47. Effects of treatments on the nematode population as percentage of non-treated check during the growing season of pepper.

| Date | Treatments and control rates (%) | | | | |
|---|---|---|---|---|---|
| | Solarization + Basamid | Solarization + *Trichoderma* | Solarization + Manure | Solarization + Straw | Methyl bromide |
| 03-10-2000 | 100.0 | 100.0 | 100.0 | 100.0 | 100.0 |
| 19-10-2000 | 98.8 | 100.0 | 100.0 | 100.0 | 100.0 |
| 07-11-2000 | 100.0 | 100.0 | 100.0 | 100.0 | 100.0 |
| 28-11-2000 | 89.9 | 100.0 | 80.0 | 100.0 | 97.8 |
| 19-12-2000 | 100.0 | 100.0 | 100.0 | 100.0 | 100.0 |
| 09-01-2001 | 96.3 | 99.2 | 79.9 | 99.3 | 100.0 |
| 30-01-2001 | 99.7 | 100.0 | 100.0 | 100.0 | 99.3 |
| 20-02-2001 | 100.0 | 100.0 | 100.0 | 100.0 | 97.0 |
| 13-03-2001 | 100.0 | 99.3 | 99.9 | 98.1 | 97.9 |
| 03-04-2001 | 79.7 | 98.7 | 96.2 | 94.3 | 91.6 |
| 25-04-2001 | 99.5 | 95.5 | 99.8 | 70.9 | 90.4 |

| | | | | | |
|---|---|---|---|---|---|
| 16-05-2001 | 72.9 | 88.8 | 94.4 | 93.2 | 87.4 |
| 06-06-2001 | 74.6 | 91.4 | 95.8 | 99.6 | 59.4 |
| 23-07-2001 | 40.8 | 59.3 | 84.9 | 0.0 | 0.0 |

100 is complete control and 0 shows no control

### 15.5.3. Weeds

***(i) Biosolarization:*** Biosolarization experiments were initiated using local organic amendments (sheep and chicken manure) to look for non chemical alternatives equally efficient as MB to manage weeds infecting bell pepper. The biosolarization had a positive effect on the control of weeds. The development and production of the plants were comparable with MB treatment in the second year. Physical and chemical properties of the soil were improved and lixiviates were not produced (Guerrero *et al.*, 2008).

Studies conducted in N.G. Ranga Agricultural University, Hyderabad (Andhra Pradesh) revealed that mulching with black polythene (25 μ) gave 10% increase in yield and 90% weed control.

## 15.6. ONION, *ALLIUM CEPA*

### 15.6.1. Diseases

#### (i) Damping-Off, *Pythium* sp., *Rhizoctonia solani*, *Fusarium* sp.

**(a) Symptoms:** Cold, wet soils often encourage the development of damping-off symptoms very early in the seedling's growth. Seedlings may fall over and die as a result of breakdown of plant tissues at the soil line. Sometimes damping-off occurs before the seedling even emerges. The disease is usually caused by *Pythium*, *Rhizoctonia* or *Fusarium* fungi, either alone or in combination. Damping-off can occur in the field or in the nursery if conditions are too wet.

#### (b) Management

**Solarization and Bioagents:** Seed treatment with *T. harzianum* at 5 g/kg and sowing in solarized soil helps in reducing the disease incidence.

#### (ii) White Rot, *Sclerotium cepivorum*

White rot, caused by the soil-borne fungus *Sclerotium cepivorum*, is a very destructive disease that begins in the field and can carry over into storage.

**(a) Symptoms:** The first above-ground symptoms are yellowing and dieback of the leaf tips, followed by a collapse of the affected leaves; however, these symptoms alone could easily be confused with other types of damage (i.e., onion maggot). When the bulbs and roots are examined, a white, fluffy mold and soft rot will be observed. Masses of tiny black sclerotia can also be seen within this mold (Fig. 15.27). These sclerotia remain in the soil for many years. Infected bulbs can rot in storage boxes and stain other bulbs. White rot typically develops in patches in the field and is less of a problem when soils are warm (higher than 24°C) and dry.

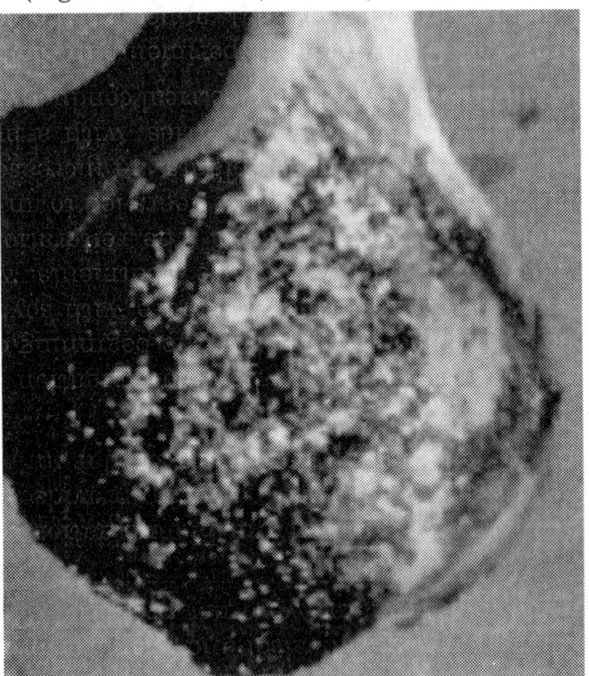

**Fig. 15.27.** White rot of onion.

### (b) Management

**Solarization:** Various experiments carried out in Israel indicated that solarization was effective against this pathogen. In the first year after solarization, growth of onions in the solarized plots was significantly improved and the white rot disease was completely controlled (Table 13.48). Disease progressed further in the non-solarized plots reaching 50 percent by the end of the season, as compared to zero in the solarized plots. A long-term effect of solarization in the second and third year was evident. The build-up of the disease in the non-treated plots was rapid, leading to a complete destruction of the plants.

Table 15.48. Effect of soil solarization on *Sclerotium cepivorum*, the causal organism of white rot disease of onion

| Treatment | Disease incidence % | | | Crop yield Fresh wt (IT/A) | | |
|---|---|---|---|---|---|---|
| | 1986/1987 | 1987/88 | 1988/89 | 1986/87 | 1987/88 | 1988/89 |
| Non-solarized | 50 | 100 | 100 | 3.15 | 0.00 | 0.00 |
| Solarized | 0 | 0 | 0.07 | 9.57 | 21.7 | 20.35 |

**Solarization and Bioagents:** Application of *Bacillus subtilis, T. harzianum, T. viride,* and *T. virens* to the solarized soil effectively controlled the disease. Incorporation of *T. harzianum* in soil after soil solarization effectively controlled *S. cepivorum* in soil and increased the control from 79 to 98%. A significant population growth of antagonist occurred attaining $10^8$ cfu/g of soil (Pereira *et al.*, 1996).

### (iii) Basal Rot, *Fusarium oxysporum* f. sp. *cepae*

Basal rot (Fig. 15.28) is caused by the soil-borne fungus *Fusarium oxysporum* f. sp. *cepae* and generally occurs when soil temperatures are very warm (optimum 29°C).

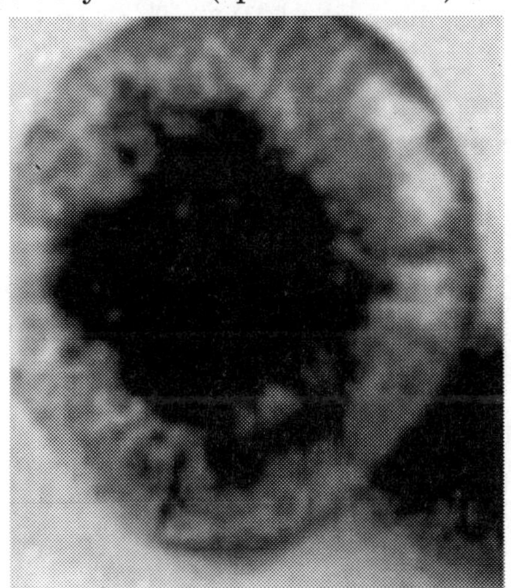

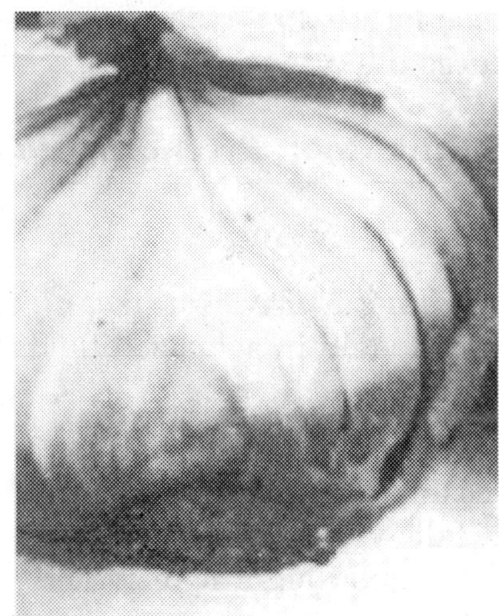

**Fig. 15.28.** Basal rot on onion bulbs.

**(a) Symptoms:** The early symptoms in the field are yellowing of leaves and tip dieback. As the disease progresses, the whole plant may collapse and, if the plant is pulled, it often comes out without any roots attached since they have decayed. The basal plate of the onion becomes pinkish-brown and secondary bacterial rots may develop in the affected area. If infection occurs late in the season, the symptoms may not show up until the onions are in storage.

**(b) Management**

**Solarization:** Solarization treatments improved growth of onions and reduced basal rot incidence by 84-96 percent and increased the yield by 151-207 percent.

Solarization increased onion seed yield by 218 percent (Table 15.49), when bulbs, previously grown in a solarized soil were planted again in the second year for seed production. In addition, seeds harvested from solarized plots were less contaminated with *Fusarium*.

Table 15.49. Onion seed production when soils were solarized by sun heating through polyethylene sheets.

| Treatment | No. of flowers / plant | Seed yield g/6m$^2$ | Mean wt of 1000 seeds (g) | % Seed germination | *F. oxysporum* % |
|---|---|---|---|---|---|
| Solarized | 5.77 | 11.58 | 11.58 | 99.00 | 0.00 |
| Non-solarized | 3.07 | 11.29 | 11.29 | 92.40 | 2.60 |

Soil solarization using polythene sheet of 750 gauge in summer season for 20-30 days reduces the infectious propagules, which in turn reduces the disease incidence.

**(iv) Pink Root, *Pyrenochaeta terrestris***

**(a) Management**

**Solarization:** Katan *et al.* (1980) reported an increase of 60-125% in onion yield in solarized fields which were earlier infected with pink root rot.

**Biofumigation:** A high glucosinalate (Ida Gold mustard), low glucosinalate (Sunrise canola), oilradish (Colonel), metham sodium and fallow treatment were tested for effectiveness against pink root. Mustard had significantly less pink root in 1 out of 3 years (Geary *et al.*, 2006).

**(v) Smut, *Urocystis cepulae***

**(a) Symptoms:** This fungus causes considerable losses at the seedling stage. The first symptom of the disease appears on the cotyledons and young leaves in the form of longitudinal blisters which are black and contain the fruiting bodies of the fungus. The seedlings often die before emergence. If the plant survives, the disease becomes systemic and they remain in a vegetative stage for the whole growing season. The bulbs also become covered with black lesions. The pathogen is soil-borne and the disease is more prevalent in temperate region. The onion is susceptible to this disease for a very short time, i.e., between germination and emergence.

**(b) Management**

**Solarization:** Seed-bed solarization (with 100µ thick clear polyethylene sheets for 6 weeks, during July and August, prior to seeding) significantly reduced onion smut pathogen (*Urocystis cepulae*) and improved seedling stand and characters. Smut disease was completely controlled by soil solarization in both seasons. In the non-treated plots disease infection ranged from 31-47% and 11-33% in both seasons, respectively. Seed germination was improved in the solarized plots, being 515 and 628 seedlings/$m^2$ compared with 264 and 456 seedlings/$m^2$ in the untreated plots, respectively.

### 15.6.2. Nematodes

**(i) Root-knot Nematode, *Meloidogyne spp.***

**(a) Symptoms:** Symptoms of root-knot on onion are very discrete, the main symptom being the presence of the protruding egg masses (Fig. 15.29).

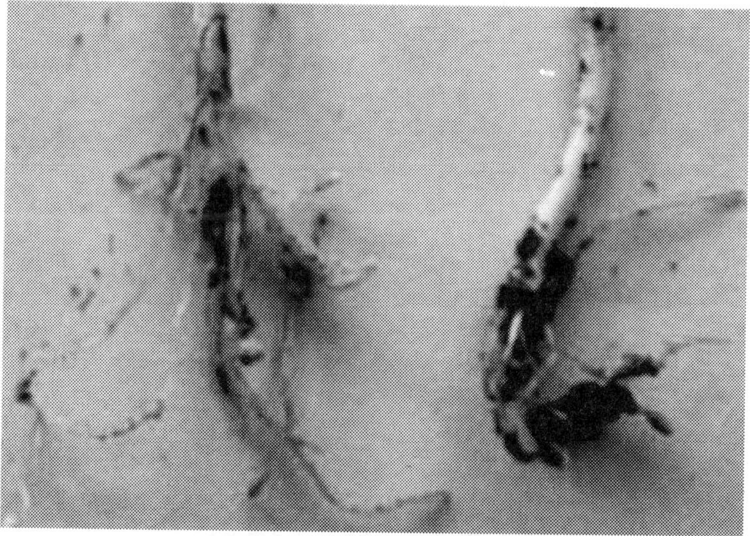

**Fig. 15.29.** Root-knot nematode on onion.

**(b) Management**

**Solarization:** Data revealed that solarization resulted in reductions of nematode populations. Table 15.50 shows that all treatments gave RF = 0, except the treatments of June (four weeks) and July (four weeks), while RF of non-solarized control plots was 2.35. Data in Table 15.51 show that solarization gave good control of nematodes. The RF in solarized treatments ranged from 0 to 0.47, while in non-solarized plots, reductions ranged from 1.18 to 1.57.

Table 15.50. Effect of solarization and timing on the control of nematodes on onion

| Treatment | Average number of nematodes per 200 g soil[1] | | Reproduction factor (RF) RF = Pf/Pi |
|---|---|---|---|
| | (Pi) | (Pf) | |
| June - 4 weeks | 280 | 10 | 0.035 |
| June-July - 6 weeks | 140 | 0 | 0 |
| July - 4 weeks | 100 | 60 | 0.60 |
| July-August - 6 weeks | 200 | 0 | 0 |
| August - 4 weeks | 180 | 0 | 0 |
| Control | 170 | 400 | 2.35 |

1 Pi = Initial nematode population; Pf = Final nematode population.

Table 15.51. Effect of solarization on the control of nematodes on onion

| Treatment | Average number of nematodes per 200 g soil[1] | | Reproduction factor (RF) RF = Pf/Pi |
|---|---|---|---|
| | (Pi) | (Pf) | |
| Onion Nursery solarized | 235 | 0 | 0 |
| Onion - solarized | 65 | 0 | 0 |
| Onion Nursery - non - solarized | 165 | 260 | 1.57 |
| Onion - non-solarized | 125 | 147 | 1.18 |

1 Pi = Initial nematode population; Pf = Final nematode population.

### (ii) Other Nematodes

#### (a) Management

**Solarization:** Table 15.52 clearly shows the effect of solarization on the control of nematodes on onion, with RF's of 0.03 for *Tylenchorhynchus* and *Trichodorus* spp. The corresponding RF values in non-solarized soil were 1.44 on the same crop and the same nematodes.

Table 15.52. Effect of solarization on the control of nematodes on onion

| Treatment | Average number of nematodes per 200 g soil[1] | | Reproduction factor (RF) RF = Pf/Pi |
|---|---|---|---|
| | (Pi) | (Pf) | |
| Onion- solarized | 580 | 16 | 0.03 |
| Onion-non-solarized | 685 | 985 | 1.44 |

Pi = Initial nematode population; Pf = Final nematode population.

Solarization was tested in a soil with *Meloidogyne, Pratylenchus. Helicotylenchus, Tylenchorhynchus* and *Hoplolaimus* spp. in Beni-Sweif, Fayoum, Qalyobia and Ismailia localities. In general, combinations of solarization and herbicides gave good control of nematodes and healthy plants, compared with non-solarized plots (Table 15.53).

Table 15.53. Effect of solarization for control of nematodes on onion

| Localities | Solarized | | | Non-solarized | | |
|---|---|---|---|---|---|---|
| | Av. No. of nemas/200 g soil | | Reproduction factor (RF) RF = Pf/Pi | Av. No. of nemas/200 g soil | | Reproduction factor (RF) RF = Pf/Pi |
| | (Pi) | (Pf) | | (Pi) | (Pf) | |
| Beni-Sweif | 80 | 3 | 0.04 | 115 | 130 | 1.13 |
| Fayoum | 125 | 0 | 0 | 170 | 135 | 0.8 |
| Qalyobia | 270 | 0 | 0 | 100 | 107 | 1.07 |
| Ismailia | 75 | 0 | 0 | 90 | 136 | 1.51 |

1 Pi = Initial nematode population; Pf = Final nematode population.

### 15.6.3. Weeds

***(i) Biofumigation:*** Biofumigation is used as an alternative for managing agricultural weeds by using Ida Gold mustard (Fig. 15.30), hairy vetch and winter wheat. These crops are grown for several weeks, and then tilled into the soil to release their bioactive compounds during decomposition, to help manage troublesome weeds such as morning glory and yellow and purple nutsedge in onion fields. Two conditions need to be present for the chemical reaction to occur-water and heat. When you incorporate the plant material into a dry soil environment, all that will happen is that the material will dry out. They will not be biologically active because there is not enough moisture.

**Fig 15.30.** Ida Gold mustard for control of onion weeds

## 15.7. GARLIC, *ALLIUM SATIVUM*

### 15.7.1. Diseases

#### (i) White Rot, *Sclerotium cepivorum*

White rot, caused by the soil-borne fungus *Sclerotium cepivorum*, is a very destructive disease that begins in the field and can carry over into storage.

**(a) Symptoms:** The first above-ground symptoms are yellowing and dieback of the leaf tips, followed by a collapse of the affected leaves; however, these symptoms alone could easily be confused with other types of damage (i.e., onion maggot). When the bulbs and roots are examined, a white, fluffy mold and soft rot will be observed. Masses of tiny black sclerotia can also be seen within this mold (Fig. 11.18). These sclerotia remain in the soil for many years. Infected bulbs can rot in storage boxes and stain other bulbs. White rot typically develops in patches in the field and is less of a problem when soils are warm (higher than 24°C) and dry.

#### (b) Management

**Solarization:** As a result of tarping moist and dry soils infested artificially with *S. cepivorum* (125 sclerotia/100 g dry soil) with four types of different colour and thickness plastic mulches during July and August, a significant increase of temperature was observed at soil surface and at 5-7 cm depth, in comparison with the control treatment. The viable number of sclerotia had decreased by 75.2-83.2 percent in the moist soils covered with clear plastic mulches. However, in the moist soil covered with black plastic mulch, the decrease in sclerotia ranged from 49.6-59.2 percent, compared with about 12 percent in the control treatment.

In contrast in dry soils, the loss of sclerotia numbers was few; 2.4-10.4 percent in tarped dry soils with clear plastic mulches, 1.6-4.8 percent in tarped treatment with black mulch, and about 1.6 percent in the control (Tables 15.54 and 15.55).

Table 15.54. The effect of solarization on the temperatures of soils covered with four types of different colour and thickness plastic sheets

| Type of plastic films | Average temperature (°C)[1] | | | |
|---|---|---|---|---|
| | Moist Soil | | Dry Soil | |
| | 0-2 cm depth | 5-7 cm depth | 0-2 cm depth | 5-7 cm depth |
| 100 mil transparent polyethylene sheet | 54.5 | 42.3 | 65.0 | 42.5 |
| 180-200 mil transparent polyethylene sheet | 54.3 | 41.5 | 65.0 | 42.0 |
| 400 mil transparent polyvinyl chloride sheet | 51.3 | 41.3 | 68.8 | 42.8 |
| 60 mil black polyvinyl chloride sheet | 48.0 | 35.0 | 56.5 | 37.3 |
| Uncovered control | 42.5 | 32.5 | 56.3 | 36.6 |

1 The average air temperature in July and August at 1:30 pm at 10 cm above the soil surface was 42.5°C.

Table 15.55. The effect of covering moist soil with plastic films on the number and characteristics of *Sclerotium cepivorum* sclerotia at 5-7 cm depth, during July and August

| Type of plastic film | Total no. of sclerotia before covering | Number of sclerotia/ 100g dry soil | | |
|---|---|---|---|---|
| | | Deca-yed | Holl-owed | % of total |
| July 18, 1989 100 mil transparent polyethylene sheet | 125 | 46 | 38 | 67.2 |
| 180-200 mil transparent polyethylene sheet | 125 | 42 | 40 | 65.6 |
| 400 mil transparent polyvinyl chloride sheet | 125 | 32 | 39 | 56.8 |
| 60 mil black polyvinyl chloride sheet | 125 | 29 | 20 | 39.2 |
| Uncovered control | 125 | 2 | 1 | 2.4 |
| Sept 5, 1989 100 mil transparent polyethylene sheet | 125 | 77 | 25 | 81.6 |
| 180-200 mil transparent polyethylene sheet | 125 | 70 | 24 | 75.2 |
| 180-200 mil transparent polyethylene sheet 400 mil transparent polyvinyl chloride sheet | 125 | 71 | 23 | 75.2 |

| | | | | |
|---|---|---|---|---|
| 60 mil black polyvinyl chloride sheet | 125 | 36 | 26 | 49.6 |
| Uncovered control | 125 | 14 | 2 | 12.8 |

Soil solarization for 8-11 weeks during summer using transparent sheet resulted in reduction of *S. cepivorum* in garlic to undetectable levels in upper 20 cm of soil.

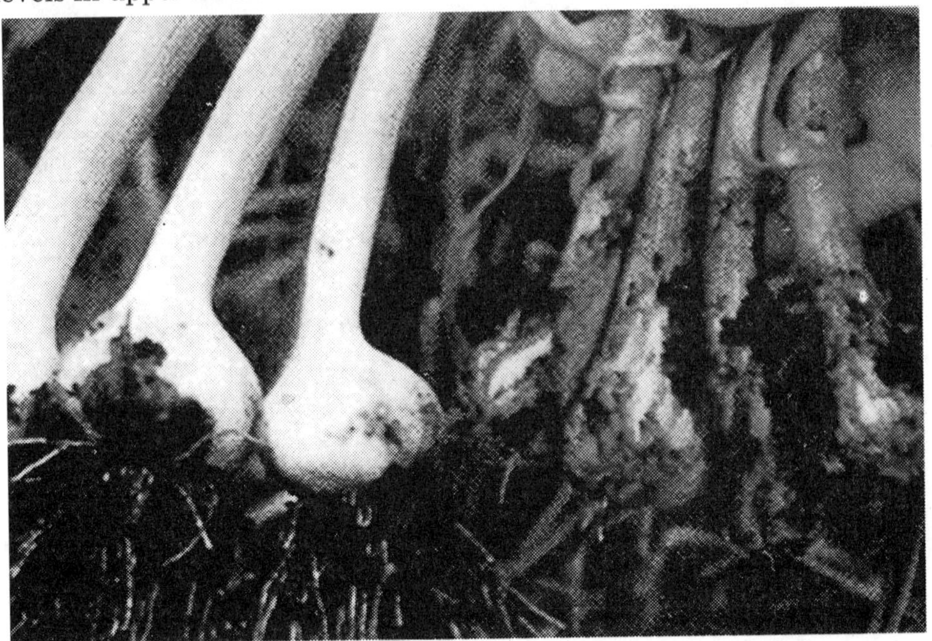

**Fig. 15.31.** The stem and bulb nematode (*Ditylenchus dipsaci*) on garlic. Left - healthy; Right - infected.

### 15.7.2. Nematodes

#### (i) Stem and Bulb Nematode, *Ditylenchus dipsaci*

**(a) Symptoms:** The leaves of garlic plants turn yellow and die but no thickening or twisting is evident (Fig. 15.31). The infected garlic bulbs become loose (the cloves separate from the base) and cannot be used for seed even if they tolerate winter storage well because the plants which form often die during the vegetative period (Fig. 15.32).

#### (b) Management

**Crop Rotation:** Onion and garlic may be rotated with wheat, barley, maize, rape, white clover, red clover, lucerne or lettuce to overcome the stem and bulb nematode damage.

**Solarization:** In fields heavily infested with *D. dipsaci* in Israel, soil solarization protected garlic bulbs throughout the growing season resulting in a greater yield increase compared with EDB and methyl bromide applications (Siti *et al.*, 1982). In Italy, control of *D. dipsaci* in a sandy soil increased with solarization periods and only 10, 6 and 2 percent nematodes were still viable after four, six, and eight weeks of solarization, respectively (Greco *et al.*, 1985).

**Fig. 15. 32.** Garlic crop in field infected with *D. dipsaci.*

## 15.8. OKRA, *ABELMOSCHUS ESCULENTUS*

### 15.8.1. Nematodes

#### (i) Root-knot Nematodes, *Meloidogyne* spp.

*M. incognita* was responsible for 28.08 to 90.90% loss in fruit yield of okra (Bhatti and Jain, 1977; Parvatha Reddy and Singh, 1981), while *M. javanica* caused 20.20 to 41.20% loss in yield (Jain *et al.*, 1986).

#### (a) Management

**Crop Rotation:** Decrease in root-knot nematode population in okra occurs following marigold, spinach and bottle gourd (Khan *et al.*, 1975). Out of 15 different cropping sequences studied, tomato-onion-resistant tomato cv. Hisar Lalit-okra was found to be the best in terms of root-knot nematode reduction and economic returns (Kanwar,

1990). Crop rotations having okra-garlic-cluster bean and okra-coriander-resistant tomato cv. Hisar Lalit effectively managed the root-knot nematode population (Anon, 1993b).

Rotation of okra with sweet potato (cv. Sree Bhadra) reduced the root-knot nematode population by 21% and increased the fruit yield by 20.57% (Sheela *et al.*, 2002). Crop rotation with cabbage, marigold and wheat significantly reduced the root-knot nematode population. Crop rotation with cereals and marigold helps in reducing the nematode population in soil. In cropping sequence involving cabbage-marigold-wheat-kochia, the population of root-knot nematode larvae reduced to such an insignificant level that growing okra following the above sequence did not bring about increase in number of larvae (Alam *et al.*, 1977).

**Intercropping:** Atwal and Mangar (1969) showed that root exudates from sesame (*Sesamum orientale*) have nematicidal properties against *M. incognita*. When okra was grown in *M. incognita* infested soil it was only slightly attacked and there were fewer nematodes compared with when okra was grown in the absence of sesame. Root exudates and extracts of sesame seedlings inhibited hatching of eggs, penetration of roots and also delayed the life cycle of *M. incognita* (Tanda and Atwal, 1988). Similarly, intercropping of okra with marigold, margosa and Persian lilac have been recorded to reduce nematode population and root galling. The root-knot development on okra was low when interplanted with *Tagetes erecta*.

**Organic Amendments:** Application of pressmud at 15 t/ha, a week prior to sowing is recommended for effective management of root-knot nematodes in *kharif* season. Crop rotation with cereals and marigold helps in reducing the nematode population in soil. Application of fresh neem or karanj cake at 1 t/ha at the time of final land preparation is recommended.

Singh and Sitaramaiah (1966) applied finely divided oil cakes to root-knot infested soil and noted a reduction in disease incidence in okra.

**Biofumigation:** Singh and Sitaramaiah (1967) reported that application of 5-10% (w/w) of leaves of neem, *Melia azedirach, Cassia fistula, Sesbania aculeata* and *Crotalaria juncea* to the infested soil reduced the incidence of root-knot nematode on okra.

## Integrated Management

**Biofumigation and Chemicals:** Application of subabool (*Leucaena leucophila)* leaves at 40 g/kg soil which are allowed to decompose for four weeks before sowing + carbofuran at 1 kg a.i./ha while sowing of okra seeds resulted in minimum galling (35.4 galls/plant) in comparison to control (64 galls/plant) and better plant growth parameters in okra (Paruthi *et al.*, 1987).

**Organic Amendments, Bioagents and Chemicals:** The combined treatment with *P. lilacinus* at 4 g/kg soil + carbosulfan 25 EC at 0.2% + poultry manure at 2.5 t/ha + FYM at 2.5 t/ha gave maximum increase in plant growth parameters and yield (9.2 MT/ha compared to 2.0 MT/ha in control), and decrease in number of galls (83.5 compared to 183.6 in control), egg masses per root system (23.6 compared to 70.5 in control) and final nematode population in soil (200 compared to 585 in control) (Das and Sinha, 2005).

### (ii) Root Rot, *Rhizoctonia solani* and Root-knot Nematode, *Meloidogyne incognita* Disease Complex

#### (a) Management

**Organic Amendments and Bioagents:** Integration of *P. lilacinus* and *T. viride* with mustard cake gave significant reduction in root galling, egg mass production, fecundity, reproduction factor and nematode population both in soil and roots coupled with increase in plant growth parameters (Tripathi and Singh, 2006).

## 15.9. CARROT, *DACUS CAROTA*

### 15.9.1. Nematodes

### (i) Root-knot Nematodes

**(a) Symptoms:** Devi (1993) reported that *M. incognita* was responsible for 56.64% loss in yield of carrots. Root-knot nematodes develop characteristic forking of the roots in carrots (Fig. 15.33).

#### (b) Management

**Crop Rotation:** The population of *M. incognita* larvae decreases in combination of spinach-marigold-carrot and spinach-radish-carrot (Hasan and Jain, 1998).

Significant carrot yield increase was observed in the plots previously rotated with *E. sativa* cv. Nemat, *B. juncea* and *R. sativus*. The beneficial effect of the rotation with mustard species was greater in biocidal crops sown in late spring rather than in late summer (Curto *et al.*, 2006).

**Fig. 15.33.** Forking of carrots incited by root-knot nematodes (Courtesy: AICRP on Nematodes, Hisar).

**Organic Amendments:** In a field trial, saw dust, neem cake, poultry manure at 1.0 and 1.5 t/ha and water hyacinth at 1 and 2 t/ha increased yield of carrot (28-66%) by reducing the infestation of *M. incognita*. The maximum reduction of *M. incognita* in soil was recorded in neem cake (38.5%) followed by saw dust (23%), poultry manure (22%) and water hyacinth (14%) at higher dose over control (Devi, 1993).

**Biofumigation:** The effectiveness of soil biofumigation treatments with *Brassica juncea* and *Eruca sativa* green manures, *B. juncea* hay, and a *Brassica* seed meal commercial pellet were investigated over two years in field experiments on a carrot - tomato rotation for the control of the root-knot nematode *Meloidogyne incognita*. Biofumigation treatments were compared with the standard nematicide fenamiphos, a quillay-derived nonchemical formulation and an untreated control. At the end of the first cycle, soil population density of *M. incognita* was significantly lower in the plots treated with green manure or hay of *B. juncea* or with green manure of *E. sativa*, either alone or combined with seed meal pellet, compared with untreated control. Nematicidal effects of biofumigation treatments did not differ from that of fenamiphos. Moreover, commercial pellet and *B. juncea* green manure and hay resulted also in carrot yield almost double the control, whereas lower increases were provided by *E. sativa* green manure, either alone or combined with the pellet.

In the following tomato crop the lowest *M. incognita* densities were recorded in soil treated with *B. juncea* hay or with *E. sativa* green manure plus pellet. Hay and green manure of *B. juncea* and integrated treatment with *E. sativa* green manure and pellet resulted in the highest tomato yields, always significantly higher than fenamiphos.

The *Brassica* crop arugula (*Eruca sativa* var. Nemat) also received interest as a potentially useful biofumigant crop to manage root-knot nematodes, as it appears to act as a trap crop for *M. hapla* (Melakeberhan *et al.*, 2006). Positive effects on yields of nematode-susceptible crops grown after arugula cultivation and biofumigation were reported for carrot.

Several chemical and non-chemical alternatives to standardized cold or hot diffusion MB (98-2) at 50 g/m$^2$ are able to maintain adequate sanitary (against *M. incognita*) and productivity levels on carrot cultivation (Table 15.56).

Table 15.56. Commercial yield of roots of carrot (kg/m$^2$).

| Treatment | Yield (kg/m$^2$) | |
|---|---|---|
| | Av. of 4 cultiva-tions | Av. relative yield |
| Control without soil fumigation | 4.46 a | 108 |
| Standardized cold diffusion MB(98-2) at 50 g/m$^2$ | 4.41 a | 107 |
| Cold diffusion MB(98-2) at 25 g/m$^2$ under VIF transparent film | 4.38 ab | 106 |
| Injection 1,3 dichloropropene + chloropicrin at 60 cc/m$^2$ | 4.25 ab | 103 |
| Injection chloropicrin at 40 cc/m$^2$ | 4.21 ab | 102 |
| Injection 1,3 dichloropropene at 20 cc/m$^2$ under VIF transparent film | 4.20 ab | 102 |
| Soil solarization alone for 6 weeks | 4.12 ab | 100 |
| Soil solarization and simultaneous cold diffusion MB(98-2) at 10 g/m$^2$ | 4.08 abc | 99 |
| Soil solarization and simultaneous emulsion Metham Sodium at 60 cc/ m$^2$ | 3.90 abc | 95 |
| Soil solarization and simultaneous injection 1,3 dichloropropene at 25 cc/m$^2$ | 3.87 abc | 94 |
| Soil solarization and simultaneous incorporation 5 kg/m$^2$ mushroom compost | 3.59 bc | 87 |
| Soil solarization and simultaneous incorporation 3 kg/m$^2$ chicken manure | 3.25 c | 79 |

### (ii) Cyst Nematode, *Heterodera carotae*

#### (a) Management

**Solarization:** Better control was achieved in Italy, where only 24-38% of eggs survived in cysts of *H. carotae* from the top 30 cm soil solarized for four to eight weeks (Greco *et al.*, 1985). In another trial the yield of carrots from plots solarized for eight weeks was 4.6 kg/m$^2$, compared with 5.5 kg in plots treated with DD at 400 1/ha and only 2.3 kg in the control.

**Biofumigation:** In a field infested by the carrot cyst nematode *H. carotae*, biofumigation with *B. juncea* hay or *Brassica* seed meal pellet was compared with commercial liquid nematicidal formulations of quillay, tagetes or neem, and with the chemical fenamiphos and an untreated control. At the end of the first year, soil populations of *H. carotae* were significantly suppressed by all the treatments, though neem and fenamiphos provided the lowest nematode populations and the highest carrot yield. At the end of carrot cycle in second year, *H. carotae* population did not increase in plots treated with hay or fenamiphos, whereas pellet and quillay resulted in significantly higher, though lower than untreated control, densities. Pellet and fenamiphos provided the highest total yield and percentage of commercial tap roots (De Mastro *et al.*, 2008d).

## 15.10. RADISH, *RAPHANUS SATIVUS*

### 15.10.1. Diseases

#### (i) Black Spot, *Verticilium dahliae*

#### (a) Management

**Biofumigation:** After cultivation of brown mustard, damage of radish black spot caused by *V. dahliae* was reduced, but it is not clear whether this is due to ITC gas effect (Hashizume *et al.*, 2008).

### 15.10.2. Nematodes

#### (i) Stunt Nematodes, *Tylenchorhynchus spp.*

#### (a) Management

**(a) Crop Rotation:** The mustard (*rabi*), radish (summer), sesame (*kharif*) rotation sequence considerably reduced the population of *Tylenchorhynchus* spp. (Haque and Gaur, 1985).

## 15.11. CABBAGE, *BRASSICA OLERACEA* VAR. *CAPITATA* AND CAULIFLOWER, *B. OLERACEA* VAR. *BOTRYTIS*

### 15.11.1. Diseases

#### (i) Wilt, *Verticillium dahliae*

*Verticillium* wilt is an increasing problem in European cauliflower production.

#### (a) Management

**Biofumigation:** Broccoli (*Brassica oleracea* var. *italica*), cauliflower (*B. oleracea* var. *botrytis*), Indian mustard (*B. juncea*), ryegrass (*Lolium perenne*) and corn (*Zea mays*) were incorporated in naturally *V. dahliae* var. *longisporum* infested soil samples collected from two cauliflower fields in Belgium, labeled Is1 and S3. In the Is1 soil, broccoli, cauliflower and ryegrass incorporation significantly reduced the inoculum level by more than 90%, while Indian mustard significantly reduced numbers of viable microsclerotia by 50%. In the S3 soil, fungicidal volatile compounds from Indian mustard killed the microsclerotia. In the S3 soil, 1% (w/w) Kraft pine lignin had to be added to observe a significant reduction on the viability of microsclerotia, whereas in the Is1 soil, a significant effect was observed when 0.1% (w/w) Kraft pine lignin was added. Incorporation of lignin-rich substrates in soil may open up new perspectives for integrated control of *Verticillium* (Debode *et al.*, 2005).

Verticillium wilt was successfully managed in a multiple-year field study by incorporating broccoli residues into infested soil. In a study conducted from 1993 to 1995 in the Salinas Valley, California, cauliflower disease incidence and severity were consistently and significantly reduced in the broccoli (*Brassica oleracea* var. *italica)* residue plots when compared with no broccoli. The commercial standard plots fumigated with methyl bromide + chloropicrin had the lowest disease incidence and severity. In both years of tarping study, *Verticillium* wilt severity was lowest in the metham sodium treatment. The cauliflower-*Verticillium* host-pathogen system therefore, can act as a model for controlling soil-borne diseases without the use of synthetic chemicals (Koike and Subbarao, 2000).

Broccoli is resistant to *V. dahliae* infection and does not express wilt symptoms. Incorporation of broccoli residues reduces soil populations of *V. dahliae*. The effects of broccoli residue were tested on the colonization of roots by *V. dahliae*, plant growth response, and disease incidence of both broccoli and cauliflower in soils with different levels of *V. dahliae* inoculum and with or without fresh broccoli residue amendments. The three soils included a low-*Verticillium* soil, a high-*Verticillium* soil, and a broccoli-rotation soil (soil from a field after two broccoli crops) with an average of 13, 38, and below-detectable levels of microsclerotia per g of soil, respectively. Cauliflower plants in broccoli-amended high-*Verticillium* soil had significantly ($P \leq 0.05$) lower wilt incidence and severity than did plants in unamended soil. Despite the absence of wilt symptoms, broccoli roots were colonized by *V. dahliae*. In high-*Verticillium* soil, the broccoli residue amendment caused a marked reduction in colonization rate of *V. dahliae* per unit of inoculum on both cauliflower and broccoli roots. In addition to its detrimental effects on the viability of microsclerotia in soil, broccoli residue may also have an inhibitory effect on the root-colonizing potential of surviving microsclerotia (Shetty *et al.*, 2000).

Although broccoli is closely related to cauliflower, wilt has not occurred on this host and *V. dahliae* isolates from cauliflower were weakly pathogenic on broccoli in greenhouse inoculations. The effectiveness of broccoli residues on propagule attrition in soil and wilt incidence on cauliflower was determined in a commercial field infested with *V. dahliae* microsclerotia. The treatments were broccoli residue with tarp, broccoli residue without tarp, chloropicrin, metham sodium, control with tarp, control without tarp, cauliflower followed by cauliflower, and cauliflower followed by fallow. Even though the pre-treatment number of *V. dahliae* propagules in broccoli-treated and control plots were similar, by the end of the cauliflower cropping season numbers either were the same or increased in control plots compared with a reduction in numbers in broccoli residue-treated plots during both seasons. There were no differences between tarped and non-tarped plots either in broccoli residue-treated or control plots. In fumigated plots, propagules declined initially but later returned to pre-treatment levels by the end of the cropping season. Continuous cauliflower or fallowing after one cauliflower crop resulted in stable or increased levels of microsclerotia. Verticillium wilt severity was lowest in metham sodium-treated plots, followed by broccoli residue-treated plots and chloropicrin in both years. Plant height, marketable heads, and head weight were higher in broccoli- treated than in control plots. These results suggest that broccoli residues reduce *V. dahliae* microsclerotia in soil and wilt of cauliflower as much as or more than chloropicrin and metham sodium, and that rotation with broccoli may be a feasible approach to manage Verticillium wilt in cauliflower and other susceptible crops (Subbarao *et al.*, 1999).

Experiments were conducted in field plots to evaluate the effects of broccoli residue on population dynamics of *V. dahliae* in soil and on Verticillium wilt development on cauliflower under furrow and subsurface-drip irrigation and three irrigation regimes in 1994 and 1995. Treatments were a factorial combination of three main plots (broccoli crop grown, harvested, and residue incorporated in *V. dahliae*-infested plots; no broccoli crop or residue in infested plots; and fumigated control plots), two subplots (furrow and subsurface-drip irrigation), and three sub-subplots (deficit, moderate, and excessive irrigation regimes) arranged in a split-split-plot design with three replications. Number of propagules in all broccoli plots declined significantly ($P < 0.05$) after residue incorporation and continued to decline throughout the cauliflower season. The overall reduction in the number of propagules after two broccoli crops was approximately 94%, in contrast to the fivefold increase in the number of propagules in infested main plots without broccoli after two cauliflower crops. Disease incidence and severity were both reduced approximately by 50% ($P < 0.05$) in broccoli treatments compared with no broccoli treatments. Differences between furrow and subsurface drip irrigation were not significant, but incidence and severity were significantly ($P < 0.05$) lower in the deficit irrigation regime compared with the other two regimes. Abundance of microsclerotia of *V. dahliae* on cauliflower roots about 8 weeks after cauliflower harvest was significantly ($P < 0.05$) lower in treatments with broccoli compared with treatments without broccoli. Rotating broccoli with cauliflower and incorporating broccoli residues into the soils is a novel means of managing Verticillium wilt on cauliflower and perhaps on other susceptible crops. This practice would be successful regardless of the irrigation methods or regimes followed on the susceptible crops (Xiao *et al.*, 1998).

Table 15.57. Soil populations of *Verticillium dahliae* (number of microsclerotia per gram of dry soil) in different treatment combinations

| Treatment, amendment* | No. of microsclerotia of *Verticillium dahliae* / gram of dry soil | |
|---|---|---|
| | Initial** | 12 weeks after planting |
| **LV soil** | | |
| Broccoli | 13 ± 1 | 4 ± 0.0 |
| No broccoli | 13 ± 1 | 8 ± 2.0 |
| **HV soil** | | |
| Broccoli | 38 ± 5..5 | 12 ± 3.7 |
| No broccoli | 38 ± 5..5 | 29 ± 4.7 |
| **BB soil** | | |

| | | |
|---|---|---|
| Broccoli | 0 | 0 |
| No broccoli | 0 | 0 |

* LV = soil with 13 *V. dahliae* microsclerotia per g of dry soil; HV = soil with 38 *V. dahliae* microsclerotia per g of dry soil; and BB = soil with undetectable *V. dahliae* microsclerotia.

** Soil sampled prior to the incorporation of broccoli residues.

In both broccoli-amended and unamended LV and HV soils, the numbers of microsclerotia at the end of the experiment were lower than were the initial numbers. The magnitude of reduction was more than 65% in broccoli-amended soils compared with that of unamended soil. The BB soil had no detectable *V. dahliae* microsclerotia throughout the experimental runs in both cauliflower- and broccoli-planted pots (Table 15.57).

Cauliflower plants were significantly taller (P = 0.05) in all soils amended with broccoli residue compared with those in controls. Cauliflower plants gained significantly (P = 0.05) higher shoot and root biomass after broccoli residue amendment on all soils compared with that of unamended controls. A significantly lower (P = 0.05) percentage of cauliflower plants grown in HV soil with broccoli amendment exhibited vascular discolouration than that of plants grown without broccoli amendment (Table 15.58).

Table 15.58. Effect of soil treatment and broccoli amendment on cauliflower growth and biomass and on *Verticillium dahliae*-induced vascular discolouration

| Treatment, amendment* | Plant height (cm) | No. of leaves | Shoot dry weight (g) | Root dry weight (g) | Vascular Discolouration** (%) |
|---|---|---|---|---|---|
| **LV soil** | | | | | |
| Broccoli | 32.9 a | 15.6 ab | 34.63 a | 8.03 a | 10 b |
| No broccoli | 27.7 b | 12.3 c | 29.73 b | 6.85 b | 0 b |
| **HV soil** | | | | | |
| Broccoli | 29.3 b | 15.9 a | 34.27 a | 7.22 ab | 20 b |
| No broccoli | 25.2 c | 15.3 ab | 19.90 c | 3.77 c | 60 a |
| **BB soil** | | | | | |
| Broccoli | 31.8 a | 16.1 a | 34.16 a | 8.38 a | 20 b |
| No broccoli | 29.0 b | 14.3 b | 26.69 b | 6.58 b | 30 ab |

Means followed by the same letter are not significantly different according to the least significant difference test (P = 0.05).

* LV = soil with 13 *V. dahliae* microsclerotia per g of dry soil; HV = soil with 38 *V. dahliae* microsclerotia per g of dry soil; and BB = soil with undetectable *V. dahliae* microsclerotia.

**Percentage of plants exhibiting vascular discoloration.

Amendment with broccoli increased the number of cauliflower leaves on LV and BB soils (P = 0.05). Broccoli amendment also significantly (P = 0.05) increased the height of broccoli plants and the number of leaves on LV and BB soils, but not on the HV soil. Significant (P = 0.05) gains in broccoli root biomass occurred only on BB soil amended with broccoli residue compared with that of the unamended controls. The differences in percent vascular discolouration in other treatments in cauliflower as well as in all treatments in broccoli plants were not significant (*P* > 0.05) (Table 15.59).

Table 15.59. Effect of soil treatment and broccoli amendment on broccoli growth and biomass and on *Verticillium dahliae*-induced vascular discolouration

| Treatment, amendment* | Plant height (cm) | No. of leaves | Shoot dry weight (g) | Root dry weight (g) | Vascular Discolouration** (%) |
|---|---|---|---|---|---|
| **LV soil** | | | | | |
| Broccoli | 43.2 a | 10.9 cd | 37.70 a | 16.96 ab | 0 a |
| No broccoli | 36.0 c | 9.0 e | 34.84 a | 15.10 b | 0 a |
| **HV soil** | | | | | |
| Broccoli | 38.5 b | 13.5 a | 36.86 a | 17.00 ab | 20 a |
| No broccoli | 38.4 b | 12.5 ab | 35.62 a | 16.03 ab | 20 a |
| **BB soil** | | | | | |
| Broccoli | 39.7 b | 11.4 bc | 25.69 a | 18.13 a | 0 a |
| No broccoli | 35.9 c | 10.0 de | 35.68 a | 14.80 b | 10 a |

Means followed by the same letter are not significantly different according to the least significant difference test (P = 0.05).

* LV = soil with 13 *V. dahliae* microsclerotia per g of dry soil; HV = soil with 38 *V. dahliae* microsclerotia per g of dry soil; and BB = soil with undetectable *V. dahliae* microsclerotia.

** Percentage of plants exhibiting vascular discoloration.

## (ii) White Rot or Stalk Rot, *Sclerotinia sclerotiorum*

**(a) Symptoms:** Leaves when infected turn light green and later yellow and are shed prematurely. Black soft rot often appears on petioles and midrib touching the soil and are covered with fluffy growth of the fungus. Stem at the ground level shows girdling. Fluffy mycelium of the fungus grows on these infected plant parts with numerous sclerotia embedded. Cool (20-25°C) and humid (95-100% RH) conditions are ideal for disease initiation and spread.

**(b) Management**

**Organic Amendments:** Application of FYM or saw dust reduces the disease incidence. Soil amendment with sunflower inflorescence, rape seeds and pine needles reduce the disease incidence (Gupta *et al.*, 1986).

**Crop Rotation:** Following cabbage/cauliflower – paddy rotation is effective against white rot.

### (iii) Club Root, *Plasmodiophora brassicae*

**(a) Symptoms:** Roots enlarge to form spindle-shaped growth (Fig. 15.34). The pathogen is responsible for temporary flagging of leaves on bright days leading to their wilting. Secondary invasion of soft rot bacteria follows. This helps in accumulation of toxic materials in plants.

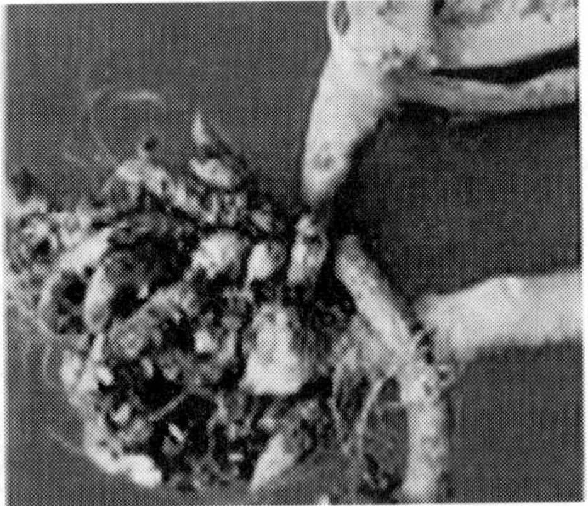

**Fig. 15.34.** Club root affected cabbage roots.

**(b) Management**

**Trap Cropping:** Trap cropping with rayosak among susceptible brassicas in the first year and tori sarson and radish among resistant brassicas in the second year with 30 days optimum trapping period was found most effective in reducing the disease intensity and increasing the yield.

**IPM**

- Deep off-season ploughing.
- Soil solarization for 2-3 weeks during summer using 400 gauge thick transparent polythene film.
- Green manuring with sunnhemp in July – August.

**(iv) Root Rot, *Rhizoctonia solani***

**(a) Symptoms:** The root rot disease caused by *R. solani* damages the cauliflower crop in curd production.

**(b) Management**

**Solarization, Green Manuring, Bioagents and Chemicals:** The seed treatment with carbendazim at 2g/kg seed, raising seedling in solarized beds, crop raising in green manure field + neem cake 25kg/ha with soil treatment by *Trichoderma viride* at 2 kg/ha gave the lowest disease intensity of 11.50 with maximum curd yield 299.58q/ha with highest percent disease control (51.00) over control treatment (Dabbas *et al.*, 2009).

**(v) Cabbage Yellows, *Fusarium oxysporum* f. sp. *conglutinans***

**(a) Symptoms:** Affected plants have a sickly appearance, lower leaves drop one by one and growth of the plant becomes stunted. The fungus is seed-borne and can also persist in soil for many years. Disease development is promoted by warm weather conditions.

**(b) Management**

**Solarization:** Cabbage and lettuce seed-bed solarization significantly reduced *Fusarium* spp. and total fungal infestations.

**15.11.2. Nematodes**

**(i) Stunt Nematode, *Tylenchorhynchus brassicae***

*T. brassicae* is the most prevalent species around the roots of cabbage and cauliflower in Uttar Pradesh.

**(a) Symptoms:** *T. brassicae* caused reduction in fresh weight of plants and suppressed root growth, resulting in reduced water absorption.

**(b) Management**

**Crop Rotation:** There was sudden decline in population of *T. brassicae* when cabbage and cauliflower were rotated with wheat (Siddiqi *et al.,* 1973). The mustard (*rabi*), radish (summer), sesame (*kharif*) sequence considerably reduced the population of *Tylenchorhynchus* spp. (Haque and Gaur, 1985).

**Intercropping:** Interculture of seedlings of marigold, neem and Persian lilac reduced multiplication rate of *T. brassicae* and improved plant growth. Intercropping of cabbage and cauliflower with neem reduced the multiplication rate of *T. brassicae* and improved the plant health (Siddiqi and Saxena, 1987a, b).

**Organic Amendments:** Amendment of soil with oil cakes of neem, groundnut, mustard and castor (equivalent to 100 kg N/ha) was effective in reducing the population of *T. brassicae* around the roots of cabbage and cauliflower (Siddiqi *et al.*, 1976). Neem cake was the best.

**Biofumigation:** Soil amendment with chopped shoot parts of jack, *Ficus carica, F. elastica, F. glomerata, Ipomea fistulosa, Nerium indicum, Tabernaemontana coronaria* gave good control of *T. brassicae* on cabbage and cauliflower (Siddiqi *et al.*, 1987).

### 15.11.3. Weeds

***(i) Solarization:*** Solarization resulted in approximately 100%, 100% and 79% weed reduction in sandy soil at Ismailia and 100%, 94% and 69% reduction in clay soil at Kalubia for annual broad leaf weeds, annual grasses and perennial weed infestation, respectively.

***(ii) Mulching:*** Increase in yield of cauliflower by 40% using black polythene film has been reported (Halappa and Srinivas, 1973). This inorganic mulch also helped in retaining moisture, prevented deterioration of soil structure and reduced the difference between maximum and minimum temperatures. Best growth and increased yield were obtained with mango leaf mulch followed by pine needles and hay mulch (Singh and Mishra, 1976).

## 15.12. PEA, *PISUM SATIVUM*

### 15.12.1. Diseases

#### (i) White Rot, *Sclerotium rolfsii*

**(a) Symptoms:** It is a ubiquitous fungus with a broad host range. Initially wet rotting of bark is observed. Entire bark of the plant near collar region rots. The characteristic symptoms include white, cottony fungus growth is observed on affected portion as well as on parts in contact with soil. Gradually this hyphal mat is converted into small, mustard-like sclerotia that survive in the soil.

#### (b) Management

**Solarization:** Summer ploughing followed by irrigation in summer helps in reducing the initial inoculum of the fungus.

**Crop Rotation:** Crop rotation with lowland paddy, wheat/barley is effective in reducing the initial inoculum.

### (ii) White Mold, *Sclerotinia sclerotiorum*

**(a) Symptoms:** The disease is commonly found on pea which is responsible for complete killing of plants. Symptoms observed include wet and soft rotting of the tissues. White fungus growth is observed on the rotted portion. Embedded sclerotia covered with white mycelium are formed on the infected portion as well as on inner portion of the pith and fruits.

### (b) Management

**Biofumigation and Bioagents:** Addition of *Eupatorium*/ *Ageratum conyzoides* plant material with wheat bran based formulation of *T. harzianum* at the time of sowing was found effective in managing root rot/wilt complex of pea (Kaur, 1999).

## 15.12.2. Nematodes

### (i) Root-knot Nematodes, *Meloidogyne spp.*

*M. incognita* was responsible for 20.00 to 50.61% loss in pod yield of peas (Parvatha Reddy, 1985; Upadhayay and Dwivedi, 1987; Sharma, 1989).

**(a) Symptoms:** The infested fields show patches in which peas are stunted, chlorotic and have few flowers which produce small and often empty pods. The root systems are reduced in size, exhibit large galls and poor nodulation. Senescence also tends to occur earlier.

### (b) Management Methods

**(a) Intercropping:** Intercropping with marigold or mustard with pea reduced the damage of root-knot nematodes.

## 15.13. BEAN, *PHASEOLUS VULGARIS*

## 15.13.1. Diseases

### (i) Root Rot, *Rhizoctonia solani*

**(a) Symptoms:** At emergence, the infection on hypocotyl and stem results in damping-off symptoms, but afterwards, elongated, sunken, reddish brown lesions are produced on the stem at ground level. On the foliage, circular to irregular brown spots having distinct borders appear. Leaf scald symptoms are common. In rainy season, the disease assumes serious proportions. Extensive damage results under continuous wet weather along with high temperature.

### (b) Management

**Solarization:** Experimentation with this crop indicated that root rot pathogens, especially *R. solani*, could be controlled to at least 86 percent by solarization compared with non-solarized plots.

## (ii) Seed and Root Rot, *Fusarium solani*

**(a) Symptoms:** Symptoms appear soon after seedling emergence. The tap root is slightly discoloured and finally becomes brown with longitudinal cracks. Bottom roots are usually killed. The fungus is soil-borne and can survive for 5 years or longer.

### (b) Management

**Solarization:** Table 15.60 shows the effect of soil solarization on maximum soil temperatures. The mulching of moist soils raised the soil temperature and the highest temperature was in the upper layer at 5 cm depth. Soil temperatures at the other depths were also increased in mulched soil compared to the unmulched soil. The differences in the maximum temperatures between mulched and unmulched soils after six weeks were 17°, 13° and 12°C at the 5, 10 and 20 cm depths, respectively. Soil solarization after four and six weeks reduced significantly the soil population of the pathogen at all depths compared to the unmulched treatment (Table 15.61). Results showed that *F. solani* population densities were considerably reduced after four and six weeks of soil mulching.

Table 15.60. Effect of soil solarization. on maximum soil temperatures at various soil depths during experiment

| Date of soil mulching* | Soil temperatures at different depths (cm) | | |
|---|---|---|---|
| | 5°C | 10°C | 20°C |
| July 1 to 15 | | | |
| Control | 38 | 31 | 24 |
| Mulched | 52 | 43 | 38 |
| July 1 to July 30 | | | |
| Control | 40 | 32 | 27 |
| Mulched | 58 | 45 | 38 |

| | | | |
|---|---|---|---|
| July 1 to August 15 | | | |
| Control | 42 | 35 | 32 |
| Mulched | 59 | 48 | 44 |

* Soil was irrigated two days before mulching

Table 15.61. Effect of soil solarization on the survival of *Fusarium solani* at different depths

| Treatment* | Fusarium propagules/g of soil at different depths (cm)** | | |
|---|---|---|---|
| | 5°C | 10°C | 20°C |
| Control | 216 a | 264 a | 338 a |
| Solarized 2 weeks | 190 a | 232 a | 289 b |
| Solarized 4 weeks | 49 b | 84 b | 93 c |
| Solarized 6 weeks | 19 c | 38 c | 55 d |

* Transparent polyethylene sheets were placed on the soil July 1.
** Values within columns followed by the same letter are not significantly different (Duncan's multiple range test, P = 0.05).

The percentage of diseased plants after 60 days from emergence dropped from 89 percent in the unmulched treatment to 38 percent in the solarized ones. Disease incidence of Fusarium root rot of broad bean plants in solarized soil was significantly reduced after 30, 45 and 60 days compared to unmulched treatments. Soil solarization improved plant growth. The increase in the growth of plants was more pronounced after 45 and 60 days from emergence. Plant dry weights indicated that plants grown in solarized soil had significantly greater growth than those in the unmulched soil.

### 15.13.2. Nematodes

#### (i) Root-knot Nematodes, *Meloidogyne spp.*

*M. incognita* and *M. javanica* appear to be the most common root-knot species of French beans and have been reported causing damage in tropics and subtropics. Lal and Ansari (1960) reported *M. arenaria* for the first time on beans from Bihar, India. *M. incognita* was responsible for 19.38 to 43.48% loss in pod yield of French bean (Das, 1994; Parvatha Reddy and Singh, 1981), while *M. javanica* caused 30-40% loss in yield (Sharma *et al.*, 2002).

**(a) Symptoms:** Gall size on roots of French bean is variable and may be nearly undetectable. The only visible symptom on the roots is the presence of large egg-masses (Fig. 15.35).

**Fig. 15.35.** Root-knot nematode on bean.

## (b) Management

**Solarization:** Root-knot nematodes can be effectively controlled by a 4-8 weeks solarization, assuming that the land will remain uncropped during summer. Data in Table 15.62 show that solarization gave good control of nematodes. The RF in solarized treatment was 0.47, while in non-solarized plots, reduction was 0.98.

Table 15.62. Effect of solarization on the control of nematodes on broad bean

| Treatment | Average number of nematodes per 200 g soil[1] | | Reproduction factor (RF) RF = Pf/Pi |
|---|---|---|---|
| | (Pi) | (Pf) | |
| Bean - solarized | 85 | 40 | 0.47 |
| Bean - non-solarized | 95 | 93 | 0.98 |

Pi = Initial nematode population; Pf = Final nematode population.

**Intercropping:** Intercropping of bean with finger millet, chilli and groundnut helps in the management of *M. incognita* Race 2.

**Crop Rotation:** Growing of chilli or groundnut or finger millet in rotation with bean significantly reduced the root-knot nematode population (Ramappa, 1988).

### (ii) Other Nematodes

#### (a) Management

**Solarization:** Table 15.63 clearly shows the effect of solarization on the control of nematodes on broad bean with RF's of 0.0 for *Tylenchorhynchus* and *Trichodorus* spp. The corresponding RF values in non-solarized soil were 0.51 on the same crop and the same nematodes.

Table 15.63. Effect of solarization on the control of nematodes on bean

| Treatment | Average number of nematodes per 200 g soil[1] | | Reproduction factor (RF) |
|---|---|---|---|
| | (Pi) | (Pf) | RF = Pf/Pi |
| Bean - solarized | 0 | 0 | 0 |
| Bean - non-solarized | 625 | 320 | 0.51 |

Pi = Initial nematode population; Pf = Final nematode population.

An experiment was conducted to study the effect of solarization and herbicides, either singly or in combination, in a soil naturally infested with *Rotylenchulus reniformis Tylenchorhynchus* spp. *Praylenchus* spp. and *Heterodera* spp. where broad bean was cultivated. Non-solarized control treatments were used. Generally, the RF in solarized treatments was 0.027, while in non-solarized it was 4.44 (Table 15.64). Solar heating of the soil by means of polyethylene mulching during hot seasons controlled nematodes in all localities.

Table 15.64. Effect of polyethylene mulching of soil and herbicide treatments on broad bean in soil naturally infested with nematodes

| Treatment | Average number of nematodes per 200 g soil | | Reproduction factor (RF) |
|---|---|---|---|
| | (Pi) | (Pf) | RF = Pf/Pi |
| Solarized | | | |
| Without herbicide | 180 | 43 | 0.24 |
| With Aradican | 840 | 10 | 0.01 |

| | | | |
|---|---|---|---|
| With Lasso | 860 | 6 | 0.01 |
| With Stomb | 740 | 14 | 0.02 |
| Non -solarized | | | |
| Without herbicide | 60 | 212 | 3.53 |
| With Aradican | 40 | 317 | 7.92 |
| With Lasso | 40 | 83 | 2.07 |
| With Stomb | 80 | 334 | 4.17 |

Pi = Initial nematode population; Pf = Final nematode population.

## 15.14. COWPEA, *VIGNA UNGUICULATA*

### 15.14.1. Nematodes

#### *(i) Root-knot Nematode, Meloidogyne incognita*

Parvatha Reddy and Singh (1981) reported that *M. incognita* was responsible for 28.60% loss in pod yield of cowpea.

**(a) Symptoms:** Symptoms of damage induced by root-knot nematode include patches of stunted and yellowed plants. Severe damage can lead to reduced numbers of leaves and buds. At high densities severe root galling occurs (Fig. 15.36). visual symptoms of damage first occurred at 1,000 and 10,000 juveniles/500g of soil.

**Fig. 15.36.** Heavy galling of cowpea roots infected with *Meloidogyne incognita*

### (b) Management

**Solarization:** Summer ploughing was found effective for the management of root-knot nematodes and increased the cowpea yield over non-ploughed field.

**Biofumigation:** The root-knot nematodes associated with cowpea can be effectively managed by the application of neem or eupatorium leaves at 15 t/ha two weeks before sowing. Neem cake incorporation in the previous crop, caused reduction in density of all nematodes in the soil on the following cowpea crop (Jain and Hasan, 1986). Cocoa pod husks incorporated at 6 t/ha caused 28% reductions in galling and 6.7% increases in yield (Egunjobi, 1985).

**Crop Rotation:** Rotation of cowpea with graminaceous crops or *Crotalaria spectabilis* is recommended. Populations of root-knot decreased greatly when compared to fallowed plots when *C. spectabilis* was grown as a weed free cover crop.

One crop of rice was sufficient to effectively reduce root-knot nematode infestations in succeeding susceptible cowpea crop. The reduction was greater than with rotations with non-host crops.

**Intercropping:** *M. javanica* populations were lower when cowpea and maize were grown under mixed rather than under sole crop cropping systems.

**Biofumigation and AMF:** Integration of chopped leaves of *Prosopis fuliflora* with *G. fasciculatum* to cowpea increased the spore production and root colonization of *G. fasciculatum* which resulted in reduced *M. incognita* population.

**Mulching:** Mulching with cowpea foliage was also highly effective in suppressing root-knot nematode populations.

## (ii) Reniform Nematode, *Rotylenchulus reniformis*

**(A) Symptoms:** Infestation of *R. reniformis* causes growth reduction and adversely affects the germination of and seedling emergence. The emergence of cowpea seedlings was delayed by 7 days and seedling stand was reduced to the tune of 6 to 11% due to *R. reniformis* at one nematode/g of soil (Nanjappa *et al.*, 1978). Sud *et al.* (1984) reported that 20 larvae of *R. reniformis* per gram of soil were more damaging on 7-day-old cowpea seedlings than older ones. *R. reniformis* was responsible for 13.20 to 32.00% loss in pod yield of cowpea (Palanisamy and Sivakumar, 1981; Hasan and Jain, 1998).

### (a) Management

**(a) Solarization:** Under field conditions, soil solarization was considered as an effective method which significantly reduced soil population of the reniform nematode up to 15 cm depth and increased the yield of cowpea.

Interesting results have also been reported using soil solarization to control the reniform nematode, *R. reniformis*. Heald and Thomas (1983) observed that the development of cowpea in plots solarized for six weeks was greatly enhanced as was the subsequent yield of lettuce. Nematode densities at the harvest of the cowpea crop were 47 specimens/100 g soil in solarized plots and 14400 in the controls. Satisfactory control was also similarly obtained by McSorley and Parrado (1986), even though nematode population densities increased during the following tomato crop.

Summer ploughing was found effective for the management of reniform nematodes and increased the cowpea yield over non-ploughed field.

**Intercropping:** The reniform nematode densities were suppressed when cowpea was grown intercropped with maize.

**Biofumigation:** The reniform nematodes associated with cowpea can be effectively managed by the application of neem or eupatorium leaves at 15 t/ha two weeks before sowing.

### (iii) *Macrophomina phaseolina* and *Meloidogyne incognita* Disease Complex

### (a) Management

**Organic Amendments and AMF:** Devi and Goswami (1992) demonstrated that *G. fasciculatum* in cowpea together with mustard cake helped in reducing the disease severity caused by *M. incognita* and *M. phaseolina*. They observed that the pre-establishment of AMF checked the entry of *M. incognita* larvae as also colonization of pathogenic fungus.

## 15.14.2. Weeds

***(i) Intercropping:*** Smother crops grown as intercrops suppress weeds without reducing total production in cowpea. Many short duration pulses such as green gram and soybean; and sorghum for fodder effectively smother weeds and increase yields of cowpea (Masood Ali, 1986; Bhat, 1989; Mutnal, 1984) (Table 15.65).

Table 15.65. Effect of intercropping of cowpea with sorghum on weed suppression and yield

| Treatment | Green fodder yield (MT/ha) | Cowpea yield (MT/ha) |
|---|---|---|
| Sole sorghum (30 cm rows) | 31.8 | --- |
| Sorghum + Cowpea (1:1) | 31.5 | 16.7 |
| Sorghum + Cowpea (2:2) | 31.4 | 16.5 |
| Sole cowpea | --- | 22.4 |

## 15.15. PIGEONPEA, *CAJANUS CAJAN*

### 15.15.1. Diseases

#### (i) Fusarium Wilt, *Fusarium udum*

The Fusarium wilt of pigeonpea is the major limitation to the growth of this crop in the Indian subcontinent.

**(a) Symptoms:** Wilting of seedlings and grown up plants as if they have suffered from water shortage although there is plenty of moisture in the field. Wilting is characterized by gradual and sometimes sudden yellowing, withering and drying of leaves, followed by drying of the entire plant or some of its branches. When the stem is split open, a purple band extends from the base of the main stem. Partial wilting of plant is a definite indication of *Fusarium* wilt distinguishing it from termite damage, drought and *Phytophthora* blight that kills the entire plant.

#### (b) Management

**Solarization:** The fields selected for solarization experiments were fusarium wilt sick plots (vertisols) of pigeonpea at ICRISAT Center. In these fields susceptible cultivars (LRG30) showed 100% mortality. Clear polythene sheeting of 400 gauge (94 g/m$^2$ and 100 μm thick) and 6 m wide was laid on the plots.

At 5 cm depth, solarization increased temperature by about 10°C. The temperatures higher than 40°C and 45°C, which could be lethal for microorganisms, were recorded for most of the duration of solarization at 5 cm and 10 cm depths.

Just before solarization pigeonpea wilt pathogen population was more than 1000 propagules/g of soil. There was a large reduction in population of the pathogen due to solarization in both dry and irrigated conditions immediately after the solarization period. In the non-solarized treatment, there was an increase in Fusarium propagules during this period. The Fusarium population remained at low levels in solarized plots throughout the growing season. The beneficial effect of solarization in reducing the population of Fusarium was confirmed in pigeonpea experiments. The effect of solarization on Fusarium propagule population extended into the second season.

Solarization reduced wilt incidence in pigeonpea under both irrigated and non-irrigated conditions.

Virtually all plants of wilt susceptible pigeonpea cultivar were killed before maturity in non-solarized plots. In wilt resistant genotype, there was no wilting in solarized and non-solarized plots. Wilt susceptible cultivar did not show mortality in solarized plots.

Stimulatory effects of solarization on plant growth were obvious for both wilt-susceptible (LRG30) and wilt-resistant (ICP 8863) genotypes in all experiments. Differences in plant height were apparent from early-growth stages (Fig. 15.37).

**Fig. 15.37.** Management of Fusarium wilt in pigeonpea by solarization. Left: Non-solarized plot, Right: Solarized plot.

Solarization reduced mean time to flowering and to maturity. The flowering period was reduced by 10 days for LRG 30 (susceptible) and by 12 days for ICP 8863 (resistant) and time to maturity reduced by 60 days for LRG 30 and 19 days for ICP 8863.

Solarization markedly increased dry-matter production and seed yield of both genotypes in all first year experiments. These increases were greater for LRG 30 than ICP 8863 and in the presence of irrigation. Improved seed yields resulted from increased pods and seeds per plant.

### 15.15.2. Nematodes

#### (i) Pigeon pea Cyst Nematode, *Heterodera cajani*

**(a) Symptoms:** Patchy appearance and unthrifty growth of the plants may be observed from the fields having heavy infestation of pigeonpea cyst nematode. It has been found to reduce bacterial nodulation which in turn affects the crop yield. Significant reduction in growth parameters was recorded at 100 larvae and above per kg of soil (Sharma, 1985; Zaki and Bhatti, 1986a). The nematode also affects the macro and micro-nutrients of pigeonpea (Zaki and Bhatti, 1986b). Under field conditions, the reduction in plant growth and yield of pigeonpea were 27.6 and 30.1%, respectively (Saxena and Reddy, 1987).

#### (b) Management

**Solarization:** The population of all parasitic nematodes of pigeonpea was markedly affected by solarization. Included in this population were *H. cajani* cysts, eggs, and larvae; *Rotylenchulus reniformis*; *Helicotylenchus retusus*; *Pratylenchus* spp; *Tylenchorhynchus* spp and *Heterodera* spp. Solarization with irrigation was most effective in reducing populations of cysts, eggs, and larvae of *H. cajani*.

**Solarization, Organic Amendments and AMF:** Soil solarization (transparent polythene sheet of 400 gauge thickness for a period of 4 weeks) and AM fungi at 100 kg/ha provided good plant growth and yield of pigeon pea by effectively controlling the detrimental effects of *H. cajani*. Maximum reduction in nematode population was observed in soil solarization + soil application of neem seed powder at 50 kg/ha + AMF, and soil solarization + seed treatment with neem seed powder at 10% w/w + AMF (Nageswari and Mishra, 2005).

### 15.15.3. Weeds

***(i) Solarization:*** Solarization markedly decreased weed growth in pigeonpea fields. Most annual weed species were effectively suppressed by solarization; however, the perennials, such as *Cynodon dactylon, Cyperus rotundus*, and *Convolvulus arvensis*, gradually recovered. Solarization for two successive years was most effective in suppressing weeds.

***(ii) Intercropping:*** Smother crops grown as intercrops suppress weeds without reducing total production in pigeon pea. Many short duration pulses such as green gram, cowpea and soybean; and sorghum for fodder effectively smother weeds and increased yields of pigeon pea (Masood Ali, 1986; Bhat, 1989; Mutnal, 1984) (Table 15.66).

Table 14.66. Effect of intercropping of pigeon pea with pulses on weed suppression and yield

| Treatment | Weed suppression (%) | Pigeon pea yield (kg/ha) |
|---|---|---|
| Pigeon pea (weed-free) | --- | 2578 |
| Pigeon pea (weedy) | --- | 1911 |
| Pigeon pea + Black gram | 20.7 | 2377 |
| Pigeon pea + Green gram | 22.2 | 2518 |
| Pigeon pea + Soybean | 22.2 | 2266 |
| Pigeon pea + Cowpea | 33.3 | 2349 |

## 15.16. CHICKPEA, *CICER ARIETINUM*

### 15.16.1. Diseases

#### (i) Fusarium Wilt, *Fusarium udum*

The Fusarium wilt of chickpea is the major limitation to growth of this crop in the Indian subcontinent. The disease is systemic in nature and the pathogen may infect the plants in any stage of growth. The pathogen is both soil-borne and seed-borne. The fungus survives in the form of chlamydospores in the hilum region of the seed. The fungus can survive for more than 5 years in the soil.

**(a) Symptoms:** Grown up plants show typical wilting *i.e.,* drooping of petioles, rachis and leaflets. All the leaves turn yellow and then light brown or straw coloured. Root rots and transverse cut of stem show blackening of xylem vessels. Sometimes, only a few branches are affected, resulting in partial wilt.

#### (b) Management

**Solarization:** The fields selected for solarization experiments were fusarium wilt sick plots (vertisols) of chickpea at ICRISAT Center. In these fields susceptible cultivars (ICCV 1) showed 100% mortality. Clear polythene sheeting of 400 gauge (94 g/m$^2$ and 100 µm thick) and 6 m wide was laid on the plots.

At 5 cm, solarization increased temperature by about 10°C. The temperatures higher than 40°C and 45°C, which could be lethal for microorganisms, were recorded for most of the duration of solarization at 5 cm and 10 cm depths.

Just before solarization chickpea wilt pathogen population was more than 1000 propagules/g of soil. There was a large reduction in population of pathogen due to solarization in both dry and irrigated conditions immediately after the solarization period. In the non-solarized treatment, there was an increase in Fusarium propagules during this period. The Fusarium population remained at low levels in solarized plots throughout the growing season. The beneficial effect of solarization in reducing the population of Fusarium was confirmed in chickpea experiments. The effect of solarization on Fusarium propagule population extended into the second season.

Virtually all plants of wilt susceptible chickpea cultivar were killed before maturity in non-solarized plots. In wilt resistant genotype, there was no wilting in solarized and non-solarized plots. Wilt susceptible cultivar did not show mortality in solarized plots.

In the Ist year of the experiment, virtually all plants of the wilt-susceptible genotype ICCV 1 were killed before maturity in non-solarized plots and no yield was obtained. In the wilt-resistant genotype JG 74, there was a 23% increase in seed yield due to solarization in the irrigated treatment and 17% in the non-irrigated treatment, mainly due to an increase in the seed mass (Fig. 15.38).

**Fig. 15.38.** Management of Fusarium wilt in chickpea by solarization. Left: Solarized plot, Right: Non-solarized plot.

### 15.16.2. Nematodes

#### (i) Pigeon pea Cyst Nematode, *Heterodera cajani*

##### (a) Management

**Solarization:** The population of all parasitic nematodes of chickpea was markedly affected by solarization. Included in this population were *H. cajani* cysts, eggs, and larvae; *Rotylenchulus reniformis*; *Helicotylenchus retusus*; *Pratylenchus* spp.; *Tylenchorhynchus* spp.; and *Heterodera* spp. Solarization with irrigation was most effective in reducing populations of cysts, eggs, and larvae of *H. cajani*.

#### (ii) Lesion Nematode, *Pratylenchus thornei*

##### (a) Management

**Solarization:** Daily (2 p.m.) soil temperatures of mulched plots were above 45°C and reached a maximum of 50°C at 5 cm depth. They exceeded 40°C at 15 cm depth (max. 46°C) and were in the range 33-35°C at 30 cm depth. Temperatures higher than 42-43°C are known to kill nematodes after a few hours exposure. Temperatures above 35°C can also injure nematodes if prolonged for several days. Therefore, according to the observed temperatures, good nematode control would have occurred in the top 20 cm soil. This assumption was confirmed by the data on percentage of *P. thornei* that survived in the soil at the end of the solarization periods (Table 15.67). Only 11.6 percent of the nematodes survived after two weeks of solarization and nearly 100 percent control was achieved by treatment for eight weeks. However, by harvest time, nematode populations had increased in all solarized plots.

Nematode numbers remained significantly lower than the control only in plots solarized for at least four weeks. This was reflected in the nematode infestation of the roots, which was significantly less than the controls in all plots solarized for more than two weeks (Table 15.67).

The total weight of the aerial plant parts increased significantly (28-34 percent; $\underline{P} < 0.01$) in all plots solarized for six to eight weeks (Table 15.68). However, although grain yield increased in all solarized plots, significant increases (23-29 percent) were only observed ($\underline{P} < 0.05$) in those solarized for six to eight weeks.

Table 15.67. Effect of soil solarization on the numbers of *Pratylenchus thornei* extracted from soil and roots of chickpea

| Treatment | Nematodes/50g soil | | | Nematodes/5g roots on 15 April |
|---|---|---|---|---|
| | Before mulching | After mulching | At harvest | |
| Control (non-solarized) | 259 | 237 | 57.5 | 2860 |
| 2 weeks solarized | 248 | 29 | 43.7 | 2519 |
| 4 weeks solarized | 297 | 11 | 19.7 | 1377 |
| 6 weeks solarized | 219 | 5 | 24.7 | 1347 |
| 8 weeks solarized | 240 | 1 | 22.5 | 1273 |
| CD ($P \geq 0.05$) | N.S. | 53.1 | 27.2 | 863.2 |
| CD ($P \geq 0.01$) | N.S. | 74.5 | N.S. | 1212.4 |

Table 15.68. Effect of soil solarization on the yield of chickpea grown in a field infested by *Pratylenchus thornei*

| Treatment | Yield of chickpea (kg/ha) | |
|---|---|---|
| | Straw and grain | Grain |
| Control (non-solarized) | 2563 | 1389 |
| 2 weeks solarized | 2889 | 1516 |
| 4 weeks solarized | 2825 | 1488 |
| 6 weeks solarized | 3286 | 1710 |
| 8 weeks solarized | 3444 | 1798 |
| CD ($P \geq 0.05$) | 476.5 | 237.3 |
| CD ($P \geq 0.01$) | 656.2 | N.S. |

The results clearly show that soil solarization is effective for the control of *P. thornei* on chickpea. This method of control has also shown promise against several soil-borne plant pathogens, including *Orobanche* spp., which could become another constraint to the production of this pulse crop.

### 15.16.3. Weeds

***(i) Solarization:*** Solarization markedly decreased weed growth in chickpea fields. Most annual weed species were effectively suppressed by solarization; however, the perennials, such as *Cynodon dactylon, Cyperus rotundus*, and *Convolvulus arvensis*, gradually recovered. Solarization for 2 successive years was most effective in suppressing weeds.

## 15.17. WATERMELON, *CITRILLUS LANATUS*

### 15.17.1. Diseases

#### (i) Fusarium Wilt, *Fusarium oxysporum f. sp. niveum*

**a) Symptoms:** The plants are prone to attack at all stages of growth. Definite wilt symptoms are visible on old plants with drooping leaves (Fig. 15.39). Affected plants turn yellow and are stunted. On nursery seedlings, damping-off symptoms appear and the seedlings die.

**Fig. 15.39.** *Fusarium* wilt disease in water melon.

The fungus is seed-borne as well as soil dweller. Optimum temperature requirement is 23°-29°C.

#### (b) Management

**Solarization:** In 1985, plots were fumigated with Metham-Sodium (Vapam at 1200 IG/ha), or solarized for one (mid-July - mid-August) or two (mid-July - mid-September) months.

Density of viable (*FO*) propagules in control and treated soils had declined, respectively, to about 19 percent and 30 percent of their initial populations by September 1985, and it further decreased in solarized soils to stabilize at about 8 percent of the initial population after December. Final disease incidence in plots solarized for one and two months averaged 28 percent and 25 percent, respectively, as compared to 41 percent for fumigated plots and 47 percent for the untreated controls.

Mean temperatures at 10 and 20 cm soil depths increased 4°C and 3.3°C, respectively, during the first month of solarization and 3.7°C and 3.5°C, respectively, over the two-month solarization period. Hourly temperatures varied within a range of 26.3° - 47.8°C and 28.9° - 43.3°C at 10- and 20-cm soil depth, respectively, during the two months of solarization; and within the range of 25.5° - 41.6°C and 26.8° - 38.8°C during the same period in the unsolarized soil. The rather low temperature increments in solarized soil, as compared to unsolarized soil, might be due to solarization within a plastic house closed environment. Although the temperature increase was low, the population of FO in soil was significantly reduced after solarization for two months, and to a lesser extent by fumigation and one month's solarization (Table 15.69).

Table 15.69. Population of *Fusarium oxysporum* in treated soil as a percentage of the initial population before treatment

| Treatment | Sampling date | | | |
|---|---|---|---|---|
| | 28/9/88 | 30/12/88 | 6/4/89 | 3/7/89 |
| Infested control | 21.1 | 51.1 | 23.3 | 24.4 |
| Solarization (1 month) | 5.8 | 15.9 | 32.7 | 7.1 |
| Solarization (2 month) | 0.9 | 3.7 | 15.6 | 18.4 |
| Metham- Sodium | 2.8 | 3.2 | 2.4 | 7.2 |

Soil solarization for two months showed the highest increases in plant dry weight, and watermelon fruit number and yield, followed by Metham-Sodium and one month's solarization (Table 15.70). While plant dry weight, and number of watermelon fruit in plots solarized for two months were significantly higher than those in the uninfested control, differences among uninfested control, solarization for one month and fumigation were not significant.

Experiments to control Fusarium wilt of watermelon under plastic houses have shown that solarization for two months provides a better control of the disease than fumigation with Metham- Sodium, and it resulted in higher increases of plant growth and yield. Solarization during one month, in July, was not highly effective in either eradicating the pathogen from soil or decreasing disease incidence. Therefore, when solarization under plastic house conditions is not feasible for more than 45 days, it should be integrated with fumigation at low dose (1) and should meet optimal environmental conditions (probably from mid-July on).

Table 15.70. Effects of soil treatment on growth and yield of watermelon in sand-mulched plastic house

| Treatment | Plant Dry weight (g) | Number of fruits/ plot | Yield (kg/plot) |
|---|---|---|---|
| Infested control | 78.2 x | 10.0 x | 6.5 x |
| Uninfested control | 111.9 x y | 16.0 x | 16.5 x y |
| Solarization (1 month) | 116.2 x y | 13.0 x | 15.1 x y |
| Solarization (2 month) | 213.9 z | 27.5 y | 32.0 z |
| Metham- Sodium | 178.2 | 17.5 x y | 21.9 y z |

Values followed by the same letter within columns are not significantly different (P=0.05) according to Student's t test.

### 15.17.2. Nematodes

### (i) Root-knot Nematodes

### (a) Management

**Biofumigation:** The effect of *Brassica* green manure, i.e. biofumigation, for the control of root-knot nematode on muskmelon and watermelon crops was investigated. *Brassica juncea* and *Eruca sativa* green manures were compared with a commercial amendment from *Brassica* meals, alone or combined with *E. sativa* green manure, and with commercial formulations of aqueous extracts of quillay and tagetes, and the nematicide cadusaphos. In the muskmelon crop cycle *E. sativa* green manure resulted in a nematicidal effect statistically similar to cadusaphos and higher than *B. juncea* amendment. *B juncea* amendment effects were not improved by combination with pellet. Muskmelon yield recorded for chemical treatment and *B. Juncea* were 79.2 and 70.8 t/ha, respectively, and were 3 times higher than untreated control.

In the following watermelon crop the lowest population densities of *M. incognita* were recorded after the combined treatment with *E. sativa* green manure and pellet, and after cadusaphos, followed by *E. sativa* green manure alone and at a significantly higher values, by *B. juncea*. Nematicidal effects of quillay formulation, though significantly higher than untreated control, was consistently lower than biofumigation treatments. Most effective treatments for nematode control provided the highest watermelon yields, 119.0 and 85.7 t/ha for

*E. sativa* manure plus pellet and cadusaphos, respectively (De Mastro *et al.*, 2008b).

The effectiveness of *Brassica* green manure biofumigation for the control of root-knot nematodes were investigated in a greenhouse experiments on a tomato - melon succession. Green manures of *Brassica juncea* and *Eruca sativa* were compared with a *Brassica* seed meal commercial pellet, alone or combined with *E. sativa* manure, two commercial formulations of quillay or tagetes, chemical nematicide cadusaphos and an untreated soil. In the tomato crop, green manure of *E. sativa* resulted in the lowest *M. incognita* population densities and gall formation, as its results were not significantly different from chemical treatment but were lower than *B. juncea* green manure, the pellet and quillay formulation. This last treatment was statistically similar to the untreated control. *E. sativa* green manure, either alone or combined with the pellet, provided a tomato yield two to three times higher than control, whereas *B. juncea* and cadusaphos almost doubled the yield of untreated soil.

In the following melon crop, the lowest nematode densities were recorded in soil treated with cadusaphos or the commercial pellet. Green manure of *E. sativa*, alone or combined with the pellet, resulted in a higher suppressivity than *B. juncea*, whereas a limited nematicidal activity emerged for the quillay formulation. The highest melon yield was provided by the seed meal pellet, followed by the combination of *E. sativa* green manure and pellet and by both single green manures (De Mastro *et al.*, 2008c).

### 15.17.3. Weeds

***(i) Mulching:*** The yields due to black, red, green, blue, white and clear (transparent) polyethylene mulches were 61.1, 50.9, 44.9, 42.5, 34.7 and 31.1%, respectively higher than control. Minimum weed growth was observed under black polyethylene mulch. The economic return from black polyethylene mulch was highest.

## 15.18. MUSKMELON, *CUCUMIS MELO*

### 15.18.1. Nematodes

#### (i) Root-knot Nematodes

#### (a) Management

**Biofumigation:** The effect of *Brassica* green manure, i.e. biofumigation, for the control of root-knot nematode on muskmelon and watermelon crops was investigated. *Brassica juncea* and *Eruca sativa* green manures were compared with a commercial amendment from *Brassica* meals, alone or combined with *E. sativa* green manure, and with commercial formulations of aqueous extracts of quillay and tagetes, and the nematicide cadusaphos. In the muskmelon crop cycle *E. sativa* green manure resulted in a nematicidal effect statistically similar to cadusaphos and higher than *B. juncea* amendment. *B juncea* amendment effects were not improved by combination with pellet. Melon yield recorded for chemical treatment and *B. juncea* were 79.2 and 70.8 t/ha, respectively, and were 3 times higher than untreated control.

In the following watermelon crop the lowest population densities of *M. incognita* were recorded after the combined treatment with *E. sativa* green manure and pellet, and after cadusaphos, followed by *E. sativa* green manure alone and at a significantly higher values, by *B. juncea*. Nematicidal effects of quillay formulation, though significantly higher than untreated control, was consistently lower than biofumigation treatments. Most effective treatments for nematode control provided the highest watermelon yields, 119.0 and 85.7 t/a for *E. sativa* manure plus pellet and cadusaphos, respectively (De Mastro *et al*., 2008b).

Ploeg and Stapleton (2001) reported that biofumigation with broccoli failed to control *M. incognita* or *M. javanica* at 20°C, but resulted in near complete control at temperatures of 30 and 35°C. In addition, they found that the time necessary to achieve control was shorter as soil temperatures increased, and concluded that biofumigation to control *M. incognita* or *M. javanica* should be done when soil temperatures of about 25°C can be achieved.

**Solarization:** In Japan and other Asian countries, several farmers growing successive crops, such as tomato and muskmelon susceptible to root-knot nematodes use solarization in plastic tunnels for 30 days in summer as an alternative to methyl bromide fumigation (Sano, 2002).

### Integrated Methods

**Crop Rotation, Intercropping and Biofumigation:** Combination of crop rotation (either sorghum or tomato in 2005 followed by melon in 2006 and 2007), biocidal intercropping (*Eruca sativa* cv. Nemat) and biofumigant treatments (pellet based on formulated defatted seed meals of *Brassica carinata*), were evaluated for the management of root-knot nematodes. The results showed that sorghum cultivation as a non host crop halved the nematode population in the soil, while the tomato host crop increased *M. incognita* infestation (reproduction factor as R = 1.33). The biofumigant treatment in spring 2006, caused a strong decrease in galling index (G.I. based on a scale from 0 to 5) on the roots of the following melon crop in rotation (G.I. = 0.3), significantly lower if compared with the untreated control (G.I. = 4.2). These positive results were further confirmed in 2007, when an important increase in quality and quantity was recorded in the yield of melon fruits. The autumn cultivation of *E. sativa* cv. Nemat and its incorporation into the soil, brought a good amount of organic matter, with positive effects on melon yield. The intrinsic characteristics of the pellet based on *B. carinata* defatted seed-meals in rotation with *E. sativa* highlighted an excellent biofumigant effect, fully comparable to chemical nematicides, and a noticeable contribution in both organic matter and nitrogen that played an important fertilizing effect. The full effectiveness of pellet in decreasing substantially the nematode population, suggest its application in alternate years with *Eruca* green manure, in the presence of low larval infestations in the soil (Curto *et al.*, 2008).

## 15.19. CUCUMBER, *CUCUMIS SATIVUS*

### 15.19.1. Diseases

#### (i) Damping-off, *Pythium aphanidermatum*

#### (a) Management

**Solarization and Biofumigation:** Fields trials were designed to evaluate solarization and biofumigation (the use of solarization following soil amendment with organic matter) for their effectiveness in reducing *P. aphanidermatum* inoculum levels and damping-off disease, and enhancing crop growth of greenhouse cucumber.

*P. aphanidermatum* colony counts (Table 15.71) showed no significant differences between treatments at the start of each season. After treatments, colony counts were significantly lower in both solarized and biofumigated soils, being reduced to approximately 20% and 50% of pre-treatment levels in the summer and winter experiments, respectively. During treatment, maximum soil temperatures in summer were 46.6, 48.1 and 40.3°C for the solarized, biofumigated and control plots respectively; in winter, the corresponding temperature maxima were 36.6, 34.6 and 30.2°C.

Table 15.71. Mean population densities of *Pythium aphanidermatum* before and after solarization and biofumigation soil treatments.

| Treatment | *P. aphanidermatum* inoculum density (cfu/g soil)* | | | |
|---|---|---|---|---|
| | Summer | | Winter | |
| | Before treatment | After treatment | Before treatment | After treatment |
| Solarization | $0.32 \times 10^3$ a | $0.08 \times 10^3$ a | $0.50 \times 10^3$ a | $0.18 \times 10^3$ a |
| Biofumigation | $0.40 \times 10^3$ a | $0.07 \times 10^3$ a | $0.50 \times 10^3$ a | $0.18 \times 10^3$ a |
| Control | $0.39 \times 10^3$ a | $0.36 \times 10^3$ b | $0.53 \times 10^3$ a | $0.37 \times 10^3$ b |

* Values in the same column followed by different letters are significantly different at ($P<0.05$) according to Duncan's multiple range test.

Disease incidence levels for each treatment during the summer and winter seasons are shown in Table 15.72. In the summer season, there were significant differences in disease incidence between treated and non-treated plots up to 35 days after planting; differences between solarized and biofumigated plots were not significant. During the winter season, disease was higher than during the summer, but as with the summer season, solarized and biofumigated plots had significantly less damping-off than control plots ($P<0.05$). Disease was significantly lower in biofumigated than in solarized plots at the final assessment, 35 days after planting.

Table 15.72. Effect of solarization and biofumigation on percent damping-off disease incidence expressed as percentage of diseased plants, in summer and winter seasons.

| Treatment | Damping-off disease incidence (%)* | | | |
|---|---|---|---|---|
| | 14 days | 21 days | 28 days | 35 days |
| Summer | | | | |
| Solarization | 1.2 a | 01.6 a | 01.6 a | 01.6 a |
| Biofumigation | 2.8 a | 03.1 a | 03.5 a | 03.5 a |
| Control | 5.5 b | 06.3 b | 06.3 b | 07.0 b |
| Winter | | | | |
| Solarization | 3.9 a | 04.7 a | 06.3 a | 11.7 b |
| Biofumigation | 6.3 b | 07.0 b | 07.0 b | 07.0 b |
| Control | 9.4 c 0 | 10.2 c | 13.3 b | 16.4 c |

* Values in the same column followed by different letters are significantly different at ($P<0.05$) according to Duncan's multiple range test.

Stem diameter and stem height measurements for cucumber seedlings are shown in Table 15.73. During the summer season there were consistent and significant differences in stem diameter between plants in solarized or biofumigated plots and plants from control plots. These differences were reflected in the stem height measurements for the summer season; at the final assessment the plants in solarized plots (141.0 cm) and biofumigated plots (134.2 cm) were substantially taller than those in the control (89.4 cm). The increases in plant height and stem diameter during the winter season are also shown in Table 15.73. Differences in plant growth were generally less than for the summer season; by day 35 there were no significant differences in either stem height or stem diameter between the three treatments.

Table 15.73. Effect of solarization and biofumigation on cucumber seedling stem diameter and stem height in summer and winter seasons.

| Treatment | Days after planting (season) | | | | | | | | | |
|---|---|---|---|---|---|---|---|---|---|---|
| | Stem diameter (cm)* | | | | | Stem height (cm)* | | | | |
| | 7 | 14 | 21 | 28 | 35 | 7 | 14 | 21 | 28 | 35 |
| Summer | | | | | | | | | | |
| Solarization | 0.28b | 0.52b | 0.91b | 1.19b | 1.26b | 2.7b | 5.9b | 14.0b | 48.9a | 141.0b |
| Biofumigation | 0.27b | 0.53b | 0.89b | 1.17b | 1.25b | 2.9b | 6.2b | 13.3b | 52.1a | 134.2b |
| Control | 0.22a | 0.43a | 0.73a | 0.92a | 0.94a | 2.3a | 4.5a | 9.4a | 45.4a | 89.4a |
| Winter | | | | | | | | | | |
| Solarization | 0.26a | 0.39a | 0.75a | 0.79a | 0.79a | 3.2a | 7.6b | 17.2b | 37.0a | 79.2a |
| Biofumigation | 0.31a | 0.40a | 0.85a | 0.91a | 0.91a | 3.3a | 7.6b | 13.9a | 89.6a | 89.6a |
| Control | 0.25a | 0.38a | 0.90a | 0.98a | 0.98a | 2.9a | 6.6a | 12.8a | 29.4a | 87.4a |

* Values in the same column followed by different letters are significantly different at ($P<0.05$) according to Duncan's multiple range test.

### 15.19.2. Nematodes

#### (i) Root-knot Nematodes

**(a) Symptoms:** The root-knot nematodes cause severe galling on cucumber roots (Fig. 15.40).

#### (b) Management

**Solarization and Organic Amendments:** Solarization alone or soil amendment alone (using poultry manure, alfalfa, cauliflower, tomato, and olive-cake residues) treatments significantly reduced densities of second stage juveniles of *M. javanica,* reduced root galling and increased yield of cucumber. Organic amendments reduced densities of *Fusariam* spp., generally increased *Aspergillus* spp., while *Trichoderma* spp. was not affected. Combinations of solarization and addition of amendments substantially augmented each other, particularly with poultry manure, alfalfa hay, and to a lesser extent with cauliflower and tomato residues.

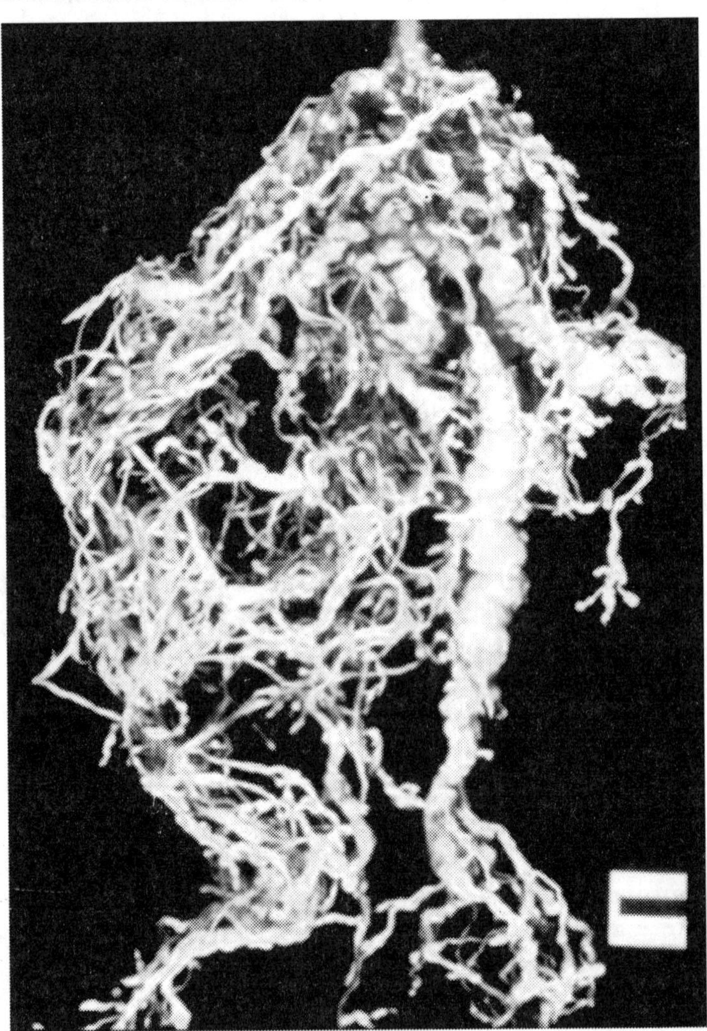

**Fig. 15.40.** Cucumber roots severely galled due to root-knot nematode infection.

**Biofumigation and Trap Cropping:** In commercial greenhouse trials in Spain an integrated management system was developed, including biofumigation with sheep manure and mushroom residue and the cultivation of short-cycle vegetables acting as trap crops.

Using this strategy, initial very high levels of *M. incognita* were reduced to near zero in the main susceptible cucumber crop (Bello, 1998).

**Solarization and Bioagents:** In a cucumber crop in glasshouse trial, the use of solarization and *Pasteuria penetrans* had an additive detrimental effect on *M. javanica* populations (Tzortzakakis and Goewn, 1994).

## 15.20. PUMPKIN, *CUCURBITA MOSCHATA*

### 15.20.1. Nematodes

#### (i) Root-knot Nematodes, *Meloidogyne spp.*

The very broad leaved plant like pumpkin which is infected with root-knot nematodes may show day time wilting and develop much larger galls. In cucurbits, the roots react to the presence of *Meloido-gyne* spp. by the formation of large, fleshy galls.

#### (a) Management

**Solarization:** Pulverizing the planting pit soil and exposing to sunlight by repeated raking and heaping 10 to 15 cm thick dry trash and burning it before application of compost helps in reducing the nematode population.

**Organic Amendments:** Application of 250g of fresh neem or karanj cake at the time of sowing reduces the nematode population in soil.

**Crop Rotation:** Crop rotation with rice, oats, wheat and taramira is also effective.

**Solarization and Chemicals:** The yield of cantaloupe was significantly increased by soil solarization with or without sulphur and by sulphur as single treatment. The thickness of the polyethylene did not affect the yields, which were significantly increased by EVA 0.150 mm, compared to EVA 0.035 mm. The addition of either 750 or 1,000 kg/ha sulphur previously to soil solarization was beneficial compared with the polyethylene tarping only. The sulphur application under EVA tarping did not statistically increased the yield with respect to 500 kg/ha sulphur alone (Table 15.74). The nematode population was significantly suppressed by either solarization or sulphur treatments: No difference was found between the two films or among the sulphur dosages and no further suppression derived by the combined use of solarization and sulphur.

Table 15.74. Effect of soil solarization and sulphur on *Meloidogyne incognita* on cantaloupe.

| Sulphur (kg/ha) | Soil solarization | Crop yield (MT/ha) | Final nema popn. (eggs & $J_2$/ml soil) |
|---|---|---|---|
| - | - | 12.6 ab | 11.7 a |
| 500 | - | 16.4 b | 3.2 c |
| 500 | Polyethylene 0.050 mm | 16.2 b | 3.1 c |
| 500 | EVA 0.035 mm | 19.0 bc | 2.9 c |
| 750 | - | 17.8 bc | 4.3 c |
| 750 | Polyethylene 0.050 mm | 20.4 c | 2.6 c |
| 750 | EVA 0.035 mm | 20.2 c | 4.3 c |
| 1000 | - | 17.0 bc | 3.7 c |
| 1000 | Polyethylene 0.050 mm | 20.4 c | 3.4 c |
| 1000 | EVA 0.035 mm | 20.4 c | 1.6 c |
| - | Polyethylene 0.050 mm | 17.0 bc | 2.1 c |
| - | EVA 0.035 mm | 16.4 b | 4.9 bc |

* Means followed by the same letters in the same column are not significantly different according to Duncan's Multiple Range Test (P = 0.01).

## 15.21. BOTTLE GOURD, *LAGENARIA SICERARIA*

### 15.21.1. Nematodes

#### (i) Root-knot nematode, *Meloidogyne spp.*

#### (a) Management

Spot application of mustard cake at 15 g/plant one week before sowing was effective in controlling root-knot nematodes and significantly improved the yield (8.30 MT/ha) followed by neem cake (8.23 MT/ha) (Paruthi *et al.*, 2010).

## 15.22. POINTED GOURD, *TRICHOSANTHES DIOIDA*

### 15.22.1. Nematodes

#### (i) Root-knot Nematode, *Meloidogyne incognita*

*M. incognita* is responsible for 30-40% loss in yield of pointed gourd. The root-knot nematode is the major limiting factor in successful production of pointed gourd.

### (a) Management

**Intercropping:** Planting of marigold, *Tagetes patula* at 3 to 5 plants/pit will manage attack of root-knot nematodes on pointed gourd.

**Intercropping, Bioagents and Organic Amendments:** Integration of *Paecilomyces lilacinus* (50 g) + *Trichoderma harzianum* (100 g) + neem cake (250 g) + marigold (3 plants/pit) increased plant growth parameters and yield (6.6 kg/plant compared to 1.4 kg/plant in control), and reduced root galling (14.3/5 g roots compared to 75.0/g roots in control), number of egg masses (40/5 g roots compared to 150/5 g roots in control) and final nematode population (66.6/100 g soil compared to 86660/100 g soil in control) (Verma *et al.*, 2005) (Table 15.75).

Table 15.75. Effect of integration of bioagents, neem cake and marigold as intercrop for the management of *Meloidogyne incognita* infecting pointed gourd

| Treatment | No. of galls / 5 g root | No. of egg masses/ 5 g root | Nema popn / 100 g soil | Yield/ plant (kg) |
|---|---|---|---|---|
| Neem cake + *P. lilacinus* (50 g) + *T. harzianum* (100g) + (250 g) + marigold (3 plants)/pit | 10.4 | 30 | 57 | 8.5 |
| *P. lilacinus* (50 g/pit) | 15.3 | 53 | 70 | 6.4 |
| Carbofuran (3 g/pit) | 22.5 | 72 | 95 | 5.8 |
| Neem cake (250 g/pit) | 22.4 | 75 | 120 | 5.0 |
| *T. harzianum* (100 g/pit) | 25.6 | 45 | 130 | 4.2 |
| Marigold (3 plants/pit) | 35.3 | 113 | 135 | 4.0 |
| Control | 80.4 | 186 | 901 | 1.3 |
| CD (P = 0.05) | 3.5 | 9.6 | 25.3 | 0.5 |

## 15.23. LETTUCE, *LACTUCA SATIVA*

### 15.23.1. Diseases

### (i) Tipburn and Bacterial Soft Rot

Bacterial rot often occurred on the inner leaves of plants that also showed tipburn symptoms. Both symptoms have been associated with plant tissue breakdown as a result of calcium deficiency. Tipburn and bacterial soft rot were the main cause of unmarketable lettuces.

**(a) Management**

**Biofumigation:** Only Fumus planted prior to the lettuce increased the percentage of marketable lettuces (Table 15.76). BQ-Mulch, broccoli and broad beans had little or no effect on tipburn and bacterial rot on marketable yield compared to the control fallow treatment. In Red Ferrosol, Fumus produced the highest level of plant biomass compared to other green manures. Therefore, the high plant biomass by Fumus may be associated with increased organic matter in soil and improved calcium uptake, hence resulting in reduced tipburn and bacterial rot.

Table 15.76. The effects of green manure crops on marketability of a subsequent lettuce crop.

| Pre-plant treatment | % Marketable lettuce | % Tipburn + soft rot |
|---|---|---|
| Control - Fallow | 71.9 a | 24.1 |
| Broccoli 8 plants/m$^2$ | 78.2 a | 18.8 |
| Broad bean 240 kg/ha | 78.5 a | 19.2 |
| Fumus 16 kg/ha | 87.8 b | 8.7 |
| BQ-Mulch 16 kg/ha | 73.5 a | 22.3 |

**(ii) White Mold, *Sclerotinia sclerotiorum***

**(a) Symptoms:** The fungus, *Sclerotinia sclerotiorum*, causes a disease called white mold in lettuce (Fig. 15.41). The cool, moist conditions in the Wiedigers' solar greenhouses in winter were perfect for white mold infection. The growing fungus would deposit hard black bodies, called sclerotia, on the soil surface (Fig. 15.42). Some sclerotia would quickly sprout into tiny mushrooms that spread millions of spores throughout the greenhouses, infecting even more lettuce plants. Others would settle into the soil, where they might sit for years before sprouting mushrooms or infecting plant roots.

**(b) Management**

**Biofumigation:** Brassica green manures that produce high Isothiocyanates (ITCs) can offer advantages over non-brassica break crops for the management of *Sclerotinia* disease of lettuce. As ITCs levels will diminish rapidly after incorporation into soil, their effects for disease suppression are expected to be relatively short term. Soil improvements by the brassica green manures have longer lasting effects, which may contribute to soil quality and soil health, in addition to soil-borne disease management. Biofumigation with Fumus gave best control (Table 15.77).

**Fig. 15.41.** A bed of young lettuce plants infested with white mould, caused by the fungus *Sclerotinia sclerotiorum.*

**Fig. 15.42.** Sclerotia sprout into tiny mushrooms that spread *S. sclerotiorum* spores. Spores infect leaves; the fungus can also grow directly from sclerotia, infecting plant roots.

Table 15.77. The effects of green manure crops on *Sclerotinia* disease incidence on a subsequent lettuce crop

| Pre-plant treatment | No. of sclerotes/ 200g soil | | Total ITCs (μmole/g) | | % *Sclerotinia* wilted plants |
|---|---|---|---|---|---|
| | Initial | Final | Root | Shoot | |
| Control – Fallow | 26 ± 6 | 10 ± 4 | Na | Na | 30.8 c |
| Oats – 20 kg/ha | 16 ± 9 | --- | 0.00 | 0.00 | 23.2 bc |
| Broad bean – 240 kg/ha | 15 ± 3 | 11 ± 4 | 0.00 | 0.00 | 24.2 bc |
| Fumus – 16 kg/ha | 16 ± 6 | 9 ± 4 | 5.92 | 6.14 | 16.5 ab |
| BQ Mulch – 16 kg/ha | 25 ± 12 | 7 ± 2 | 42.23 | 2.69 | 3.1 a |

**Biofumigation and Solarization:** On-farm trails were conducted to monitor the effects of biofumigation and soil solarization on *Sclerotinia sclerotiorum*. This study suggests that a month of summer soil solarization can control populations of *S. sclerotiorum* to a depth of 15 cm in Kentucky high tunnels. The effect was seen at both at the middle and edge of solarized plots. Biofumigation, by incorporating a mixture of *Brassicae juncea* leaves and stems at a rate of 900 g/m$^2$, did not reduce germination of *S. sclerotiorum* sclerotia (Figs. 15.43 and 15.44).

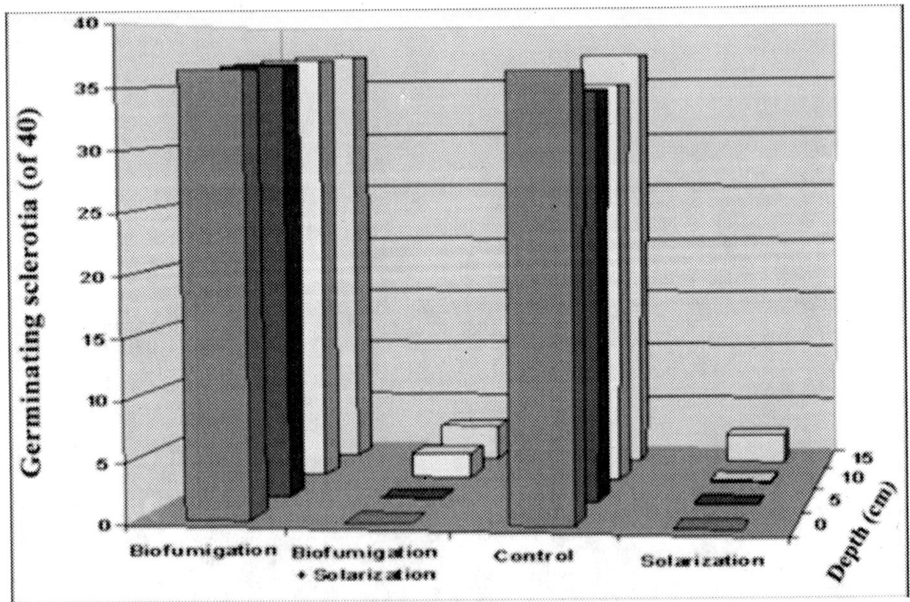


**Fig. 15.43.** The graph clearly shows that solarization was almost completely lethal to *S. sclerotiorum* to 15 cm, and that biofumigation had no effect on the fungus.

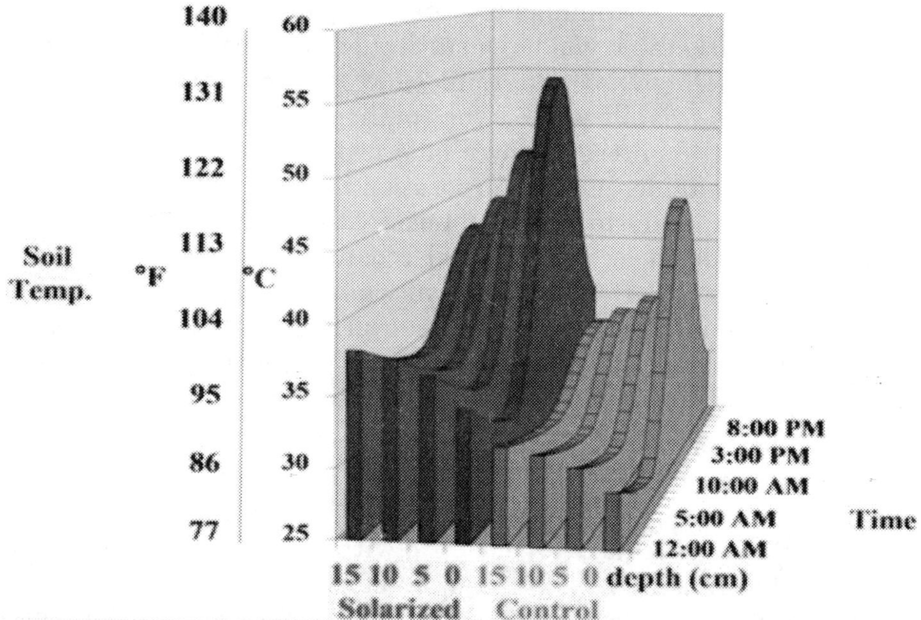


**Fig. 15.44.** The daily fluctuation in soil temperature in solarized (purple) and unsolarized (blue) plots. Soil depth, in centimeters, is shown above each treatment label. Fluctuations were less pronounced deeper in the soil than at the soil surface. Soil temperatures in solarized plots consistently stayed above 30°C each night.

## (iii) Lettuce Drop, *Sclerotinia minor*

### (a) Management

**Biofumigation:** The effects of biofumigation (mulching and subsequent irrigation) with five incorporated cover crops (*Sinapis alba* 'Ida Gold,' *Brassica juncea* 'Pacific Gold,' *Brassica napus* x *Brassica campestris* 'BQ mulch' *Vicia faba*, and *Triticum aestivum* x *Secale cereale*) on sclerotia of *S. minor* (lettuce drop), seed of five weed species, and yield of Romaine lettuce were evaluated. Disease incidence in lettuce was low; however, disease severity was reduced by 25% when biomass amendments were doubled, regardless of cover crop. Lettuce heads were 15% larger after 'Ida Gold' compared to 'BQ mulch,' which corroborates previous findings that 'Ida Gold' resulted in improved agronomic performance of subsequent lettuce. This may be attributed to the presence of benzyl glucosinolate in above-ground tissues and 4-methylthiobutyl and 4-hydroxybenzyl glucosinolates in roots of 'Ida Gold' and their absence in the other crops. (Daugovish *et al.*, 2006).

In field studies, biofumigant *Brassica* green manure crops could increase the marketable yield of a subsequent lettuce crop by reducing plant loss from *S. minor* infection. However, the brassicas did not reduce sclerotia viability (i.e. the pathogen's inoculum level). It appears they are suppressing mycelial growth of the pathogen. As ITC levels diminish rapidly after incorporation into soil, effects for disease suppression are likely to be short term. Other important benefits of green manures such as reduced soil crusting, improved infiltration, increased organic matter and the subsequent increase in beneficial microflora and microfauna are often overlooked. Soil improvement following a *Brassica* may produce longer lasting effects on disease suppression and crop health, as well as soil health.

The effect of tissue mulching and irrigation on survival of *S. minor* was measured. Oriental mustard (*Brassica juncea*), yellow mustard (*Sinapis alba)*, faba bean (*Vicia faba*) and a mixture of cereals (rye and triticale) were chopped with a mower at late flowering and sprinkler-irrigated with 30 mm of water within 0.5 h. The pathogen was buried in gas and water-permeable bags at 20 cm prior to irrigation, recovered 7 days later and analyzed for viability.

Biofumigation did not affect *S. minor* sclerotia, except when oriental mustard was used in combination with plastic (75% reduction compared to control). No sclerotia grew in aqueous extracts from oriental mustard tissue and sclerotia growth was reduced by about 60% in tissue extracts from oriental mustard + plastic, and cereal. Oriental mustard exceeded other crops in inhibiting sclerotial growth in both field and lab tests. Abundant presence of *Trichoderma* spp. (fungi with useful biocide potential) was observed in plates with sclerotia of both field and lab isolates from oriental mustard plots. This observation agrees with Australian results that documented prolific growth of *Trichoderma* spp. at ITC concentrations well beyond what killed a range of pathogenic fungi. This suggests that dynamic changes in the microbiological community may be partially responsible for suppressive effects of *Brassica* spp. on certain soil-borne pathogens.

Infection of lettuce with *S. minor* was 91 and 68% less after 'biofumigation' with yellow and oriental mustard, respectively, than after faba bean. Lettuce after yellow mustard/plastic was nearly infection-free (7.6 times lower infection score than control). Yellow mustard/plastic resulted in the largest lettuce heads. Control celery were lightest and heaviest after mustards. This study and literature reports show beneficial effects of mustards on subsequent crops.

Experimental work in the Salinas Valley, California, reported that broccoli (B. *oleracea* var. *italica)* rotations were effective in controlling certain soil-borne fungal pathogens, such as *S. minor* in lettuce crop rotations (Hao *et al.*, 2003).

Field trial studies on green manure crops (mustard, rape and broad beans) have shown that biofumigant green manure crops could increase marketable yields of subsequent lettuce crops by reducing *Sclerotinia* disease.

The biofumigants BQ-Mulch and Fumus significantly reduced the percentage of plants with *Sclerotinia* wilt (Table 15.78). BQ-Mulch appeared more effective than Fumus in reducing wilted plants. Oats and broad beans had little or no effect.

Table 15.78. Effect of green manures on *Sclerotinia minor* sclerotia and disease in subsequent lettuce crop

| Treatment | Mean no. of sclerotia/200 g soil | | |
|---|---|---|---|
| | Initial count 5/7/01 | Final count 1/3/02 | % plants with *Sclerotinia* wilt* |
| Untreated control | 26 | 10 | 31c |
| Oats | 16 | nt** | 23bc |
| Broad bean | 15 | 11 | 24bc |
| Fumus | 16 | 9 | 17ab |
| BQ-Mulch | 25 | 7 | 3a |

* Value with same letter are not significantly different at 5%, ** nt – not tested.

Apart from any biofumigation effects, the crop residues also increase organic matter and nutrients, and improve soil structure. Fumus is a variety of mustard, while BQ-Mulch is two rapes.

Fumus matured and flowered approximately two weeks earlier than BQ-Mulch. Analysis of the biofumigant plant samples collected prior to green manure incorporation into soil, showed that Fumus produced higher levels of ITCs on their shoots, whereas most of the ITCs for BQ-Mulch were in the roots (Table 15.79).

Table 15.79. Total ITC concentration in green manure plants (μmol/g)

| Variety | Tissue | ITC concentration |
|---|---|---|
| BQ-Mulch | Shoot | 2.7 |
| Fumus | Root | 16.6 |
| BQ-Mulch | Shoot | 6.1 |
| Fumus | Root | 5.9 |

The better *Sclerotinia* disease control with the BQ-Mulch plants may be due to the higher ITC levels in their roots.

### (iv) *Fusarium* spp.

### (a) Management

**Solarization:** Cabbage and lettuce seed-bed solarization significantly reduced *Fusarium* spp. and total fungal infestations.

## 15.23.2. Nematodes

### (i) Root-knot Nematode, *Meloidogyne spp.*

The root-knot nematode (Fig. 15.45) cause large decreases in yield of lettuce.

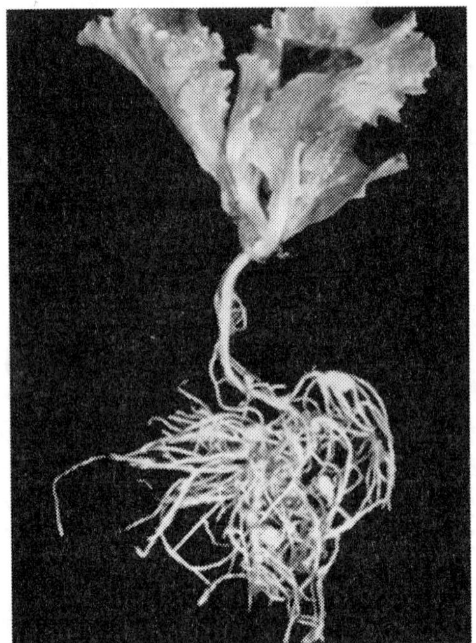

**Fig. 15.45.** Root-knot nematode on lettuce

### *(a) Management*

**Intercropping:** Planting of 4 and 5 seedlings of *Tagetes erecta* in a pot around a lettuce plant reduced root galling by 56 and 72% and nematode population by 73 and 94%, respectively (Pervez *et al.*, 1988).

**Biofumigation:** The influence of organic amendments applied as green manures and composts were tested against root-knot nematode *M. incognita* in resistant and susceptible lettuce cultivars. Chopped leaves, stems and roots of six plants namely, broccoli and cauliflower (*Brassica oleracea* var. *botrytis*), cabbage (*B. oleracea* var. *capitata*), mustard (*B. juncea*), radish (*Raphanus sativus*), and *Tithonia diversifolia*, including processed chicken manure were incorporated at 20g per kg of soil previously inoculated with 10,000 second stage juveniles of root-knot nematode. A treatment using fallow for 3 weeks was included as control. Significant increase ($P \leq 0.05$) in plant height and top weight were recorded in the two lettuce cultivars incorporated with organic amendments. Highest top weight and plant height were noted in broccoli- and cabbage-amended soil. Root weight was highest in broccoli- and cauliflower- amended soil. This was significantly different from the other treatments. Significantly lower number of galls was recorded in resistant cultivar Great Lakes as compared to that of susceptible cultivar Tyrol. Highest number of galls was observed in fallow and processed chicken manure which was significantly different from the rest of the treatments. No egg masses were produced in cultivar Great lakes while in cultivar Tyrol, fallow gave the highest number of egg masses significantly different from the amended soil. Significant reductions in second stage juveniles in the soil were found in mustard and radish amendment but comparable with the other treatments. (Pedroche *et al.*, 2008).

### (ii) Reniform Nematode, *Rotylenchulus reniformis*

### (a) Management

**(a) Solarization:** Under field conditions, soil solarization significantly reduced soil population of the reniform nematode up to 15 cm deep and increased the yield of lettuce.

## 15.23.3. Weeds

***(i) Solarization:*** Solarization resulted in approximately 100%, 100% and 79% weed reduction in sandy soil at Ismailia and 100%, 94% and 69% reduction in clay soil at Kalubia for annual broad leaf weeds, annual grasses and perennial weed infestation, respectively.

***(ii) Biofumigation and Solarization:*** The effect of tissue mulching and irrigation on survival of seeds of bur clover, annual ryegrass and pigweed was measured. Oriental mustard (*Brassica juncea*), yellow mustard (*Sinapis alba)*, faba bean (*Vicia faba*) and a mixture of cereals (rye and triticale) were chopped with a mower at late flowering and sprinkler-irrigated with 30 mm of water within 0.5 hr.

Yellow mustard/plastic resulted in the largest lettuce heads. Control celery were lightest and heaviest after mustards. This study and literature reports show beneficial effects of mustards on subsequent crops.

Biofumigation with rapeseed alone killed 95 percent of milkweed seeds, 68 percent of foxtail seeds and 51 percent of dock seeds. Rapeseed plots that were tarped for solarization killed 94 percent of dock and 100 percent of milkweed. However, high temperatures under tarps apparently stimulated foxtail. Twice as many foxtail seeds germinated after solarization as in the control.

## 15.24. SPINACH, *BETA VULGARIS* VAR. *BENGALENSIS*

### 15.24.1. Diseases

#### (i) Wilt, *Fusarium oxysporum* f. sp. *spinaciae*

#### (a) Management

**Biofumigation:** Healthy spinach was obtained after cultivation of brown mustard and cleome in infested soil with *F. oxysporum* f. sp. *spinaciae* (Mitsuhori) (Hashizume *et al.*, 2008).

## 15.25. BABY CORN, *ZEA MAYS*

### 15.25.1. Weeds

***(i) Mulching:*** Application of sugarcane trash (10 cm thickness) proved best in improving the yield of baby corn and suppression of weed flora. Among the bio-mulches, coriander gave satisfactory results (Table 15.80).

Table 15.80. Effect of organic and bio-mulches on weed count, biomass and control efficiency (Mean of 2001 and 2002).

| Treatment | Weed count/m$^2$ | Weed biomass ( kg/ha) | Weed control efficiency |
|---|---|---|---|
| Control | 32.95 | 47.23 | --- |
| Water hyacinth residue at 5 cm thickness | 9.90 | 22.41 | 69.92 |
| Sugarcane trash at 10 cm thickness | 7.00 | 16.20 | 78.71 |
| Coir pith at 2 cm thickness | 8.67 | 19.47 | 73.80 |
| Saw dust at 2 cm thickness | 17.67 | 35.82 | 46.16 |
| Mint ( bio-mulch) | 11.57 | 23.26 | 64.70 |

| | | | |
|---|---|---|---|
| Coriander ( bio-mulch) | 11.37 | 22.73 | 66.80 |
| Fenugreek ( bio-mulch) | 15.65 | 26.20 | 52.30 |
| CD at 5% | 1.09 | 2.93 | 3.29 |

## 15.26. MUSHROOM, *AGARICUS* SPP.

### 15.26.1. Nematodes

#### (i) *Aphelenchoides sacchari*

*A. sacchari* was reported by Sharma *et al.* (1981) and found to be highly pathogenic to white button mushroom, *Agaricus bisporus*. An average of 50-100% yield loss in sporophores in white button mushroom has been reported due to this nematode species (Sharma *et al.*, 1984). It is quite prevalent on mushrooms in India (Thapa *et al.*, 1981). Singh *et al.* (2003) reported that *A. sacchari* was responsible for 40.6 to 100% loss in yield of mushroom.

**(a) Symptoms:** Initially whiteness of spawn slowly starts turning brown followed by sparse and patchy appearance of the mycelium. This mycelium is stingy in nature. The compost surface sinks. Alternately, medium and low quantity of sporophores (fruiting bodies) is obtained in succession of flushes. There are extremely poor yields in the later flushes finally leading to total failure of sporophore production thus reducing the total yields as well as duration of the crop.

**(b) Losses:** Economic losses due to decline in yields are generally very high. Extent of losses mainly depends upon the time of nematode infestation, nematode species involved and population of the nematodes. Losses caused by *A. sacchari* are only so far worked out and have been given in Table 15.81 (Sharma *et al.*, 1984).

Table 15.81. Yield losses in mushroom due to *Aphelenchoides sacchari*

| Time of inoculation | Initial inoculum | Yield loss (%) |
|---|---|---|
| Spawning | $1 \times 10^1$ | 98.2 |
| | $1 \times 10^2$ | 100.0 |
| | $1 \times 10^3$ | 100.0 |
| | $1 \times 10^4$ | 100.0 |
| Casing | $1 \times 10^1$ | 90.6 |
| | $1 \times 10^2$ | 92.1 |
| | $1 \times 10^3$ | 87.3 |
| | $1 \times 10^4$ | 100.0 |

### (c) Management

**Organic Amendments:** Incorporation of neem/pongamia/coconut cake at 2 kg/100 kg of compost during first or second turning is recommended. Dry leaf powder of neem can be mixed with compost at 2-5% concentration. This has been found useful not only to check nematode multiplication in mushroom growing trays but also helps in proper mycelial growth and improved sporophore yields (Khanna and Sharma, 1994). Addition of oil cakes like neem, karanj, coconut, castor and groundnut have been reported to control nematodes and increase button production (Rao *et al.*, 1992). Further, incorporation of fresh leaves of Pedilanthus, eucalyptus and pongamia in the compost during first or second turning was observed to reduce the population of mushroom nematodes and to increase the yield of button mushrooms. The incorporation of dried leaves of neem, castor and *Cannabis sativa* resulted in the suppression of *A. composticola* below economic injury levels and increased mushroom yields by 6.5 to 19% (Grewal, 1989).

Incorporation of *Pedilanthus tithymaloides* leaves both at 2 and 4% in paddy straw compost was found to be more effective than carbofuran at 0.04% for the management of *A. sacchari* and increased the yield of mushroom, *Agaricus bisporus. Glyricidia maculata* leaves incorporated at 4% was as effective as carbofuran at 0.02% for the management of *A. sacchari* and increased the yield of mushroom (Rao *et al.*, 1991).

*Chapter - 16*

# ORNAMENTAL, MEDICINAL, AROMATIC AND TUBER CROPS

## 16.1. ORNAMENTAL CROPS

India is known for growing of traditional flowers such as jasmine, marigold, chrysanthemum, tuberose, crossandra and aster. Commercial cultivation of cut flowers- rose, orchids, gladiolus, carnation, gerbera, anthurium and lilies-has also become popular.

With the liberalization of Indian economy, the domestic floricultural trade, which hitherto remained an unorganized sector, became an organized sector with the active participation of corporate giants. The establishment of state-of-art hi-tech floricultural units in early 90's for intensive cultivation of cut flowers contributes about Rs. 300 crores of valuable foreign exchange today. Various government agencies are actively involved in the promotion of this sector and are highly committed to resolve the teething problems. The world demand for floricultural products is approximately US $ 40 billions and cut flowers contribute nearly 60% (US$ 25 billions) to the global trade. India's share in the world flower market is about 0.7% which is contributed mainly from open cultivation.

Floriculture is estimated to cover an area of 161,000 ha with production of 870,000 MT of loose flowers during 2008. The area under cut flowers has increased significantly in recent years and the production has reached 4341 million numbers (Tables 16.1 and 16.2). Another 500 ha of climatic control greenhouse are available for growing quality flowers for export. Many corporate houses (more than 100) have set up 100% export oriented units with an investment of more than Rs. 1000 crores. India is bestowed with natural advantages like favourable weather conditions, soil, cheap labour which are likely to make a real mark in international trade. With this, India is emerging as one of the powerful forces in world flower trade. India exported floriculture products worth US $ 84.12 million during 2007.

Table 16.1. Area and production of ornamental crops in India (2007-08) (Bijay Kumar, 2009)

| Ornamental Crops | Area ('000 ha) | Production ('000 MT) | Productivity (MT/ha) |
|---|---|---|---|
| Flowers Loose | 161 | 870 | 5.4 |
| Flowers Cut* | --- | 4341* | --- |

* Million numbers

Table 16.2. State-wise area and production of flowers (2007-08)

| State / UT's | Area ('000 ha ) | Production | | | |
|---|---|---|---|---|---|
| | | Loose ('000 MT) | % of total production | Cut (million nos.) | % of total production |
| Karnataka | 22.34 | 169.12 | 19 | 555.000 | 13 |
| Maharashtra | 16.74 | 69.45 | 8 | 572.800 | 13 |
| Uttar Pradesh | 8.41 | 12.36 | --- | 375.200 | 9 |
| Uttaranchal | 0.90 | 0.70 | --- | 145.545 | 3 |
| Gujarat | 9.74 | 49.50 | 6 | 506.300 | 12 |
| Delhi | 5.50 | 5.70 | --- | 103.800 | 2 |
| West Bengal | 27.42 | 48.45 | 6 | 1968.000 | 45 |
| Tamil Nadu | 26.74 | 214.38 | 25 | --- | --- |
| Andhra Pradesh | 23.52 | 126.27 | 14 | 6.782 | --- |
| Punjab | 1.00 | 77.90 | 9 | --- | --- |
| Haryana | 6.11 | 61.76 | 7 | 1.053 | --- |
| Others | 12.30 | 34.78 | 6 | 7.266 | 3 |
| Total | 160.72 | 870.37 | --- | 4341.746 | --- |

Floriculture is an intensive type of agriculture wherein the income per unit area is much higher than any other agricultural product if it is done in a scientific way. The area used for production of ornamental plants is low, but the cash returns are high because of the intensive cropping. The flowers grown in large areas include jasmine, rose, marigold, chrysanthemum, crossandra, champak, tuberose, gladiolus and aster. New crops like lilies, tulips, anthurium, carnation, gerbera, etc., are also finding potential place in the foreign market. With the rapid development in urban planning and hotel industry, there has been an increasing demand for cut flowers and ornamentals. The cut flowers followed by live plants, cut foliage, seeds and bulbs have got an immense potential for export to Western countries where the cost of production is comparatively high. India exports roses, gladioli, carnations, chrysanthemums, lilies and jasmines to Gulf, Germany, France, England, Netherlands and Singapore. Our share in the export market has registered a sharp increase from Rs.14.45 crores in 1991-92 to Rs.299.41 crores in 2005-06.

### 16.1.1. CONTAINER GROWN ORNAMENTALS

#### (i) Weeds

**(a) Biofumigation:** Mustard seed meal (MSM) was applied to soil surfaces to evaluate its effect on several common weeds in container-grown ornamentals. MSM applied to the soil surface of containers at 113, 225, and 450 g/m$^2$ reduced the number of annual bluegrass seedlings by 60%, 86%, and 98%, respectively, and the number of common chickweed seedlings by 61%, 74%, and 73%, respectively, at eight weeks after treatment. Surface-applied MSM may control weeds in container-grown ornamentals without injuring the crop. In addition, MSM provides about five percent nitrogen and contributes to the nutrient needs of the crop.

### 16.1.2. ROSE, *ROSA* SPP.

#### *(i) Nematodes*

#### (a) Root-knot Nematode, *Meloidogyne* spp.

*M. incognita* was recorded on rose (*Rosa indica*) from Aligarh (Alam *et al.*, 1973) and *M. javanica* from Delhi (Prasad and Dasgupta, 1964).

**Symptoms:** Nematodes of the genus *Meloidogyne* attack the roots of a wide range of plants including roses. The damaged root system develops root knots (Fig. 16.1), which cause slow growth, wilting, and yellowing of leaves. Nematodes are spread by introducing infested plants or soil to the field, and also can be carried on garden tools. Symptoms of *Meloidogyne* infestation in roses is stunting, slow-growth, pale green leaves and wilting in mild weather.

#### Management

**Intercropping:** Mexican marigold, also known as Tagetes, has been successfully used in the control of root-knot nematode in roses by a Kenyan Farmer (Report on ToT for Alternatives to the Use of Methyl Bromide for Soil Fumigation in Brazil and Kenya).

### (b) Lesion Nematode, *Pratylenchus* sp.

The lesion nematodes are the major problems in rose cultivation. *P. zeae* was associated with poor growth of rose plants.

**Symptoms:** The affected plants showed chlorotic symptoms, poor and stunted growth with necrotic lesions on rose roots. *P. zeae* was responsible for 69.6, 36.4, 59.6 and 33.3% reduction in shoot length, shoot weight, root length and root weight, respectively. Affected roots show patchy discolouration in cortical region, which coalesce and lead to root rotting.

**Fig. 16.1.** *Meloidogyne* sp. damage on rose roots.

### Management

**Intercropping:** Interplanting of roses with African marigold helps in reducing the population of *P. penetrans*.

### *(ii) Weeds*

**(a) Mulching:** Mulching with 5-8 cm layer preferably of organic material to the surface of a rose bed in late spring as soon as the soil begins to warm up keep the soil moist by preventing evaporation, smother the majority of annual weeds during summer, breakdown into humus and provide plant foods particularly if FYM is used (Gibson, 1990; Randhawa and Mukhopadhyay, 1986). Adequately thick mulch gave better weed control. Saw dust controlled weeds better than bark, rock wool or pine straw mulch.

**(b) Solarization:** Soil solarization for two months in the absence of major known pathogens increased the percentage of rooted cuttings and plant vigour of Indica rootstock, reduced the weed infestation, and increased the gross value production of grafted plants by 37 percent of that in untreated control. Those effects were highest in plots fumigated with methyl bromide, which increased the gross value production of grafted plants by 35 percent of that in untreated control (Table 16.3).

Table 16.3. Effects of soil treatments on growth of *Rosa indica* rootstock[a]

| Treatment | Incidence of dead cuttings[a] | Plant growth vigour[b] | Incidence of weed infestation (Number/1.25 $m^2$) | Gross value production ($ US) |
|---|---|---|---|---|
| Control | 5.6[c]x | 1.8[c]x | 112.5 x | 1 377 x |
| Solarization | 4.0 xy | 6.0 y | 32.5 y | 1 881 y |
| Methyl bromide | 3.1 y | 8.1 z | 3.5 z | 1 863 y |
| Metham-sodium | 3.1 y | 3.4 x | 24.3 y | 1 783 z |

a Determined four months after planting.
b Based on 1-10 scale (1= least vigour, 10= greatest vigour).
c Mean of four replications. Values followed by the same letter within columns are not significantly different (P= 0.5) according to Student's t test.

**(c) Biofumigation:** Three biofumigant crops (white mustard, brown mustard, and rapeseed) were evaluated against small-seeded weeds—redroot pigweed and barnyard grass under greenhouse conditions. When mixed into soil, the three mulches reduced weed germination by 20 percent to 95 percent. Reductions of 5 percent to 95 percent in growth of seedlings from surviving seeds were also observed.

### 16.1.3. GERBERA, *GERBERA JAMESONII*

#### *(i) Weeds*

**(a) Mulching:** The highest yield of gerbera flowers (2.782 million/ha) was recorded in weed-free check which was on par with pendamethalin at 1.0 kg a.i./ha (2.465 million/ha) followed by alachlor (2.310), black polythene sheet (2.270 million/ha) and paddy straw mulch (2.219) (Shalini and Patil, 2004) (Table 16.4).

Table 16.4. Effect of weed control treatments on flower yield in gerbera

| Treatment | Flower yield (million/ha) |
|---|---|
| Weed free | 2.782 |
| Unweeded control | 1.125 |
| 2 hoeings + 2 hand weedings at 30 and 60 DAP | 1.860 |
| Pendimethalin at 1.0 kg a.i./ha | 2.465 |

| | |
|---|---|
| Alachlor at 1.5 kg a.i./ha | 2.310 |
| Metalachlor at 1.0 kg a.i./ha | 2.122 |
| Butachlor at 1.0 kg a.i./ha | 1.730 |
| Atrazine at 1.0 kg a.i./ha | 1.370 |
| Oxyfluorfen at 0.15 kg a.i./ha | 1.832 |
| Oxadiargyl at 90 g a.i./ha | 1.960 |
| Paddy straw mulch | 2.219 |
| Black polythene sheet | 2.270 |
| CD (P = 0.05) | 0.431 |

## 16.1.4. GLADIOLUS, *GLADIOLUS* SPP.

### *(i) Diseases*

### (a) Wilt, *Fusarium oxysporum* f. sp. *gladioli*

**Symptoms:** This disease is caused by the fungus *Fusarium oxysporum var. gladioli* and is often referred to as Fusarium yellows or Fusarium Wilt. The symptoms are yellowing of the leaves starting with the older outside ones first and this is accompanied by the apparent stunting of newer leaves. In addition, the spike itself will often be stunted and faded in colour. The plant may also have blackened areas at the base which spread onto the corm (Fig. 16.2). Eventually, if left, the whole plant will wilt and when lifted for destruction, (for this is the only course of treatment), the interior of the corm when opened will appear marbled with a brown colour.

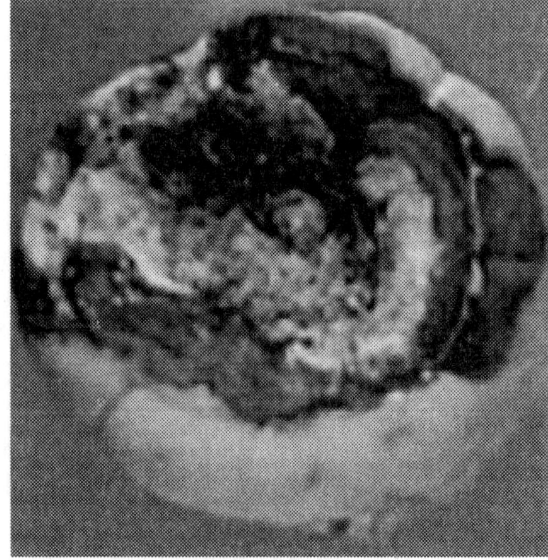

**Fig. 16.2.** Fusarium infection on gladiolus.

**Management**

**Solarization:** Solarization of field soil by mulching with transparent polythene sheets (100 gauze) raised the soil temperature which was lethal to wilt pathogen.

### (b) Root Rot, *Rhizoctonia solani*

**Biofumigation:** In gladiolus, incorporation of *Brassica juncea* gave significant control of the Rhizoctonia disease (van Os *et al.*, 2004).

### *(ii) Nematodes*

### (a) Root-knot Nematodes, *Meloidogyne* spp.

**Management**

**Crop Rotation:** Crop rotation or intercropping with marigold helps in reducing the nematode population.

**Intercropping:** Intercropping with marigold helps in reducing the nematode population.

**Organic Amendments and Bioagents:** Soil application of neem cake enriched with *T. harzianum* at 1 t/ha is also effective.

### *(iii) Weeds*

**(a) Mulching:** Mulching is used to check weeds in this crop. Plastic mulching benefitted the gladiolus crop with black and photo selective plastic being preferable to the transparent type. It gave good weed control, conserved soil moisture well and kept the temperature around the roots comparably low.

## 16.1.5. TUBEROSE, *POLIANTHES TUBEROSA*

### *(i) Nematodes*

### (a) Root-knot Nematode, *Meloidogyne incognita*

**Symptoms:** Affected plants exhibit yellowing, drying up of leaves and retarded growth. In severely affected plants the emergence of spikes is also suppressed. Sundarababu and Vadivelu (1988) found that at higher inoculum level of *M. incognita*, 65% reduction occurred in top growth of plant.

**Management**

**Organic Amendments and Bioagents:** Treatment of tuberose bulbs with neem and calotropis leaf extracts mixed with *P. lilacinus* spores significantly reduced *M. incognita* infection and multiplication besides stimulating the plant growth (Nagesh *et al.,* 1996).

***(ii) Weeds***

**(a) Mulching:** Mulching is done by the growers. Mulching the plots with fresh manure, strips of black polythene, dried grass clippings, chopped straw, saw dust, husk, peat, bark is effective in controlling weeds, besides helping in conserving soil moisture.

## 16.1.6. CHRYSANTHEMUM, *DANDRATHEMA GRANDIFLORA*

***(i) Diseases***

### (a) Root Rot, *Pythium* sp., *Phytophthora* sp.

**Symptoms:** Infected plant parts (root, stem and leaves) show sudden wilting.

**Management**

**Solarization:** Soil solarization by covering with black polythene mulch during summer helps in reducing the inoculum level.

### (b) Stem Rot, *Fusarium solani*

**Solarization:** Soil solarization by covering with black polythene mulch during summer is effective.

***(ii) Weeds***

**(a) Mulching:** Covering of black plastic or green woven cloth prevented weed growth in soil under pot grown chrysanthemums. Mulching with bark of 5 cm thickness effectively controls weeds and conserve moisture.

## 16.1.7. CROSSANDRA, *CROSSANDRA UNDULAEFOLIA*

***(i) Nematodes***

### (a) Root-knot Nematode, *Meloidogyne incognita*

In India, *M. incognita* was first reported on crossandra by Rajendran *et al.* (1976).

**Economic Importance and Losses:** The nematode is responsible for 25.62% and 21.64% loss in number of flowers and weight of flowers, respectively (Khan and Parvatha Reddy, 1994).

**Symptoms:** Infected plants are stunted with dried peripheral branches bearing smaller chlorotic leaves almost turning to white in later stages (Rajendran *et al.*, 1976). Roots exhibit severe galling. Inflorescences are small and sometimes fail to produce flowers.

### Management

**Solarization:** Deep ploughing and fallowing in summer for about a month helps in reducing the nematode population.

**Intercropping:** Intercropping of crossandra with enemy plant like marigold significantly reduced the root-knot nematode population both in soil and roots (Khan and Parvatha Reddy, 1994).

**Organic Amendments and Intercropping:** Incorporation of FYM in soil and intercropping of crossandra with marigold or pangola grass would reduce the root-knot nematode infection.

### (b) Lesion Nematode, *Pratylenchus delattrei*

**Economic Importance and Losses:** Heavily infested plants do not produce tertiary spikes and thereby the flower yield is reduced by nine-fold (Srinivasan and Muthukrishnan, 1975).

**Symptoms:** *P. delattrei* was frequently associated with crossandra crop exhibiting stunting, chlorosis of leaves and wilting. The leaves showed mottled appearance which turned to brown and becoming pinkish eventually. Later, the leaf tips dried and entire leaves turned yellow and shed. The plants were devoid of side shoots and remained completely defoliated. The roots exhibited brown to black lesions. Heavily infested plants did not produce tertiary spikes and thereby flower yield is reduced (Srinivasan and Muthukrishnan, 1975).

### Management

**Intercropping:** Intercropping of crossandra with enemy plant like marigold significantly reduced the lesion nematode population both in soil and roots (Khan and Parvatha Reddy, 1994). Application of farm yard manure has been found to reduce the lesion nematode population in crossandra.

## 16.1.8. DAHLIA, *DAHLIA* SPP.

### *(i) Weeds*

**(a) Mulching:** Mulching is an important operation in dahlia cultivation. This can be done with grass clipping, old hay and saw dust showing several beneficial effects. It helps to maintain an even soil temperature and causes earlier and better bloom. Mulch slows down water evaporation and reduces the irrigation requirements (Bhattacharjee and Swamy Vivekananda, 1989). It also keeps weeds under control. Using black plastic mulch for growing dahlia is also beneficial as it improves growth, flowering, tuberous root formation and suppresses weed growth.

## 16.1.9. NARCISSUS

### *(i) Nematodes*

#### (a) Lesion Nematode, *Pratylenchus penetrans*

**Intercropping:** Marigold effectively reduced *P. penetrans* on narcissus and reduced root rot disease.

## 16.1.10. LILY, *LILIUM* SPP.

### *(i) Diseases*

#### (a) Bare Patch Disease, *Rhizoctonia solani*

**Management**

**Biofumigation:** In an experiment with lily, *Brassica juncea* was compared to untreated (fallow) and other green manure crops, like *Tagetes patula* and *Trifolium* spp. In non-infested field plots, incorporation of all crops resulted in yield increase compared to the fallow treatment, probably caused by fertilization effects. In infested field plots, incorporation of *B. juncea* also reduced stem and bulb infection by *R. solani* AG2-2IIIB in lily, resulting in a significant increase of bulb yield compared to the other green manure crops (van Os *et al.*, 2004).

In a field experiment with lily, application of defatted seed meal from *Brassica carinata* (2.5 ton/ha) resulted in a better crop stand compared to the non-treated control, probably due to increased availability of nitrogen from the meal. Bulb yield was increased by

70%, and total nitrogen uptake was increased by 80% after seed meal application. In infested soil with *R. solani*, application of seed meal resulted in significant control of stem infection in lily. Percentage of healthy plants was 20% in the infested control treatment and 60% in the infested seed meal treatment. The efficacy of seed meal was equal to the chemical treatment with azoxystrobine (6 l/ha) (van Os and Lazzeri, 2006).

## 16.1.11. TULIP

### *(i) Diseases*

#### (a) Root Rot, *Rhizoctonia solani*

In tulip, incorporation of *Brassica juncca* decreased infection by *R. solani* AG2-T and increased bulb yield in one field experiment (van Os *et al.*, 2004).

## 16.1.12. PROTEAS, *LEUCADENDRON SALIGNUM LAUREOLUM*

### *(i) Diseases*

#### (a) Sudden Death, *Phytophthora cinnamomi*

In Western Australia, approximately 40 per cent of the properties that were surveyed had suffered from Sudden Death, and autopsies of the plants demonstrated that the major causal agent was *P. cinnamomi*, the Jarrah dieback fungus.

### Management

**Biofumigation:** Biofumigants to protect Protea plants through mulching the Brassicas (Indian mustards, fodder rapes) and incorporating it through the top soil was developed. Biofumigation is a great alternative to the use of synthetic chemical fumigants that are frequently used in agriculture and horticulture, such as metham sodium and methyl bromide. Therefore, integrating biofumigation with a series of other 'green' control measures such as solarization, composting and the sterilization of irrigation water will provide a first step for Protea farmers to reduce the impact of this devastating pathogen.

*P. cinnamomi* is sensitive to the suppressive nature of biofumigants. Biofumigants can suppress the *in vitro* growth, sporulation, infective ability of *P. cinnamomi* and reduce the incidence of the disease caused by the pathogen in the glasshouse. Of the two *Brassica* species investigated, *B. juncea* was superior in its ability to control *P. cinnamomi* compared to *B. napus*. When applied in the field, biofumigation using *B. juncea* was found to be more suppressive than soil solarization, but not as effective as metham sodium.

## 16.1.13. ASTER, *CALLISTEPHUS CHINENSIS* AND LISIANTHUS, *LISIANTHUS RUSSELLIANUS*

### *(i) Nematodes*

### (a) Root-knot Nematode, *Meloidogyne incognita*

The root-knot nematodes are a major limiting factor both in nursery and main field (Khan and Parvatha Reddy, 1992).

### Management

**Biofumigation:** Soil biofumigation with green manures of *Brassica juncea* or *Eruca sativa*, and a *Brassica* seed meal commercial pellet, alone or combined with *E. sativa* manure, was compared with treatments of quillay or tagetes formulations, the chemical nematicide fenamiphos and an untreated control for suppression of the root-knot nematode *M. incognita*. In the greenhouse cultivated with *Lisianthus russellianus*, nematode population densities in plots treated with *B. juncea* and *E. sativa* green manures were reduced by 50 - 60% compared to untreated soil and resulted in significantly lower populations than the quillay treatment. Combination of *E. sativa* green manure with a reduced dose of the pellet treatment improved, though not significantly, the nematicidal effect of single green manure or pellet treatments. Moreover, *E. sativa* green manure, alone or combined with pellet, resulted in the lowest root gall indices. Compared to untreated soil, all treatments resulted in a significant increase of stem length and diameter and in a higher number of flowers per stem.

In the following *Aster* crop the highest suppression of *M. incognita* population and the lowest gall formation on *Aster* roots were achieved after *E. sativa* green manure and pellet, alone or in combination, which was also significantly lower than fenamiphos. A consistent but lower reduction of nematode population was found after *B. juncea* green manure. Control plots showed significantly lower stem length and diameter, number of flowers per stem and dry matter compared to soil treated with *E. sativa* and *B. juncea* green manures or pellet (De Mastro *et al.*, 2008a).

## 16.1.14. IRISES

### *(i) Nematodes*

#### (a) Root-knot Nematode, *Meloidogyne* spp.

**Biofumigation:** This greenhouse study evaluated the biofumigant effects of crushed seed meal from brown mustard and field pennycress against the root-knot nematode. Early results revealed a 70 percent to 80 percent nematode decline in pots containing the seed meals.

### *(ii) Weeds*

**(a) Biofumigation:** The biofumigant effects of crushed seed meal from brown mustard and field pennycress on chickweed and prickly lettuce were evaluated under greenhouse conditions. In pennycress-treated pots, about 80 percent fewer chickweed seeds and 55 percent fewer prickly lettuce seeds sprouted. Brown mustard seed meal reduced chickweed emergence by 65 percent and prickly lettuce by half. The irises were unaffected.

## 16.2. MEDICINAL CROPS

In recent times, due to increasing realization of health hazards and toxicity caused by synthetic drugs (side effects), there has been a renewal of interest in the use of plant based drugs throughout the world. A campaign is going on worldwide to utilize more and more plant derived chemicals in the human health care/personal care system. As a result, the demand for plant-based raw materials has increased enormously in both national and international markets. Herbal plants are finding diverse uses as raw materials, not only for medicines but also as biopesticides.

In medicinal plant sector, the WHO has estimated that about 80% of the population in developing countries rely on traditional medicines mostly plant drugs for primary health-care needs. Even modern medicines contain about 25% drugs derived from plants. A large number of people are earning their livelihood from different activities of medicinal plants such as collection, cultivation, marketing, processing, etc. The area under cultivation of the above crops is low, but the cash returns are high because of intensive cropping.

Medicinal crops form one of the important groups due to their demand in various pharmaceutical industries and also to earn foreign exchange by way of export. The potential for earning foreign exchange by India from the exports of medicinal and aromatic plants is estimated over US$ 3,000 million per annum. The export of these plants and their products has a tremendous potential in advanced countries like Europe, USA and Japan. The international market for medicinal plants related trade is estimated at US$ 60 billion/year having a growth rate of 7% per annum. Destination of foreign tourists is now aimed at India for health treatment. Kerala and Gujarat have become their favourite destinations since Ayurvedic treatment is very popular in these two states.

Unless we start cultivating the important medicinal plants, it will not be possible to meet the increased demands. However, cultivation of medicinal plants has to be taken up in polyculture models rather than in monoculture models. There is a huge task to be carried out at every level to make the Indian system of medicine a viable, sustainable and modern. For this task, urgency is felt which can only be established through regular and systematic scientific research on collection, conservation, evaluation and development of good agricultural practices to compete in the world market.

### 16.2.1. COLEUS, *COLEUS FORSKOHLII*

#### *(i) Nematodes*

#### (a) Root-knot Nematode, *Meloidogyne incognita*

The root-knot nematode infestation was reported on coleus from Kerala and Orissa. The dry weight of the tubers was reduced by 20% due to root-knot nematodes. The percentage of starch on fresh weight basis showed drastic reduction (16%) in the infested tubers. Senthamarai *et al.* (2006) reported that *M. incognita* was responsible for 70.2% loss in tuber yield of coleus.

**Symptoms:** The galls on coleus roots are very big and pronounced. The root-knot nematode damage often leads to crop failure. The infested tubers swell in size with irregular surface and cracking of the skin. When the infestation is severe, rotting sets in even before harvest. Infested tubers rot after harvest and rarely reach market.

#### Management

***(i) Trap Cropping:*** Sree Bhadra, a high yielding variety of yam released by Central Tuber Crops Research Institute, Trivandrum, is identified as a resistant trap crop for the root-knot nematode. The nematodes were able to penetrate the root but giant cell formation was not induced and as a result the nematodes die inside the roots. Planting this variety in root-knot nematode infested field helped in clearing the field free of nematodes and giving good tuber yield. Subsequently growing of susceptible crop like coleus escaped nematode damage (Mohandas, 2001).

## Integrated Management

**Solarization, Organic Amendments and Bioagents:** Integration of soil solarization in the nursery for 15 days with 150 gauge LDPE film and application of *Paecilomyces lilacinus* + neem cake or *P. lilacinus* + *Bacillus macerans* in the main field are the best treatments in increasing plant height (64.3 and 60.3 cm compared to 40.0 cm in control), number of leaves (593.3 and 583.3 compared to 310.0 in control), weight of tubers/plant (560.0 and 546.6 g compared to 350.0 g in control), yield (11.5 and 11.3 kg/plot compared to 6.9 kg/plot in control) and in reducing root galls (0.3 and 1.0 compared to 50.6 in control), nematode population in soil (25.0 and 30.0/100 ml soil compared to 196.6/100 ml soil in control) and roots (1.0 and 1.6/5 g roots compared to 79.0/5 g roots in control) (Nisha and Sheela, 2006).

**Organic Amendments, Intercropping and Bioagents:** Integration of dipping stem cutting in 0.1% *Pseudomonas fluorescens* + soil application of neem cake at 400 kg/ha + growing marigold as an intercrop followed by their biomass incorporation during earthing up increased the root tuber yield by 22.7 to 30.0% and reduced the root-knot nematode (*M. incognita*) population by 71.2 to 73.8%. However, integration of *P. fluorescens* + marigold intercrop proved to be more economical (with benefit: cost ratio of 6.4 to 8.8) and effective management practice for the management of root-knot nematodes in medicinal coleus (Seenivasan and Devarajan, 2008) (Table 16.5).

Table 16.5. Effect of integration of bioagents, neem cake and marigold intercrop for the management of *Meloidogyne incognita* infecting Coleus

| Treatment | Gall index (1-5 scale) | | Yield (MT/ha) | | Benefit: cost ratio | |
|---|---|---|---|---|---|---|
| | *Kharif* | *Rabi* | *Kharif* | *Rabi* | *Kharif* | *Rabi* |
| Cutting dip in 0.1% *P. fluorescens* + neem cake at 400 kg/ha | 4.3 | 3.6 | 10.41 | 9.86 | 1.8 | 1.3 |
| *P. lilacinus* at 2.5 kg/ha + neem cake at 400 kg/ha | 4.3 | 3.6 | 10.21 | 9.81 | 1.4 | 1.2 |

| | | | | | | |
|---|---|---|---|---|---|---|
| Cutting dip in 0.1% *P. fluorescens* + marigold intercrop | 2.3 | 2.1 | 12.08 | 11.07 | 8.8 | 6.4 |
| *P. lilacinus* at 2.5 kg/ha + marigold intercrop | 3.6 | 3.3 | 10.92 | 10.20 | 4.6 | 3.2 |
| *P. fluorescens* + neem cake + marigold | 2.3 | 2.1 | 12.11 | 11.12 | 3.4 | 2.5 |
| *P. lilacinus* + neem cake + marigold | 3.6 | 3.3 | 11.06 | 10.26 | 2.0 | 1.4 |
| Carbofuran at 1 kg a. i./ha | 4.3 | 3.6 | 10.36 | 9.85 | 2.1 | 1.6 |
| Control | 4.6 | 4.3 | 9.31 | 9.06 | --- | --- |
| CD (P = 0.05) | 0.36 | 0.36 | 0.83 | 0.80 | --- | --- |

### 16.2.2. ASHWAGANDHA, *WITHANIA SOMNIFERA*

#### *(i) Nematodes*

#### (a) Root-knot Nematode, *Meloidogyne incognita*

The root-knot nematode, *M. incognita* is an important limiting factor in the successful cultivation of ashwagandha.

**Symptoms:** The infected plants were chlorotic, stunted, less branched with fewer and smaller leaves. Roots of such plants were severely galled.

#### Integrated Management

**Organic Amendments and Bioagents:** Integration of neem cake with *T. harzianum,* vermicompost with *T. harzianum*, and cow urine with *T. harzianum* considerably reduced the root-knot nematode development and enhanced plant growth and yield. Maximum root-knot suppression was noticed in vermicompost with *T. harzianum* followed by mentha distillate with *Glomus aggregatum*. Highest increase in plant yield was recorded when the soil was amended with mentha/curry leaf distillates along with *T. harzianum* / *G. aggregatum* (Pandey and Kalra, 2005) (Table 16.6). Vermicompost apart from providing a complete nutrition to the plant may also support the growth of *T. harzianum* thus helping it perforate in the rhizosphere and improve the soil health and thus sustainability to restrict the nematode population.

Table 16.6. Effect of organic materials and bioagents on plant growth of Ashwagandha infected with *Meloidogyne incognita*

| Treatment | Dry weight of plant (g) | Root-knot index (RKI) | Total nema popn (soil + roots) |
|---|---|---|---|
| Untreated – Uninoculated | 8.5 | --- | --- |
| Untreated – Inoculated | 5.2 (-38.8) | 4.00 | 6460 |
| Carbofuran | 8.0 (-5.9) | 1.66 | 3012 |
| Neem comp + Davana | 23.8 (+62.4) | 1.66 | 2820 |
| Neem comp + Curry leaf | 7.4 (-12.9) | 1.33 | 2261 |
| Neem comp + Vermicompost | 12.0 (+41.2) | 1.66 | 2649 |
| Neem comp + *T. harzianum* | 12.8 (50.6) | 1.66 | 2674 |
| Neem comp + *G. aggregatum* | 14.3 (+68.2) | 3.00 | 2964 |
| Neem comp + *Mentha* distillate | 14.2 (67.1) | 1.66 | 2409 |
| Davana + Curry leaf | 10.8(+27.1) | 1.33 | 2030 |
| Davana + Vermicompost | 14.1 (65.9) | 3.00 | 2594 |
| Davana + *T. harzianum* | 9.6 (+12.9) | 1.66 | 2110 |
| Davana + *G. aggregatum* | 10.3 (+21.2) | 1.33 | 2050 |
| Davana + *Mentha* distillate | 13.2 (+55.3) | 1.33 | 2150 |
| Curry leaf + Vermicompost | 14.0 (+64.7) | 1.66 | 2440 |
| Curry leaf + *T. harzianum* | 15.0 (+76.5) | 3.33 | 3252 |
| Curry leaf + *G. aggregatum* | 13.4 (+57.6) | 3.00 | 2635 |
| Curry leaf + *Mentha* distillate | 15.3 (+80.0) | 1.66 | 2404 |
| Vermicompost + *T. harzianum* | 14.4 (+69.4) | 0.66 | 1400 |
| Vermicompost + *G. aggregatum* | 13.6 (+60.0) | 1.99 | 2480 |
| Vermicompost + *Mentha* distillate | 13.9 (+63.5) | 1.33 | 2000 |
| *G. aggregatum* + *Mentha* distillate | 15.6 (+83.5) | 1.00 | 1784 |
| CD (P = 0.05) | 0.71 | 0.01 | 465.70 |

## 16.2.3. SARPAGANDHA, *RAUWOLFIA SERPENTINA*

### *(i) Weeds*

**(a) Intercropping:** Growing of intercrops such as maize, cowpea or brinjal during *kharif* and radish, wheat or a cole crop during *rabi* gave largest overall profits. Intercropping with soybean, garlic or onion gave higher net returns than when grown alone. Even alkaloid content was more when it was intercropped. Suppression of weeds by intercrops could be the reason for enhanced alkaloid yield in sarpagandha (Maheswari *et al*., 1985).

### 16.2.4. OPIUM POPPY, *PAPAVER SOMNIFERUM*

#### *(i) Weeds*

*Phalaris minor, Polypogon* spp. and *Cynodon dactylon* among monocots, and *Anagalis arvensis* and *Heliotropium* spp. among dicots were the predominant weeds of poppy fields.

**(a) Intercropping:** Intercropping of poppy with garlic was more productive than single crop and that latex yield was as good as of poppy grown alone. Intercropping with garlic in black soils is beneficial in controlling weeds (Nigam *et al*., 1986).

### 16.2.5. KACHOLAM, *KAEMPFERIA GALANGA*

#### *(i) Nematodes*

#### (a) Root-knot Nematode, *Meloidogyne incognita*

The root-knot nematode was reported to cause severe damage to the medicinal plant Kacholam (Sheela *et al*., 1996). *M. incognita* was responsible for 64% loss in yield of kacholam (Sheela and Rajani, 1998).

Table 16.7. Effect of mulching of green leaves on root-knot nematodes and yield of kacholam.

| Treatment | Root-knot index | Nema popn. (100g soil) | % redn. over control | Nema popn. (5g root) | % redn. over control | Rhizome yield (kg/4 m$^2$) | % increase over control |
|---|---|---|---|---|---|---|---|
| Neem leaf | 1 | 45.44 (6.74)* | 81.80 | 7.00 (2.76) | 89.60 | 5.6 | 160.7 |
| Glyricidia | 2 | 52.48 (7.24) | 79.05 | 14.67 (2.84) | 78.21 | 3.9 | 79.6 |
| Mangium | 5 | 178.26 (13.35) | 28.50 | 33.67 (5.87) | 49.99 | 2.6 | 21.6 |
| Clerodendron | 2 | 95.35 (9.92) | 60.60 | 22.70 (4.82) | 6.33 | 3.4 | 56.4 |
| Calotropis | 2 | 85.86 (9.27) | 65.60 | 16.00 (4.03) | 76.24 | 3.6 | 69.8 |
| Cromolaena | 1 | 48.96 (6.99) | 80.40 | 9.33 (3.18) | 86.14 | 5.3 | 143.3 |

| | | | | | | | |
|---|---|---|---|---|---|---|---|
| Control | 5 | 249.37 (15.79) | --- | 67.33 (8.26) | --- | 2.2 | --- |
| CD (P=0.05) | --- | (1.02) | --- | (1.45) | --- | 1.28 | --- |

*Figures in parentheses are values after square root transformation.

## Management

**Biofumigation:** Mulching of kacholam with green leaves of neem and chromolaena at 5 kg/m$^2$ 15 days before planting rhizomes, increased the rhizome yield by 160.7% and 143.3% over control, respectively. The above treatments also reduced root galling (1 as compared to 5 in control), the nematode population both in soil (81.8% and 80.4%, respectively) and roots (89.60% and 86.14%, respectively) (Table 16.7).

## 16.3. AROMATIC CROPS

In recent years, there has been an increased interest in the cultivation of aromatic plants to meet the requirements of cosmetic, flavouring and perfumery industries. Aromatic crops are estimated to cover an area of 386,100 ha with production of 325,000 MT during 2008 (Bijay Kumar, 2009) (Table 16.8). The aromatic plants provide raw material for the production of flavours, herbal cosmetics, perfumery etc. Aromatic plants and their products, particularly the essential oils, are now becoming one of the more important export items from many developing countries of Asia. Chemically essential oils are terpenes, which act as carriers of the aromatic substances. Most essential oils also contain camphors, and the more odiferous compounds present in them consist of oxygen derivatives of terpenes, alcohols, esters, aldehydes and ketones. India has enjoyed a prominent position in the manufacture of superior perfumes and aromatics by using essential oils. The important aromatic plants are lemon grass, vetiver, patchouli, palmarosa, citronella, mints, geranium, lavender, basil, jasmine etc.

Table 16.8. State-wise area and production of aromatic crops (2007-08) (Bijay Kumar, 2009)

| State | Area ('000 ha) | Production ('000 MT) | Productivity (MT/ha) |
|---|---|---|---|
| Madhya Pradesh | 19.6 | 117.6 | 6.0 |
| Rajasthan | 198.2 | 94.1 | 2.1 |
| Chhattisgarh | 11.5 | 65.6 | 5.6 |
| Andhra Pradesh | 9.6 | 13.8 | 1.4 |

| | | | |
|---|---|---|---|
| Uttar Pradesh | 133.7 | 13.4 | 0.1 |
| Tamil Nadu | 7.6 | 11.2 | 1.4 |
| Karnataka | 2.0 | 7.2 | 3.6 |
| Punjab | 2.4 | 1.1 | 0.5 |
| Haryana | 1.4 | 1.0 | 0.7 |
| Jharkhand | 0.1 | 0.0 | --- |
| Total | 386.1 | 325.0 | 0.8 |

In the world market, India stands in second position. Most of the essential oils produced though marketed within the country, but a sizeable amount has also been exported. Among the essential oils exported from India are Japanese mint oil, ambrette seed essence, sandal wood oil, citronella oil, lemon grass oil, palmarosa oil, ginger oil, clove oil, eucalyptus oil, vetiver oil, pepper oil, etc. The annual exports of the derivatives of aromatic plants are to the tune of Rs. 600-700 million.

### 16.3.1. MINT, *MENTHA* SPP.

#### *(i) Nematodes*

#### (a) Root-knot Nematodes, *Meloidogyne incognita, M. javanica, M. hapla*

Root-knot disease of Menthol/Japanese mint, Spearmint, Scotch Spearmint, Peppermint and Bergamot mint caused by *M. incognita* and *M. javanica* was observed for the first time in Lucknow and Terai region of Uttar Pradesh which reduce the herbage yield and oil content (Haseeb and Pandey, 1989). *M. incognita* dominates over *M. javanica* in mixed infection. The quality of mint oil was also adversely affected due to nematode infection. The root-knot nematode, *M. incognita* caused 40.2% loss in herbage yield and 46.6% loss in oil yield in menthol mint (Pandey, 2001).

**Symptoms:** Initial symptoms of the disease include occasional yellowing of leaves which in a month time spread to a large portion of the foliage. Growth ceases soon after yellowing. Leaves turn yellow and thin, scorching easily and eventually turning brown. Initially symptoms appear in patches as the reduced plant growth with smaller leaf size and temporary wilting under slight stress of water during hot sun. As the crop grow, especially after first harvest, symptoms become more severe and appears as stunted growth with yellowing of leaves, while veins remain green.

The below-ground symptoms are the large numbers of small sized galls with large egg masses on the roots. Under severe infection, initiation of lateral roots and rootlets on suckers is checked. As a result, uptake of nutrients is inhibited, which in turn produces deficiency symptoms on the aerial portion of the plants (Haseeb and Pandey, 1989).

## Management

**Crop Rotation:** Cultivation of some non-host crops in rotation with Japanese mint, water logging, rice cultivation during rainy season caused reduction in the population of root-knot nematodes. Crop rotation sequences of Menthol mint with maize, potato, mustard, pigeon pea, paddy and wheat were effective in reducing the root-knot nematode population and in increasing the yield (Table 16.9).

Table 16.9. Effective crop rotation sequences for the management of root-knot nematode in Menthol mint.

| Crop rotation sequence | Root-knot index | Net benefit ( in rupees) |
|---|---|---|
| Maize – potato – menthol mint | Mild | 83,000 |
| Paddy – potato – menthol mint | Mild | 80,000 |
| Maize – mustard – menthol mint | Mild | 76,000 |
| Pigeon pea – menthol mint | Mild | 72,000 |
| Paddy – menthol mint | Mild | 75,000 |
| Paddy – wheat – menthol mint | Mild | 68,000 |

**Organic Amendments:** The application of hydro-distillation wastes of *Mentha arvensis, M. piperita, Cymbopogon martini* (palmarosa) and *C. winterianus* (citronella) in the soil (1 month before transplantation) was found effective in reducing the nematode population under field conditions.

Maximum reduction in *M. incognita* population in Menthol mint was recorded in neem cake treated soil followed by mustard cake and *Trichoderma harzianum*. Higher herbage yield (21 MT/ha) was observed in the soil treated with oil cakes in comparison to control (14.66 MT/ha). The crop produced had significantly higher oil content (2.7%) and yield (151 kg/ha) in neem cake treatment followed by mustard cake (2.7% oil and 143.6 kg/ha yield), *T. harzianum* (2.5% oil and 116.4 kg/ha yield) and *Glomus aggregatum* (2.4% oil and 112 kg/ha yield). The untreated crop gave 1.8% oil and 70.4 kg/ha yield (Pandey *et al.*, 1998) (Table 16.10).

Table 16.10. Effect of bio-organics and a pesticide on herb and oil yield of menthol mint and root-knot indices of *Meloidogyne incognita.*

| Treatment | Herbage yield (MT/ha) | % oil content | Oil yield (kg/ha) | Root-knot indices |
|---|---|---|---|---|
| Untreated control | 1.466 | 1.8 | 70.4 | 3.66 |
| Carbofuran | 1.880 (+28.4) | 2.3 | 114.2 (+62.2) | 2.33 (-34.7) |
| Mustard cake | 1.998 (+36.3) | 2.7 | 143.6 (+104.0) | 2.00 (-45.4) |
| Neem cake | 2.100 (+43.2) | 2.7 | 151.0 (+114.5) | 2.00 (-45.4) |
| *Glomus aggregatum* | 1.758 (+19.9) | 2.4 | 112.2 (+59.4) | 2.66 (-27.3) |
| *Trichoderma harzianum* | 1.750 (+19.4) | 2.5 | 116.4 (+65.3) | 2.66 (-27.3) |
| CD (P = 0.05) | 0.284 | 0.5 | 40.4 | 0.1 |

### *(ii) Weeds*

**(a) Mulching:** Mulching with citronella distillation waste and sugarcane leaves can be used to control weeds.

**(b) Crop Rotation:** Crop rotation may also help to maintain a reasonable control on weed growth. Paddy as preceding crop in rotation with mint has been found to minimize weed competition by 30-40%. The following crop rotations may be followed: (i) Mint – Maize – Potato, (ii) Mint – early Paddy – Potato, (iii) Mint – Maize – Rapeseed/ Mustard, (iv) Mint – Paddy.

## 16.3.2. GERANIUM, *PELARGONIUM GRAVEOLENS*

### *(i) Weeds*

**(a) Intercropping:** Black gram, garlic, onion or peas can be taken as intercrops during the initial stages of the crop to keep down the weed population. One row of black gram or pea and 2 rows of garlic or onion are sown between 2 rows of geranium, subsequent to transplanting of geranium.

## 16.3.3. DAVANA, *ARTEMISIA PALLENS*

### *(i) Nematodes*

### (a) Root-knot Nematode, *Meloidogyne incognita*

The root-knot nematode is a major problem in the cultivation of davana. *M. incognita* is responsible for more than 30% reduction in oil yield of davana (Haseeb and Pandey, 1990).

**Symptoms:** The main symptoms were chlorotic and stunted plants with less number of flower buds (which are the major source of essential oil) showing patchy appearance in the field. Their roots were severely galled by root-knot nematodes (Fig. 16.3). One larva / 2 g soil has been found as economic threshold level of *M. incognita* on this crop.

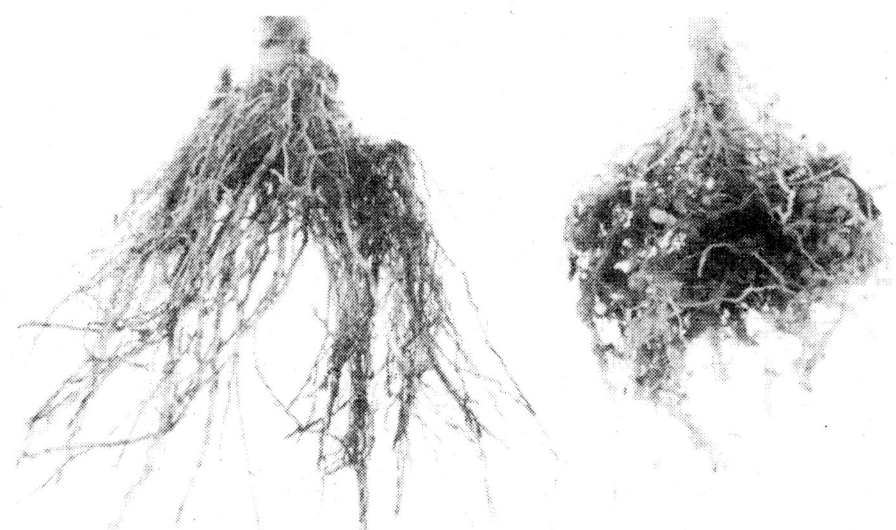

**Fig. 16.3.** Root-knot nematode on davana.
Left – Healthy roots; Right – Infected roots (Courtesy: Haseeb, 1992).

## Management

**Organic Amendments:** A considerable enhancement of herbage biomass, flower buds and oil yields were noticed when davana plants were treated with the distillation waste of palmarosa, citronella, curry leaf, menthol mint and vermicompost of pyrethrum, marigold and menthol mint. Least nematode infections were recorded in distillation waste of curry leaf, palmarosa, lemon grass and vermicompost of pyrethrum, marigold and menthol mint (Pandey, 2005) (Table 16.11).

Table 16.11. Effect of distillation waste and vermicompost on growth and root-knot development of davana infected with *Meloidogyne incognita*

| Distillation waste/ vermicompost | Fresh wt. of herbage & flower buds (g) | Total nematode population (soil + root) | Reproduction factor | Root-knot index |
|---|---|---|---|---|
| Untreated-Uninoculated control | 40.0 | --- | --- | --- |
| Untreated-Inoculated control | 24.7 | 3480 | 3.48 | 4.00 |
| Carbofuran control | 36.8 | 1800 | 1.80 | 2.66 |

| | | | | |
|---|---|---|---|---|
| Lemon grass (Distillation waste) | 41.8 | 1260 | 1.26 | 2.00 |
| Palmarosa (Distillation waste) | 48.8 | 1100 | 1.10 | 1.66 |
| Citronella (Distillation waste) | 44.2 | 1040 | 1.04 | 1.66 |
| Menthol mint (Distillation waste) | 42.7 | 1640 | 1.64 | 2.33 |
| Curry leaf (Distillation waste) | 58.5 | 1020 | 1.02 | 1.00 |
| Geranium (Distillation waste) | 32.3 | 1680 | 1.68 | 2.66 |
| Patchouli (Distillation waste) | 31.3 | 1800 | 1.80 | 2.66 |
| Marigold (Distillation waste) | 41.6 | 1600 | 1.60 | 2.33 |
| Menthol mint (Vermicompost) | 51.3 | 1260 | 1.26 | 2.00 |
| Citronella (Vermicompost) | 51.0 | 1380 | 1.38 | 2.00 |
| Chrysanthemum (Vermicompost) | 44.0 | 1240 | 1.24 | 1.66 |
| Marigold (Vermicompost) | 37.9 | 1040 | 1.04 | 1.66 |
| CD (P=0.05) | 2.176 | --- | --- | 0.118 |

### 16.3.4. LEMON GRASS, *CYMBOPOGAN FLEXUOSUS*

#### *(i) Weeds*

**(a) Organic Amendments:** Distillation waste applied as organic mulch at 3-5 t/ha has been found effective for controlling weeds in the crop. Mulching was the most economical method of weed control and gave good yield of essential oil in lemon grass. There was practically no weed growth in the mulched plots (Majumdar and Kahali, 1986).

### 16.3.5. PALMAROSA, *CYMBOPOGAN MARTINI*

#### *(i) Weeds*

**(a) Mulching:** Mulching the plantation with distillation waste at 5.0 to 7.5 t/ha has been found to be most appropriate for weed management and improving the productivity of palmarosa.

**(b) Intercropping:** Intercropping with black gram is effective in reducing the weed population.

## 16.4. TUBER CROPS

Tuber crops form an important part of the diet in many areas of the tropics and subtropics. They are most important after cereals and grain legumes. They are mostly grown for the carbohydrates stored in underground storage organs viz., roots or rhizomes. In addition, the leaves or leaf stalks of some of the tuber crops are rich in protein, vitamins and minerals and are used for either human consumption or cattle feed. Tuber crops are a major source of food and calories in many tropical countries. The annual tropical root crop production (sweet potato and cassava) is in the range of 10.200 million MT in 0.396 million ha with productivity of 20.5 MT/ha (Table 16.12). Root and tuber crops are predominantly grown in the upland or rain fed areas, mainly by resource poor small and marginal farmers. The major tuber crops include cassava (tapioca), sweet potato, colocasia, amorphophyllus, yams and coleus. Cassava and sweet potato are grown to a larger extent in southern and eastern India. Their cultivation has spread to Maharashtra, Gujarat and North-Eastern states as well. The Salem belt in Tamil Nadu and the Samalkot belt in Andhra Pradesh are known for cassava as an industrial crop. Yams and Colocasia are popular in Andhra Pradesh, Tamil Nadu, West Bengal, Uttar Pradesh and Orissa states.

Table 16.12. Area, production and productivity of tuber crops in India (2007-08) (Bijay Kumar, 2009)

| Tuber crop | Area ('000 ha) | Production ('000 MT) | Productivity (MT/ha) |
|---|---|---|---|
| Sweet potato | 126 | 1146 | 9.1 |
| Cassava (Tapioca) | 270 | 9054 | 33.5 |
| Total | 396 | 10200 | 20.5 |

Tuber crops, especially cassava can be successfully substituted as alternative source of feed for livestock in both tropical and subtropical countries. Cassava has also been used as a substrate for the production of ethyl alcohol. The tropical root crops, in general, have a great potential in meeting basic food and energy needs of the developing world. Tuber crops are very efficient in solar energy transfer, having a clear superiority over cereals in biological efficiency and are comparatively free from severe pests and diseases. Cassava and sweet potatoes account for about 30% of the total production of root crops from developing countries. The potential of root crops as nutritionally rich sources of β-carotene, anti-oxidants, dietary fibre and minerals like calcium have begun to be recognized as a result of the multifarious research programmes worldwide.

## 16.4.1. CASSAVA, *MANIHOT ESCULENTA*

### *(i) Nematodes*

### (a) Root-Knot Nematodes, *Meloidogyne* spp.

*Meloidogyne incognita, M. javanica* and *M. arenaria* have been reported from cassava. Nirula (1963) reported *M. javanica* for the first time from India. Attack by root-knot nematodes was considered serious when cassava was intercropped with other susceptible hosts like brinjal and roselle. Jatala (1988) reported considerable yield reduction in cassava due to *M. incognita* and *M. javanica* infestation. *M. incognita* race 2 reduced fresh storage root weight by 49% and dry top weight by 22%.

**Symptoms:** Root-knot nematodes produce the typical swelling or knotting of cassava roots. *M. incognita* race 2 and *M. javanica* significantly reduced stalk length, stalk weight and storage root weight of cassava after 15.5 months growing period (Caveness, 1981).

## Management

**Summer Fallowing:** The summer fallowing which is already being practiced by farmers in many parts of cassava growing areas in Kerala due to non-availability of water is an excellent means of controlling nematode population.

**Organic Amendments:** Cassareep, a by-product of the cassava industry, was apparently effective in controlling *M. incognita* and *M. javanica* on cassava.

### *(ii) Weeds*

**(a) Cover Cropping:** Early season weed suppression with the low growing 'Egusi' melon (*Citrullus lanatus*) followed by sweet potato gave yields of maize, cassava and yam equal to those from 3 hand weedings. This method of weed control also protects the soil from erosion in high rainfall areas.

Perennial legume green covers gave effective and long lasting weed control than mulches (Leihner, 1980).

**(b) Intercropping:** Intercropping with legumes (cowpea, field bean and black gram) reduced the weed intensity in cassava. Intercropping saved 77% of weeding cost, besides obtaining additional income. Intercropping with cowpea and 'Egusi' melon which suppressed weed growth gave highest economic returns under cassava (Unamma *et al.*, 1986). The reduction in cassava yield due to bean competition (20%) was offset by dry bean yield.

**(c) Mulching:** Mulching with maize stalks, sugarcane leaves or *Stylosanthes* straw controlled weeds.

## 16.4.2. YAM, *DIOSCOREA* SPP.

### *(i) Nematodes*

### (a) Root-Knot Nematode, *Meloidogyne* spp.

**Symptoms:** Yam seedlings infested with root-knot nematodes can be severely stunted and chlorotic; very young seedlings may be killed by severe infestation. The roots swell in size at the site of infection producing the characteristic and typical knots on the roots. Tuber formation was severely affected, often infested tubers are smaller and most of the tuber forming roots does not develop. Rotting of tubers has been reported.

### Management

**Trap Cropping:** The root-knot nematode can be controlled by growing sweet potato cv. Shree Bhadra as a trap crop.

### (b) Yam Nematode, *Scutellonema bradys*

The most important nematode problem on yams is the dry rot of tubers caused by *S. bradys*. Nadakal and Thomas (1967) reported the occurrence of dry rot of *Dioscorea alata* and the association of *S. bradys* from Kerala. *Dioscorea alata, D. cayenensis, D. esculenta* and *D. rotundata* are good hosts of the yam nematode. Sesame and cowpea support high root populations and melon can increase soil populations.

**Economic Importance:** Dry rot of yams alone causes a marked reduction in the quality, marketable value and edible portions of tubers, and these reductions are more severe in stored yams. When dry rot is followed by wet rot in stored yams, losses of whole tubers can be as high as 80-100%. Weight differences between healthy and diseased tubers harvested from the field have been estimated to be 20-30%.

**Symptoms:** *S. bradys* causes a disease of tubers commonly known as 'dry rot'. In early stages of infection, small yellow lesions can be observed just below the skin; these become dark brown and eventually form a continuous dark, dry rot layer just beneath the surface in heavily infested tubers. Outwardly the infected tubers have cracks in the skin, which are often deep, can be malformed, or the skin flakes off exposing dark brown, dry rot tissue underneath. *S. bradys* reproduces and builds up large populations in stored yam tubers and causes severe damage during storage.

Nematode damage to the tubers results in considerable reduction in the edible portion and marketable value of the tubers. Moisture loss is also significantly greater from diseased tubers during storage.

## Management

**Crop Rotation:** Soil populations of *S. bradys* will be reduced if a non-host or poor-host crop, such as peanut, chilli pepper, tobacco, Indian spinach, cotton, maize or sorghum are grown prior to yams.

## 16.4.3. ELEPHANT FOOT YAM, *AMORPHOPHALLUS CAMPANUTATUS*

### *(i) Diseases*

### (a) Storage rot, *Sclerotium rolfsii*

**Mulching:** Mulching with paddy straw/polythene sheet reduce collar rot.

*Chapter 17*

# PLANTATION AND SPICE CROPS

## 17.1. PLANTATION CROPS

Plantation crops serve a variety of human needs such as food, oil, industrial raw materials, beverages and confectionary items. They generate huge employment opportunities directly or indirectly to several million people in their production, processing, marketing and international trade sectors. Plantation crops are a large group of crops. These are grown over an area of 3.226 million ha with production of 12.045 million MT in India (Bijay Kumar, 2009) (Tables 17.1 and 17.2). The plantation crops contributed 5.82% of total horticulture production, in 2007-2008. India is the largest producer and consumer of areca nut, cashew nut, tea, third largest producer of coconut, fourth largest producer and consumer of rubber and sixth largest producer of coffee in the world. While the major plantation crops include coconut, areca nut, oil palm, cashew, coffee, tea and rubber, minor plantation crops are cocoa and betel vine. They play an important role in export as well as domestic requirement, employment generation and poverty alleviation particularly in rural sector. The annual income generated is more than Rs.5,52,590 million and export earning is approximately Rs.93,350 millions.

Table 17.1. Crop-wise area and Production of plantation crops in India (2007-08) (Bijay Kumar, 2009)

| Plantation Crops | Area ('000 ha) | Production ('000 MT) | Productivity (MT/ha) | % Share of India production |
|---|---|---|---|---|
| Coconut | 1939.9 | 19893.9 | 5.6 | 90.4 |
| Cashew nut | 868.0 | 665.0 | 0.8 | 5.5 |
| Areca nut | 386.6 | 476.0 | 1.2 | 4.0 |
| Cocoa | 31.8 | 10.6 | 0.3 | 0.1 |
| Total | 3226.4 | 12045.5 | 3.7 | |

Table 17.2. State-wise area, production and productivity of plantation crops in India (2007-08) (Bijay Kumar, 2009)

| State | Area in '000 ha | Production in '000 MT | Productivity in MT/ha | % Share of India production |
|---|---|---|---|---|
| Kerala | 1073.7 | 43.59.9 | 4.0 | 36.2 |
| Tamil Nadu | 503.9 | 3810.6 | 7.5 | 31.6 |
| Karnataka | 679.2 | 1401.0 | 2.0 | 11.6 |
| Andhra Pradesh | 289.0 | 1021.2 | 3.5 | 8.5 |
| Maharashtra | 190.2 | 334.1 | 1.7 | 2.8 |
| Orissa | 182.0 | 279.8 | 1.5 | 2.3 |
| West Bengal | 44.1 | 279.1 | 6.3 | 2.3 |
| Assam | 104.3 | 186.1 | 1.7 | 1.5 |
| Goa | 82.2 | 120.8 | 1.4 | 1.0 |
| Gujarat | 20.4 | 99.2 | 4.8 | 0.8 |
| Andaman & Nicobar | 25.5 | 66.9 | 2.6 | 0.6 |
| Others | 71.8 | 86.7 | 1.2 | 0.7 |
| Total | 3226.3 | 12045.5 | --- | |

There is a good scope for switching over the production of plantation crops viz., coffee, tea, coconut and areca nut to organic farming, as it is possible to realize higher returns from the unit quantity exported when grown organically. The plantation crops are generally grown in ecologically fragile hilly tracts, adopting organic farming methods would protect environment and also prevent contamination thereby producing products of high nutritional quality. Further, the plantation crops are amenable for organic farming as they produce huge amount of waste biomass for recycling, which can meet major portion of nutrients required.

## Export of Plantation Produce

India is the largest producer in the world of cashew nuts, coconuts and tea. India is largest producer, importer, processor, exporter and second largest consumer of cashew. Indian exports during 2006-2007 were US $ 866 million in tea and coffee (Table 17.3).

Table 17.3. Export of plantation crops (in US $)

| Commodity | Apr-Mar 2006 | Apr-Mar 2007 | % Growth | % Share |
|---|---|---|---|---|
| Tea | 390.92 | 432.00 | 10.51 | 0.34 |
| Coffee | 358.83 | 434.77 | 21.16 | 0.34 |
| Cashew | 585.76 | 553.41 | -5.52 | 0.44 |

India is the largest producer and consumer of tea in the world. During 2006-07, India produced 0.95 million MT of tea of which 0.218 million MT was exported. Tea accounts for Rs. 80 billion in turnover in India.

India produced 0.29 million MT of coffee accounting for 4% of the global production. Over 80% of the coffee produced in India is exported.

India is the fourth largest producer of natural rubber with a share of 8.8% in world production in 2006. The yield/ha in India at 1879 kg/ha, was amongst the highest in the world. India produced 0.852 million MT of natural rubber during 2006-07 and exported 56545 MT.

## 17.1.1. COCONUT, *COCOS NUCIFERA*

### *(i) Nematodes*

#### (a) Burrowing Nematode, *Radopholus similis*

The burrowing nematode was reported from coconut palms in Kerala, India by Weischer (1967). *R. similis* is the most important nematode pest of coconut and is responsible for considerable amount of root rotting (Koshy *et al.*, 1978). It causes 30% yield loss in coconut (Koshy and Geetha, 1992). The threshold inoculum density required to cause significant reduction in various growth parameters of coconut is 100 nematodes/625 ml sandy loam soil over a period of 5 years under field conditions (Koshy, 1986).

**Symptoms:** The burrowing nematode infested coconut palms exhibit general decline symptoms like yellowing, button shedding, reduction in number and size of leaves and leaflets, delay in flowering and reduced yield which are non-specific. Symptoms on the root are more specific. *R. similis* on infestation produces isolated elongate orange coloured lesions on tender and semi-hard roots. Consequent to nematode parasitization and multiplication, these lesions enlarge and coalesce to cause extensive rotting of roots. Tender roots on heavy infestation become spongy in texture. On semi-hard orange coloured roots, surface cracks are commonly seen. As high as 4,000 nematodes were recovered from one gram (one inch length) of main roots. The drastic reduction in the number and mass of tertiary feeder roots on parasitization by the nematode limits plant growth (Koshy and Sosamma, 1987).

**Management**

**Green Manure Cropping:** *Crotalaria juncea* may be grown as a green manure crop in the basins and interspaces and its incorporation helps in the inhibition of nematode multiplication.

**Intercropping:** Growing of intercrops like cocoa that enriches the soil with sizeable quantities of shed foliage which help in the build up of beneficial organisms which may inhibit nematode multiplication.

**Organic Amendments:** Application of oil cakes, FYM and green foliage to the basins, which help in the build up of beneficial organisms which may inhibit nematode multiplication.

Application of 50 kg cow dung/FYM, 2 kg neem/marotti cake and 25 kg green manure to the coconut basins in November is quite effective in bringing down the nematode population. Thirty per cent increase in yield has been recorded by application of *Hydnocarpus* sp oil cake at 8 kg/palm/year in June-July and October-November to the burrowing nematode infested coconut palms (Koshy, 1986).

***(ii) Weeds***

**(a) Mulching:** Mulching coconut basins with mechanically removed or hand weeded material will help not only conserve soil moisture but also serves as source of organic matter improving soil health. Highest survival of coconut seedlings (86.7%) was obtained with coir pith mulch laid 10 cm thick. Thick mulch of green or dry leaves in planting pits of young palms reduce weed growth considerably.

**(b) Intercropping:** Since the interspaces of coconut are to be utilized by adopting inter/mixed cropping system, weeding in such coconut plantations do not arise. Growing cover crops such as *Calapogonium mucunoides* (legume) in coconut gardens suppressed the weeds very effectively within 3 months when grown as a green manure – cum cover crop. This method of weed control is more practical and economical than other methods (Albuquerque and Mohamed Ibrahim, 1956). Cover crops like *Mimosa invisa* and *Stylosanthes* aid in suppressing weed growth due to fast growing nature. Growing competitive crops like sunflower, soybean, green or black gram, cowpea and other legumes had a suppressing effect on weeds in coconut gardens. Intercropping and mixed cropping of coconut gardens with arecanut, ginger, turmeric, dioscorea, banana, tapioca, cocoa and cinnamon was found useful for keeping that coconut plantation free from weeds to a great extent. All weeds except *Rottboellia exaltata* were suppressed by winged bean (*Psophocarpus tetragonolobus*) when it was fully established 4 months after sowing in coconut gardens.

**(c) Green Manuring:** Green manuring crops like *Indigofera, Tephrosia* and sunnhemp grow fast and provide organic manure besides competing with weeds.

## 17.1.2. ARECANUT, *ARECA CATECHU*

### *(i) Weeds*

**(a) Green Manuring:** The raising of green manure-cum cover crops such as *Calapogonium mucunoides, Mimosa invisa* and *Stylosanthes gracilis* was found to be advantageous. Growing of green manure crops on the onset of monsoon will help to suppress the weed growth, prevent soil erosion and add large quantities of organic matter to the soil.

**(b) Intercropping:** Intercrops such as elephant foot yam, black pepper, arrowroot (*Manihot* spp.) and banana, and mixed cropping with cocoa, pepper and betel vine can be taken up to increase the income of growers without any detrimental effect to the yield of main crop (Singh *et al.*, 1982).

**(c) Cover cropping:** Cover crops like guinea grass can also be grown to suppress the weeds.

## 17.1.3. COFFEE, *COFFEA* SPP.

### *(i) Nematodes*

#### (a) Root-knot Nematodes, *Meloidogyne exigua, M. coffeicola*

**Economic Importance:** A 5% loss in yield due to *M. coffeicola* damage has been reported in Hawaiin coffee plantations. The yield of non-infected plants was more than that of artificially inoculated plants (104 and 50 kg, respectively). Losses due to *Meloidogyne* spp. in Brazil are estimated at 15% of that country's production

**Symptoms:** Affected trees look unthrifty and have chlorotic, wilting and dying leaves. *M. exigua* produces small elongated galls that are located primarily at the root tips (Fig. 17.1). In addition, roots may show cracks, necrosis and sloughing of cortical tissue. The effectiveness of root system is greatly diminished, with rootlets and root hairs practically eliminated. Trees may die, particularly during periods of drought and cold temperatures. Affected plants lack vigour and are unable to withstand adverse conditions.

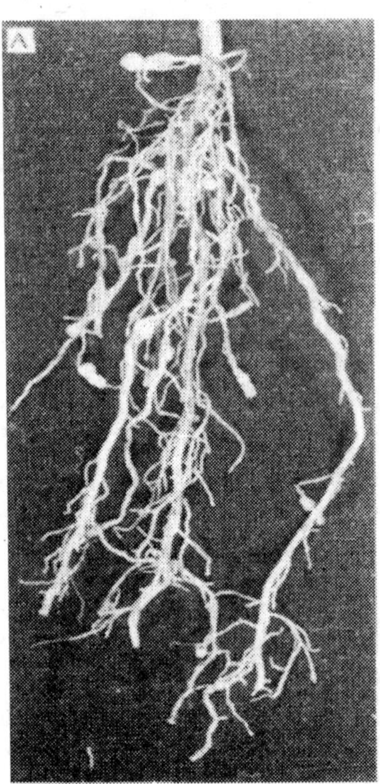

**Fig. 17.1.** Arabica coffee roots infected with *Meloidogyne exigua.*

## Management

**Cover Cropping:** The decision of growing forage crops and turning back again the area to coffee will certainly reduce drastically the population of *Meloidogyne* species in 6-12 months.

**Crop Rotation:** Rotation of coffee with cotton, soybean and maize in *M. exigua* infested areas for one year significantly reduced the nematode population, after which the farmer can return to coffee cultivation. Coffee can also be rotated with groundnut, castor (both immune), *Styzolobium deeringiatum* and *Crotalaria spectabilis* (both resistant).

## (b) Reniform Nematode, *Rotylenchulus reniformis*

The reniform nematode has been observed on coffee in India, Puerto Rico, Brazil and Hawaii.

**Symptoms:** *R. reniformis* infection of coffee trees results in an almost complete absence of feeder roots, a poorly developed tap root, and the yellowing and wilting of above-ground parts. In soils heavily infested with reniform nematode, Arabica coffee cannot be grown.

### Management

**Crop Rotation:** In Puerto Rico, crop rotation with Pangola grass and sugarcane have effectively controlled *R. reniformis*. Rotation with sugarcane has also provided effective control in Brazil.

### *(ii) Weeds*

**(a) Green Manuring:** Cultivation of green manure crops/cover crops/grain legume crops would help in smothering of weeds. Growing cover crops like lima bean, cowpea, soybean, groundnut and pigeon pea are preferred which reduce the density of weeds.

**(b) Mulching:** Soil must be mulched properly for good growth of coffee. Mulching with weed slashings and shade tree leaf litter, etc. would help in smothering of weeds.

## 17.1.4. TEA, *CAMELLIA SINENSIS*

### (i) Diseases

### (a) Black Root, *Rosellina arcuata*

The black root disease generally occurs in soils that are rich in organic litter, in cool or cold habitat. It is a major problem in southern India at elevations of 1500 m above MSL and Darjeeling (West Bengal). Occasionally, it is also noticed at mid and low elevations.

**Symptoms:** *Rosellina arcuata* produces black mycelium on affected roots and white star shaped mycelium between bark and wood. The fungal mycelium is cob-webby in appearance, whitish grey or black in colour, spreading rapidly on the surface of the soil on organic debris. When the mycelium comes in contact with tea roots, it penetrates the bark and produces typical star-shaped mycelial structures on the surface of the wood beneath the bark. Very rarely black spherical perithecia are seen at the collar region of the plant. Apart from tea, the disease also affects certain green manure and shade plants commonly grown in tea plantations.

### Management

**Cover Cropping and Bioagents:** Rehabilitating the soil with Guatemala grass and use of biocontrol agents at 200 g/plant of *T. viride* effectively check the disease.

### (b) Brown Root, *Fomes noxius*

**Symptoms:** The brown root is the most common root disease of tea. This pathogen mainly spreads by root contact. Hence, the disease occurs in distinct concentric patches in tea fields. The fungal mycelium is fawny in colour. *Fomes noxius* affected roots become soft and spongy. The infected root disintegrates in advanced stages of the disease and becomes spongy. When such roots are split longitudinally, typical brownish honeycomb-like reticulations are seen. The surface of the root is heavily encrusted with soil. The fructifications are typical bracket- shaped, but occur rarely.

#### Management

**Cover Cropping and Bioagents:** Rehabilitating the soil with Guatemala grass and use of biocontrol agents at 200 g/plant of *Trichoderma harzianum / T. viride / T. hamatum / T. resei / T. koningii / T. virens* effectively check the disease.

### (c) Red Root, *Poria hypolateritia*

**Symptoms:** The red root disease is extremely common in all tea areas of southern India. *Poria hypolateritia* produces blood red mycelium on affected roots. The fungal mycelium is initially white, then turns pink to pale red and in later stages it is dark red, almost black in colour. The mycelium is generally present as a sheet on the root surface and the typical red colour is invariably seen, when the root is rubbed vigorously under running water and held to bright light. The disease spreads both by root contact as well as by free mycelial spread to a limited extent. Roots of the affected bushes are encrusted with soil. In advanced stages, the affected root becomes spongy. It takes almost 5 years for a mature bush to succumb following the attack.

#### Management

**Cover Cropping and Bioagents:** Rehabilitating the soil with Guatemala grass and use of biocontrol agents at 200 g/plant of *Trichoderma harzianum / T. virens* effectively check the disease.

### (d) Root Splitting, *Armillaria mellea*

**Symptoms:** The root splitting disease occurs in isolated areas, mainly in South India. It derives its name from longitudinal fissures observed on the bark of the invaded tea roots. It can also be observed readily by the characteristic black rhizomorphs which resemble a shoe lace and are seen in large numbers. The fungal mycelium which is white in colour grows densely under the bark and due to the pressure the bark splits. The fungus invades the host plant either by root contact or rhizomorphs which are capable of traversing long distances through the soil from the food base. The fruiting body is typical mushroom-like basidiocarp, and although it is rarely recorded. When it occurs, it is found in abundance around the collar of the diseased plant.

### Management

**Cover Cropping and Bioagents:** Rehabilitating the soil with Guatemala grass and use of biocontrol agents at 200 g/plant of *Trichoderma harzianum* / *T. virens* effectively check the disease.

**Solarization and Bioagents:** Talam and Otieno (2002) reported that soil solarization and application of *T. harzianum* were effective in controlling root splitting disease caused by *A. mellea*. They also found that the control achieved with *T. harzianum* was excellent when applied after soil solarization.

### *(ii) Nematodes*

### (a) Root-knot Nematodes, *Meloidogyne incognita* (in nurseries), *M. brevicauda* (in field)

### Symptoms

**On Young Tea:** The first report of root-knot nematode in young tea was from South India, where large number of seedlings were found infected (Barber, 1901). *Meloidogyne incognita, M. javanica* and *M. hapla* have been found associated with tea in India.

Young nursery plants, both seedlings and vegetatively propagated clonal plants, are severely damaged by root-knot nematodes. Seedling plants in which both the tap-root and lateral rots are severely attacked suffer greater damage than do clonal tea plants of equivalent age, probably because seedling tea plants possess less than half the root bulk of clonal plants of similar age. A marked increase in resistance is observed between 8 and 15 months.

**On Mature Tea:** *M. brevicauda* is the only root-knot nematode that is a serious pest of mature tea. Affected bushes are unthrifty, have severely galled main roots and lack feeder roots. *M. brevicauda* occurs in southern India at altitudes of 3000 to 3500 ft. in the Wynaad area (Venkata Ram, 1963). It can attack both seedlings and mature tea.

Root-knot nematodes (*M. brevicauda*) cause severe galling on the roots of mature tea bushes with resultant decline (Fig. 17.2). Tea plants of all ages are susceptible to infection by this nematode which has been recorded in only South India and Sri Lanka.

**Fig. 17.2.** Heavy root galling of tea with *Meloidogyne brevicauda* (Sivapalan, 1972).

The infected bushes grow more slowly and the leaves tend to become smaller, dull and yellow in appearance. The roots show the characteristic swelling and pitting that is typical of *M. brevicauda.*

The effect of infection on the growth of tea bushes is most evident during the period of recovery from pruning. Healthy bushes are usually harvested about 16 weeks after pruning, but the harvesting of severely infected bushes is necessarily delayed up to a further 6 months. The production of new branches by infected bushes is very restricted, and as a consequence the cropping capacity of the bush is reduced. The most severely affected bushes fail to recover from pruning.

### Management in Nurseries

**Solarization:** Soil solarization in nurseries is adequate to manage root-knot and other nematodes as the plants develop complete resistance by about 9-15 months (Gnanapragasam *et al.*, 1989).

**Organic Amendments:** Incorporation of organic matter into nematode infested soil is known to suppress the incidence of root-knot nematodes on tea.

**Intercropping:** The cultivation of marigold as a pre-plant crop considerably reduces the population of *Meloidogyne* spp. The reduction in the nematode population is significantly greater than when the land is left fallow. The interplanting of marigold in tea has resulted in a 7% increase in yield compared to 10% increase that followed soil fumigation with DBCP.

**Cover Cropping and Bioagents:** The rehabilitation crops such as Mana grass (*Cymbopogon confertiflorus*) or Guatemala grass (*Tripsacum laxum*) (which help to improve the soil) are resistant to *Meloidogyne* spp. The soil population of parasitic nematodes decline rapidly when these grasses are grown for one or more years before replanting tea.

### Management under Field Conditions

**Cover Cropping:** In Sri Lanka, uprooted old tea lands are routinely reconditioned by planting deep-rooted grass species including Guatemala grass and Mana grass which are non-hosts to parasitic nematodes and bring down the residual nematode population (Kerr and Vythilingam, 1966).

**Intercropping:** Plants such as marigolds, *Tagetes erecta* and *T. patula*, *Vetiveria zizanoides* and *Eragrostis curvula* have been found to help reduce nematode populations.

### (b) Lesion Nematode, *Pratylenchus coffeae, P. loosi*

The most destructive and most intensively investigated nematode pest of tea is the root lesion nematode *P. loosi*, which is equally destructive to both young and mature tea plants and is a problem in nurseries and in mature tea fields. Crop loss in tea infested with *P. loosi* has been estimated to average between 225-350 kg made tea/ ha/ year (Sivapalan, 1972). Sasser (1989) estimated the average crop loss on a worldwide scale to be 8.2%.

**Symptoms:** The infested bushes show unthrifty tea with weak maintenance foliage. The leaves appear pale yellowish and dull mainly due to nutritional imbalances caused by altered rate of uptake of essential nutrients. The heavily infested plants show tendency for premature flowering and fruiting.

The roots of infested plants show typical marked reduction in the feeder root system. The feeder roots appear brownish in colour and dark brown necrotic patches appear on both the feeder roots, as well as the storage roots. Under greenhouse conditions, the damage threshold level of *P. loosi* to tea was estimated tobe 40 nematodes/100 g soil at 24°C in a clayey loam soil (Gnanapragasam & Manuelpillai, 1984).

**Management**

**Green Manuring:** The incorporation of large quantities of green manure such as loppings from dadaps, *Tephrosias* and marigolds, resulted in population decline of *P. loosi*. Mulching with Guatemala grass at 20 to 30 t/ha gave effective control of *P. coffeae* on tea.

**Intercropping:** The cultivation of marigold (*Tagetes erecta* and *T. patula*) as a pre-plant crop considerably reduces the population of *P. loosi*. The interplanting of marigold in tea has resulted in a 7% increase in yield. *P. loosi* is attracted to marigold roots, but upon entry the nematodes die within a week. Roots of marigold in the vegetative stage are effective in this way, but roots of plants in the full bloom or post-bloom are less effective. Consequently, it is preferable to maintain the plants in a non-flowering state by periodic lopping. Growing of grasses like Guatemala (*Tripsacum laxum*), Manna (*Cymbopogon confertiflorus*) and *Eragrostis curvula* helps to bring down the population of *P. loosi*.

***(iii) Weeds***

*Mikania micrantha* is a serious weed of tea plantations.

**(a) Cover Cropping:** Weeds are controlled by cultivation of green manure crops/cover crops/grain legumes.

**(b) Mulching:** Mulching helps in conservation of soil moisture, suppression of weeds, induction of feeder root proliferation and addition of organic matter to the soil. Weeds are controlled by mulching with weed slashing, pruning litter and shade litter.

### 17.1.5. BETEL VINE, *PIPER BETEL*

***(i) Nematodes***

**(a) Root-knot Nematode, *Meloidogyne incognita***

*M. incognita* was responsible for 21-38% loss in leaf yield of betel vine (Saikia, 1992; Jonathan *et al.*, 1990).

**Symptoms:** Blackening and drooping of the growing tip and yellowing of leaves occurs due to root-knot nematode infestation. Galls of varying sizes and shapes develop in roots leading to quick root decay.

## Management

**Solarization:** White polyethylene mulching for 15 days on beds prepared for planting betel vine revealed high reduction in nematode populations due to increased soil temperature of 44.1°C (Sivakumar and Marimuthu, 1987).

Summer ploughing, summer ploughing and solarization have been found beneficial.

**Organic Amendments:** Application of FYM at 25 t/ha or neem cake at 1 t/ha before planting helps to reduce the nematode population in soil.

Soil amendment with sawdust at 2 t/ha + NPK (3638 leaves) and neem cake at 2 t/ha is effective in reducing nematode numbers and increasing yields (Sivakumar and Marimuthu, 1986).

**Crop Rotation:** A pre-rice crop reduces populations of *M. incognita, Rotylenchulus reniformis* and *Helicotylenchus incisus* in the subsequent betel vine crop. Crop rotation with non-host crops like marigold, sesame and groundnut also suppress root-knot nematode populations. Support plants such as *Erythrina indica, Moringa oleifera* and *Bombax malbaricum* which are not susceptible to root-knot nematodes can be used in betel vine gardens.

**Intercropping:** In pots having 5 seedlings of *Tagetes erecta* around a betel vine plant, the population of *M. incognita* was reduced by 41.4% and root galling by 54% (Medhana *et al.*, 1985).

**Biofumigation:** Green leaf manuring with African marigold, neem or *Xanthium* have also been quite effective against root-knot, lesion and other nematodes. Significant reduction (60%) in the nematode population has been observed in beds amended with chopped and shade dried leaves of *Calotropis gigantea* at 2.5 t/ha followed by neem oil cake and poultry manure (44.4 and 40.9%, respectively). Beds amended with *C. gigantea* leaves yielded 14.2 kg/4840 leaves and with neem oil cake yielded 12.1 kg/4220 leaves.

**Organic Amendments and Bioagents:** *Trichoderma viride* mixed with mustard oil cake at 100 g/10 kg cake applied in three split doses reduced the nematode infestation and enhanced the leaf yield.

### 17.1.6. RUBBER, *HEVEA BRASILIENSIS*

#### *(i) Weeds*

*Imperta cylindrica, Chromolaena odorata* and *Mikania cordata* are some of the serious weeds of rubber plantations.

**(a) Cover Cropping:** The general practice to control weeds in rubber is by growing cover crops. Legume cover crops like *Calapogonium mucunoides, Pueraria phaseoloides* and *Centrosema pubescens* are grown in between rows of young rubber trees (Mangoensoekarjo and Kadnan, 1978; Vernon, 1981).

### 17.1.7. OIL PALM, *ELAEIS GUINEENSIS*

#### *(i) Weeds*

**(a) Cover Cropping:** Cover cropping with *Centrosema pubescens* and *Calapogonium caeruleum* in oil palm plantations helps in reducing the incidence of weeds.

## 17.2. SPICE CROPS

From time immemorial, India is regarded by the rest of the world as the "magic land of spices". Spices are natural plant products used to improve flavour, aroma, taste and colour of food products, beverages, liquors, pharmaceutical, cosmetic and perfumery products. No other country in the world has such a diverse variety of spice crops as in India. Indian spices are also noted for their excellent aroma, flavour and pungency not so easily matched by any other country. Even in minute quantities, spices are a real delight to all our sensory organs, making the food dishes more palatable. Usage of spices in food processing industries is increasing quite fast in all the countries. Hence, the world demand for spices in recent years is continuously on the rise.

India is the largest producer, consumer and exporter of spices in the world. India is the largest producer in the world of ginger, turmeric and black pepper. The annual production of spices is 4.103 million MT valued at US$ 1,500 million. India's share of the world spices trade is estimated at 45 to 50% by volume and 25% by value. The present annual production of spices in the country is 4.103 million MT from over 2.603 million hectares (Bijay Kumar, 2009) (Tables 17.4 and 17.5). The spices contributed 1.98% of total horticulture production in 2007-2008. About 8.5% of India's export earnings from agricultural and allied products come from spices. Indian spices flavour foods in over 134 countries. The important spice crops include black pepper, cardamom, ginger, garlic, turmeric and chilli.

Table 17.4. Crop-wise area and Production of spice crops in India (2007-08) (Bijay Kumar, 2009)

| Spice crop | Area in '000 ha | Production in '000 MT | Productivity in MT/ha | % Share of India production |
|---|---|---|---|---|
| Chilli | 765 | 1244 | 1.6 | 30 |
| Turmeric | 180 | 830 | 4.6 | 20 |
| Garlic | 168 | 810 | 4.8 | 20 |
| Ginger | 106 | 376 | 3.5 | 9 |
| Coriander | 398 | 243 | 0.6 | 6 |
| Tamarind | 60 | 192 | 3.2 | 5 |
| Cumin | 429 | 172 | 0.4 | 4 |
| Black pepper | 242 | 68 | 0.3 | 2 |
| Fennel | 49 | 62 | 1.3 | 2 |
| Fenugreek | 55 | 55 | 1.0 | 1 |
| Others | 152 | 51 | 0.3 | 1 |
| Total | 2603 | 4103 | 1.6 | |

Table 17.5. State-wise area, production and productivity of major spices in India (2007-08) (Bijay Kumar, 2009)

| State | Area in '000 ha | Production in '000 MT | Productivity in MT/ha | % Share of India production |
|---|---|---|---|---|
| Andhra Pradesh | 317.8 | 1235.2 | 3.9 | 30.1 |
| Rajasthan | 556.4 | 520.6 | 0.9 | 12.7 |
| Gujarat | 299.8 | 356.8 | 1.2 | 8.7 |
| Karnataka | 235.2 | 344.9 | 1.5 | 8.4 |
| Tamil Nadu | 126.8 | 279.2 | 2.2 | 6.8 |
| Madhya Pradesh | 194.6 | 249.8 | 1.3 | 6.1 |
| Orissa | 147.0 | 199.2 | 1.4 | 4.9 |
| Uttar Pradesh | 56.8 | 166.9 | 2.9 | 4.1 |
| Kerala | 312.3 | 159.3 | 0.5 | 3.9 |
| West Bengal | 79.0 | 114.4 | 1.4 | 2.8 |
| Maharashtra | 114.3 | 100.2 | 0.9 | 2.4 |
| Meghalaya | 18.4 | 80.9 | 4.4 | 2.0 |
| Others | 144.6 | 295.3 | 2.0 | 7.2 |
| Total | 2603.0 | 4102.7 | 1.6 | |

Since spices form a part of traditional/ethnic medicines, the demand for organically produced spices is growing. Export of organic spices from India has started in right earnest. The country at present exports around 50 MT of different varieties of organic spices. Export will get a significant boost in the coming years as more and more farmers switch to organic methods.

**Export of Spice Produce**

India is the largest producer, consumer and exporter of spices and spice products in the world and produces more than 50 spices. Spice export figures for the year 1996-97 and 2006-07 shows the phenomenal growth of the Indian spice industry. The Indian spice export for the year 1996-97 was 225,000 MT which was worth $350 million; in the year 2007-08 the spice exports touched the figure of 444,250 MT which was worth US$ 1102 million. Thus, these figures show the exponential growth of the Indian spice industry in a very short time. Spice exports from India have touched US $ one billion mark during 2007-08. The major destination of Indian spice exports is USA followed by European Union, Sri Lanka, Japan and Middle East Countries.

## 17.2.1. BLACK PEPPER, *PIPER NIGRUM*

### *(i) Diseases*

#### (a) Foot Rot, *Phytophthora capsici, P. palmivora*

**Symptoms:** The pathogen causes black spots on leaves with fine fibre like projection from advancing margins which rapidly enlarge and cause defoliation (Fig. 17.3). Tender leaves and succulent shoot tips of new runner shoots trailing on the soil turn black. When the main stem is affected, the entire vine wilts followed by shedding of spikes with or without black spots.

**Management**

**Solarization and Bioagents:** Raising plants in solarized nursery mixture, fortified with *Trichoderma* spp. and endomycorrhizae, has been found effective in suppressing root rot. Potting mixture may be sterilized through solarization. To the solarized mixture, biocontrol agents such as *Glomus fasciculatum* at 100 g/kg mixture and *T. harzianum* (1 x $10^{10}$ cfu/g) may be added at the time of filling of nursery mixture in polybags.

**Fig. 17.3.** Foot rot infected plants of black pepper.

### *(ii) Nematodes*

#### (a) Burrowing Nematode, *Radopholus similis*

Van der Vecht (1950) first reported the association of the burrowing nematode with the "Yellows disease" of black pepper and also proved the pathogenicity of *R. similis*. "Yellows disease" was reported to have wiped out 22 million pepper vines within 20 years in Bangka Island in Indonesia. *R. similis* is found associated with "Slow-wilt" disease of black pepper in Kerala and Karnataka, India (Mohandas and Ramana, 1987; Ramana *et al.*, 1987a).

A population level of 250 nematodes/g of root was constantly recorded with slow wilt affected pepper vines in Kerala (Ramana, 1986). Menon (1949) reported mortality up to 10% of the vines due to the disease. Mohandas and Ramana (1991) reported that *R. similis* was responsible for 59% loss in yield of black pepper.

**Symptoms:** Primary symptoms of slow-wilt (yellows) disease is the appearance of pale yellow drooping leaves on the vines whose number gradually increases and finally the entire foliage becomes yellow. Yellowing is followed by shedding of leaves, cessation of growth and die-back symptoms. Within 3-5 years of the initiation of the primary symptoms, all the leaves are shed and the death of the vine takes place. Young and old plants are susceptible. In bearing vines, shedding of the spikes is another major symptom. At the base of the affected vines, large numbers of shed spikes can be seen. In plantations, barren standards that have lost the vines or standards supporting dead vines are of common occurrence. The foliar symptoms become more pronounced when the soil moisture is depleted.

The symptoms on the roots are the presence of characteristic orange to purple coloured lesions on the thin, white feeding roots which are not clearly seen on old brown roots. The main roots remain devoid of fine feeder roots and extensive root rotting takes place. Subsequently, extensive necrosis of the larger lateral roots takes place (Fig. 17.4).

**Fig. 17.4.** Burrowing nematode on black pepper. A-Infected; B-Healthy.

## Management

**Soil Solarization:** Soil solarization of nursery sites or potting mixture and random rotation of nursery sites are reported to be effective in black pepper.

**Intercropping:** *R. similis* was not encountered when black pepper was intercropped with coffee (Venkitesan, 1976). Growing of non-host cover crops like Siratro (*Macroptilium atropurpureus*) in the interspaces is recommended to reduce nematode population.

**Mulching:** Mulching (with Guatemala grass) has an ameliorating effect on the symptoms of slow wilt/pepper yellows (*R. similis*).

**Organic Amendments:** Application of organic amendments like neem cake (200 g), green foliage (3-5 kg) or FYM (1 kg) per vine and earthing up in September-October are recommended.

**Trap Cropping:** Growing marigold as a trap crop and uprooting the trap crop at flowering stage and burning reduce the incidence of nematodes.

### (b) Root-knot Nematodes, *Meloidogyne incognita, M. javanica*

Butler (1906) reported root-knot nematodes from black pepper in Wynaad, Kerala, India. *M. incognita* and *M. javanica* have been reported from India. The root-knot infestation is a serious problem in Government nurseries in Kerala. Up to 91% root-knot infestation was reported from Para, Brazil (Ichinhoe, 1975) and Kerala, India (Ramana *et al.*, 1987).

An initial inoculum level of 10 $J_2$ per rooted cutting was found to reduce growth by 15%, while at 1,00,000 $J_2$ level 50% reduction in growth was observed over one year period (Koshy *et al.*, 1979). More than 50% death of transplants occurs in field planting of infected cuttings. *M. incognita* was responsible for 46% loss in yield of black pepper (Mohandas and Ramana, 1991).

**Symptoms:** Prominent symptoms of root-knot infestation on black pepper are unthrifty growth and yellowing of leaves. Inter-veinal yellowing of the foliage is also noticeable. The leaves exhibit dense yellowish discolouration of the interveinal areas making the leaf veins prominent with deep green colour. Heavy galling of the root system is also present.

### Management

**Intercropping:** Growing of non-host cover crops like Siratro (*Macroptilium atropurpureus*) in the interspaces is recommended to reduce the root-knot nematode (*M. incognita*) population on black pepper in the Amazonian region (Ichinohe, 1980).

**Mulching:** Mulching with Guatemala grass (*Imperata cylindrica*) is recommended to reduce the root-knot nematode (*M. incognita*) population on black pepper in the Amazonian region (Ichinohe, 1980).

### (c) Nematodes, *Meloidogyne incognita, Radopholus similis* and Foot Rot, *Phytophthora capsici* Disease Complex

Increased susceptibility of *M. incognita* and *M. javanica* infested cultivars of black pepper to *Phytophthora* infestation has been reported.

**Management**

**Solarization, Organic Amendments and Bioagents:** Integrated management of foot rot (*P. capsici*) and nematodes (*M. incognita* and *R. similis*) on black pepper was achieved by (i) mixing AMF and *Trichoderma harzianum* in solarized nursery mixture to raise healthy and robust seedlings, (ii) application of *T. harzianum* and FYM in planting pit, (iii) field application of neem cake at 1 kg/vine mixed with 50 g of *T. harzianum* during August.

### *(iii) Weeds*

**(a) Mulching:** Mulching is an important cultural operation practiced in pepper to suppress weeds and conserve soil moisture. Saw dust mulch was found to give highest yield in pepper (Sukumaran Pillay *et al.*, 1987). Cover crops (*Calapogonium mucunoides, Mimosa invisa*) are grown to provide effective soil cover and to prevent soil erosion.

## 17.2.2. CARDAMOM, *ELETTARIA CARDAMOMUM*

### *(i) Diseases*

### (a) Damping-off, *Pythium vexans*

**Symptoms:** Occurs during SW monsoon. The symptoms include pale yellow colour in the foliage and premature death of older leaves. Collar portion of the aerial shoots becomes brittle and tiller breaks easily. Rotting develops in the collar region emitting a foul smell.

**Management**

**Solarization and Bioagents:** Solarization of nursery beds before sowing seeds and incorporation of *T. harzianum* has resulted in prevention of damping-off and production of pathogen-free healthy seedlings.

Soil solarization can be done for sterilizing the nursery mixture. To the sterilized mixture, biocontrol agents such as arbuscular mycorrhizal fungus at 100 g/kg and *T. harzianum* ($10^{10}$ cfu/g) may be added at the time of filling of nursery mixture in polybags.

### (b) Rhizome Rot, *Pythium vexum, Rhizoctonia solani*

**Symptoms:** The disease occurs during South West monsoon. The symptoms include pale yellow colour on the foliage and premature death of older leaves. Collar portion of the aerial shoots becomes brittle and tiller breaks easily from the rhizome at the bulbous base (Fig. 17.5). Water soaked lesions appear first on leaves. The pathogen causes rotting and shredding of leaves along the veins. Rotting develops in the collar region emitting a foul smell. Capsules turn dull greenish brown emitting a foul smell.

**Fig. 17.5.** Clump rot of cardamom.

## Management

**Solarization and Bioagents**: Solarization of nursery beds before sowing seeds and incorporation of *T. harzianum* has resulted in prevention of damping off and production of pathogen-free healthy seedlings. Soil solarization can be done for sterilizing the nursery mixture. To the sterilized mixture, biocontrol agents such as arbuscular mycorrhizal fungus at 100 g/kg and *T. harzianum* ($10^{10}$ cfu/g) may be added at the time of filling of nursery mixture in polybags.

Soil solarization of the nursery beds combined with AMF and biocontrol agents would be a better approach to suppress the disease in nurseries.

### (c) Capsule Rot/Azhukal, *Phytophthora meadii, P. nicotianae* var. *nicotianae*

**Symptoms:** This is a fungal disease caused by *Phytophthora* spp. occurring during the rainy season. It affects the leaves, tender shoots, panicles and capsules. On the infected leaves, water soaked lesions appear first and rotting and shedding of leaves along the veins occur there after. The infected capsules become dull greenish brown and decay. This emits a foul smell and subsequently capsules are shed. Infection spreads to the panicles also.

## Management

**Solarization and Bioagents:** Solarization of nursery beds and subsequent soil application of *T. harzianum* reduced root rot infection and also nematode infection caused by *Meloidogyne incognita* (Eapen and Ramana, 1996).

## *(ii) Nematodes*

### (a) Root-knot Nematodes, *Meloidogyne* spp.

*M. incognita* and *M. javanica* have been reported as widely occurring in the cardamom nurseries and plantations in Kerala, Karnataka and Tamil Nadu (Kumar *et al*., 1971; Koshy *et al*., 1976; Ali, 1982, 1986). An yield loss of 32-47% due to root-knot infestation has been reported (Ali, 1984, 1986). An initial population of 100 nematodes/ plant causes discernible damage to cardamom (Eapen, 1987).

**Symptoms:** Heavy infestation of root-knot on mature plants in a plantation causes stunting, reduced tillering, yellowing, premature drying of leaf tips and margins, narrowing of leaf blades, delay in flowering, immature fruit drop and reduction in yield. Infested roots are characterized by excessive branching. Galling of roots is not conspicuous on mature plants.

In primary nursery, the second stage juveniles infest the radicle and plemule as a result of which even 50% of the germinating seeds do not emerge. At the two leaf stage itself the infested seedlings show marginal yellowing and drying of leaves and severe galling of roots. Up to 40% of such seedlings do not establish in the secondary nursery. The infested secondary nursery plants exhibit stunting, yellowing, poor tillering, drying of leaf tips and margins and heavy galling of roots (Ali and Koshy, 1982).

## Management

**Solarization:** Soil solarization for 40-45 days using 400 gauge transparent LDPE was proved ideal pre-sowing treatment for the management of nematodes and other soil-borne diseases in cardamom nurseries. Mean soil temperature during the solarization period increased by 8.7°C in solarized beds and the germination was enhanced by 25.5%, while weed growth was suppressed by 82%. *Pythium vexans* was totally eliminated, while populations of *Rhizoctonia solani, Phyllosticta elettaria* and nematodes were suppressed to varying levels. Solarization was also found to stimulate the growth of cardamom seedlings (Eapen, 1995).

**Solarization and Bioagents**: Soil solarization of nursery beds and application of bioagents such as *P. lilacinus* or *Trichoderma* spp. improved growth of cardamom seedlings by suppressing root-knot nematode population (Eapen and Venugopal, 1995). Application of *T. harzianum* multiplied on decomposed coffee husks (7 day old) at the time of sowing at 2.5 kg/bed (4.5 m x 1.0 m) and repeated after 3 months is recommended for the control of root-knot nematodes and damping-off in nurseries.

Besides soil solarization of nursery beds, subsequent soil application of *T. harzianum* reduced root rot infection and also infection caused by the root-knot nematode, *M. incognita* (Eapen and Ramana, 1996).

### (b) Root-knot Nematode, *Meloidogyne* sp. and Rhizome Rot, *Rhizoctonia solani* Disease Complex

*M. incognita* was found to predispose cardamom seedlings to *R. solani* infection, which causes damping off and rhizome rot in the primary nursery.

### Management

**Solarization, Bioagents and Chemicals**: Soil solarization alone enhanced the germination by 25.5% and suppressed weed growth by 82.0%. Solarization also enhanced the growth and vigour of cardamom seedlings. The disease complex was suppressed by incorporation of *P. lilacinus*/*T. harzianum* and phorate into the solarized nursery beds. This approach is being adopted on a large scale for the production of nematode-free cardamom seedlings.

## 17.2.3. GINGER, *ZINGIBER OFFICINALE*

### *(i) Diseases*

### (a) Bacterial Wilt, *Ralstonia solanacearum*

**Symptoms:** At first, symptoms occur on a single plant or on plants in a small patch. There is a downward curling of the leaves, without any marked yellowing, and quickly the leaves turn orange-brown or golden-brown. They then collapse and hang down around the stem. This gives the appearance of wilt. If these stems are placed in water, a white substance streams from the cut ends. Stems become infected at the base, near the rhizome. Dark, water-soaked patches occur, and soon after the stems can be pulled easily from the rhizomes.

Usually, the rotten stems have a foul smell. The young, tender, sprouts of the rhizomes are also infected. The vascular tissues become dark brown or black. Water-soaked areas occur and a milky liquid oozes out when rhizomes are cut. At harvest, there is nothing left of the plant or only the skin remains; everything else has been destroyed by *Ralstonia* and other organisms. Biotype 4 was found to produce the most severe symptoms.

### Management

**Solarization**: Solarization of ginger rhizomes for 2 to 4 hr from 9.00 a.m. to 1.00 p.m. (which raised the rhizome temperature to 55°C during summer) just before planting, eliminated the seed-borne bacterial infection from ginger rhizomes.

**Biofumigation and Bioagents:** Lowest disease incidence (15.63) was recorded in the treatment of *P. fluorescens* mass cultured in the mixture of vermicompost (VC) and mustard oil cake (MOC) applied as seed treatment and soil application. It was followed by *T. harziamum* mass cultured in MOC (21.88%), which was statistically at par with the application of copper oxychloride (22.50%) (Bora and Bora, 2009).

As a seed disinfection procedure, a novel technique called rhizome solarization has been devised at the Indian Institute of Spices Research, Calicut. The disinfected rhizomes when treated with *T. harzianum* and rhizobacterial strain consortia as seed treatment and soil application resulted in higher yields and growth promotion and soft rot suppression and minimal bacterial wilt. In addition, in combination with *Glomus* spp., the disease was absent probably through growth mediation.

### (b) Rhizome Rot, *Pythium aphanidermatum, P. myriotylum*

**Symptoms:** Soft rot is the most serious disease of ginger in India and in some other countries. It is caused by *Pythium* spp., of which *P. aphanidermatum* is the principal species in India, although *P. butleri*, *P. gracile*, *P. myriotylum*, *P. nigriotilum*, and *P. vexans* have also been recorded. The bases of the aerial shoots become soft, watery and then rot. The affected plants become pale; the tips of the leaves turn yellow, followed by complete yellowing and drying up of the leaves (Fig. 17.6). The shoots fall and cease to produce rhizomes. The infection extends to the rhizomes; the inner tissues being reduced to a soft and black, putrefying mass. Losses can be high. The disease is favoured by high moisture content of the soil with insufficient drainage.

**Fig. 17.6.** Rhizome rot of ginger.

## Management

**Solarization and Bioagents**: Usman *et al.* (1996) successfully utilized the native strains of *Trichoderma* and *Gliocladium* against rhizome rot of ginger caused by *P. aphanidermatum, P. myriotylum* and *Fusarium solani* complex. The disease suppression was enhanced when biocontrol was succeeded by soil solarization.

Soil solarization prior to planting followed by application of biocontrol agents showed synergistic effects in disease reduction and increased yield (Sarma *et al.*, 1996) (Table 17.6).

Table 17.6. Effect of the biocontrol agents on germination, rhizome rot incidence and yield of ginger

| Biocontrol agent | Solarized soil | | | Non- solarized soil | | |
|---|---|---|---|---|---|---|
| | Germination (%) | Disease index (%) | Yield (kg/plot) | Germination (%) | Disease index (%) | Yield (kg/plot) |
| *T. viride* | 77.69 | 13.70 | 2.846 | 71.52 | 35.25 | 1.225 |
| *T. harzianum* 1 | 79.61 | 15.80 | 3.552 | 79.43 | 32.07 | 2.673 |
| *T. harzianum* 2 | 77.87 | 19.38 | 2.910 | 72.91 | 43.61 | 1.439 |
| *T. hamatum* | 75.00 | 16.16 | 2.744 | 73.23 | 39.26 | 1.602 |
| *G. virens* | 74.48 | 18.25 | 2.641 | 74.43 | 36.63 | 1.705 |
| Mancozeb | 79.23 | 36.22 | 2.260 | 73.89 | 46.70 | 0.818 |
| Control | 81.53 | 57.58 | 1.692 | 79.86 | 53.80 | 0.992 |

| | | | | | | |
|---|---|---|---|---|---|---|
| Mean | 77.92 | 25.30 | 2.278 | 74.90 | 41.08 | 1.485 |
| CD (P=0.05) | NS | 6.26 | 0.430 | NS | 6.26 | 0.430 |

### (c) Soft Rot, *Pythium aphanidermatum, P. graminicolum*

**Symptoms:** Soft rot is the most serious disease of ginger in India and in some other countries. It is caused by *Pythium* spp., of which *P. aphanidermatum* is the principal species in India, although *P. butleri*, *P. gracile*, *P. myriotylum*, *P. nigriotilum*, and *P. vexans* have also been recorded. The bases of the aerial shoots become soft, watery and then rot. The affected plants become pale; the tips of the leaves turn yellow, followed by complete yellowing and drying up of the leaves (Fig. 9.3). The shoots fall and cease to produce rhizomes. The infection extends to the rhizomes; the inner tissues being reduced to a soft and black, putrefying mass. Losses can be high. The disease is favoured by high moisture content of the soil with insufficient drainage.

#### Management

**Solarization, Organic Amendments and Bioagents**: Solarization of beds before planting and addition of *T. harzianum* formulation (with 10 cfu/g) at 50 g/3 m$^2$ bed along with neem cake/ FYM reduced the disease incidence.

### *(ii) Nematodes*

### (a) Root-knot Nematodes, *Meloidogyne* spp.

*M. incognita, M. javanica, M. arenaria* and *M. hapla* have been reported to be associated with ginger from various countries. Under pot conditions, an initial inoculum level of 10,000 nematodes per plant over a period of six months caused 74% reduction in rhizome weight. A population level of one juvenile/30 g of soil was found to cause significant reduction in yield (Sukumaran and Sundararaju, 1986a). Kaur (1987) estimated 41-59% yield loss in ginger when the crop was raised using apparently healthy rhizomes in nematode infested fields.

**Symptoms:** Heavily infested plants exhibit stunting and chlorotic leaves with marginal necrosis. The root-knot nematodes cause galling and rotting of roots and underground rhizomes. Infested rhizomes show brown, water-soaked lesions in the outer tissues, particularly in the angles between shoots. The J2 of *M. incognita* invade the rhizome through the axils of leaf sheaths in the shoot apex. In fibrous roots, penetration occurs in the area of differentiation and in fleshy roots.

The entire length of root is invaded. In the rhizomes and fleshy roots extensive internal lesions develop.

**Management**

**Solarization:** Summer ploughing at an interval of 15 days and use of hot water treated ginger rhizomes (at 50°C for 10 min.) as seed proved promising in reducing the nematode population and in increasing the yield up to 65%.

**Organic Amendments:** Application of well decomposed FYM/compost at 25-30 t/ha or neem cake at 2 t/ha helps in reducing nematode multiplication (Kaur, 1987).

**Mulching:** Mulching with green leaves at 10-12 t/ha at planting and repeating the mulching during the growth period helps in reducing nematode multiplication (Kaur, 1987). Mulching with FYM is effective in reducing the nematode population.

**Crop Rotation:** Crop rotation with cereals and millets should be practiced, preferably with rice at least once in 3 years. For ginger cultivation in Fiji, a ginger-taro-fallow rotation has been recommended.

**Intercropping:** Intercropping of ginger with maize, capsicum is effective in reducing the nematode population.

**Biofumigation:** When Brassica family plants (Sunnhemp *Crotolaria juncea* and fodder radish, *Raphanus sativa*) are incorporated into soil, microbial decomposition of glucosinolates release 2-phenylethl isothiocyanate lowering soil populations of wilt bacteria and RKN.

There were low root knot nematode populations after growing sunnhemp. There were also large populations of beneficial saprophytic bacteria observed in the soil during the sunnhemp residue decomposition. There were low numbers of RKN in soil near growing ginger 30 days after planting where fodder radish had time to decompose and biofumigate the soil.

### (b) Fusarium Wilt, *Fusarium oxysporum* f. sp. *zingiberi* and Lesion Nematode, *Pratylenchus coffeae* Disease Complex

**Management**

**Solarization and Bioagents**: The efficacy of biocontrol agents like *Aspergillus niger*, *Trichoderma harzianum, T. hamatum* and *Gliocladium virens* either alone or in combination with soil solarization has been reported for the management of yellows disease (*Fusarium* + *Pratylenchus coffeae)* in Himachal Pradesh (Dohroo, 1995). Similar results have been reported by Eapen and Ramana (1996).

### *(iii) Weeds*

**(a) Mulching:** Mulching with sisham leaves produced tallest plants with high number of litters per plant. Mulching with green leaf, sugarcane trash and polythene film gave highest yield of green ginger and could conserve moisture up to 50%. Suppression of weed growth, increase in crop emergence and enhanced yields were noticed in plots mulched with dry leaves of straw.

## 17.2.4. TURMERIC, *CURCUMA LONGA*

### *(i) Nematodes*

#### (a) Root-knot Nematodes, *Meloidogyne* spp.

*M. incognita* and *M. javanica* have been reported on turmeric of which *M. incognita* is more important. In pot experiments, 100,000 nematodes/plant resulted in 76.6% reduction in the rhizome weight after 6 months (Sukumaran and Sundararaju, 1986b).

**Symptoms:** Turmeric plants infested with *M. incognita* exhibit stunting, yellowing, reduced tillering and marginal and tip drying of leaves. Galling and rotting of roots can also be noticed. Infested rhizomes have brown, water soaked areas in the outer tissues and lose their bright yellow colour (Mani *et al.*, 1987). High populations of *M. incognita* in field cause stunting, yellowing and withering of plants in large patches. Premature death of plants takes place leaving a poor crop stand.

#### Management

**Solarization:** Summer ploughing at an interval of 15 days and use of hot water treated turmeric rhizomes as seed proved promising in reducing the nematode population and in increasing the yield up to 65%.

**Crop Rotation:** Crop rotation with cereals and millets should be practiced, preferably with rice at least once in 3 years. Deep summer ploughing with furrow turner ploughs should be practiced.

**Biofumigation:** Application of well decomposed FYM/compost at 25-30 t/ha or neem cake at 2 t/ha and mulching with green leaves at 10-12 t/ha at planting and repeating the mulching during the growth period helps in reducing nematode multiplication (Kaur, 1987).

### *(ii) Weeds*

The predominant weeds are *Trianthema portulacastrum, Eclipta alba, Euphorbia hirta and Phyllanthus niruri among dicots, and Cyperus spp., Echinochloa colonum, E. crusgalli* and *Digitaria marginata* among monocots.

**(a) Mulching:** Mulching the beds with green leaves at 10-12 t/ha 40 and 50 days after planting is done in turmeric to suppress weeds (Srirama Rao, 1978; Nair, 1982; Rathaiah, 1986). Wheat and paddy straw can be used as dry mulch both for weed control and soil moisture conservation (Panigrahi *et al.*, 1987). Cover cropping and mulching with leaves and straw applied just after planting gave earliness in sprouting of turmeric and less number of weeds (Mohanty *et al.*, 1991).

## 17.2.5. CUMIN, *CUMINUM CYMINUM*

### *(i) Weeds*

**(a) Intercropping:** Pea and beans were found to be the most suitable intercrops for cumin as they are leguminous crops which fix atmospheric N in soil and improves soil fertility status besides giving better returns per unit area (Kaith, 1981). The other intercrops like cabbage, cauliflower, root crops or *Fagopyrum* spp. reduced the seed yield of cumin.

## 17.2.6. MUSTARD (RAPE SEED)

### *(i) Nematodes*

#### (a) *Tylenchorhynchus* and *Trichodorus* spp.

#### Management

**Solarization:** Table 17.7 clearly shows the effect of solarization on the control of nematodes on rape with RF's of 0.14 for *Tylenchorhynchus* and *Trichodorus* spp. The corresponding RF values in non-solarized soil was 1.45 on the same crops and the same nematodes.

Table 17.7. Effect of solarization on the control of nematodes on rape seed

| Treatment | Average number of nematodes per 200 g soil[1] | | Reproduction factor (RF) RF = Pf/Pi |
|---|---|---|---|
| | (Pi) | (Pf) | |
| Rape - solarized | 285 | 180 | 0.14 |
| Rape - non-solarized | 1980 | 2880 | 1.45 |

Pi = Initial nematode population; Pf = Final nematode population.

*Chapter 18*

# OTHER CROPS

## 18.1. COTTON, *GOSSYPIUM* SPP.

### 18.1.1. Diseases

#### (i) Black Root Rot, *Thielaviopsis basicola*

Black root rot of cotton (Fig. 18.1) has been observed on most farms in the Macquarie, Namoi, Gwydir and Macintyre valleys of Australia. It is causing high yield losses in the cooler areas where continuous cultivation of cotton is followed. The disease develops over time when the fungus *Thielaviopsis basicola* is present in adequate amount where cotton is grown in an environment favourable for infection. This pathogen is cosmopolitan and is a serious pathogen of cotton in Australia.

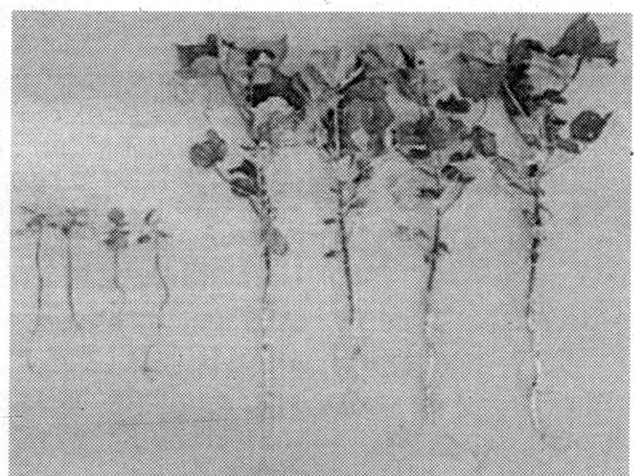

**Fig. 18.1.** Cotton black root rot. Left – Infected, Right – Healthy.

*Thielaviopsis basicola* does not grow on crop residues and survives as long-lived spores in the soil. Each crop of infected cotton deposits more spores in the soil and the severity of black root rot increases according to the number of cotton crops

### (a) Management

**Biofumigation:** Hairy vetch (*Vicia villosa*) has been used successfully as a biofumigation crop against black root rot, caused by *T. basicola*, in cotton in the USA. In soils used to grow cotton in Australia, green manure crops with hairy vetch and *Brassica juncea*, demonstrated useful biofumgation potential against *T. basicola*. Biofumigation clearly has potential to reduce the severity of black root rot (Nehl, 2008).

Canola, vetch or mustard can be used as biofumigants for the management of *T. basicola*. When vetch is incorporated into soil, its breakdown can release ammonia levels sufficient to kill *T. basicola* while it also provides nitrogen to the subsequent cotton crop.

In eight out of eleven field trials with vetch or mustards in Australia, disease severity in the following cotton crop was decreased significantly (by 24 to 70%). The success of biofumigation depends upon the growth of the biofumigation crop and good incorporation (at least four weeks before cotton). Vetch also provides a nitrogen bonus to the soil (as much as 200 units of N per ha).

Woolly pod vetch and Indian mustard have a 'biofumigation' effect on *Thielaviopsis* (i.e. toxic to spores) when grown as 'green manures'.

**Crop Rotation:** In preliminary research, the population of *T. basicola* in a severely infested field was reduced dramatically by three consecutive years of wheat, but a single rotation had no effect. The disease severity was very low in three years of continuous wheat even after two years of continuous cotton.

In a preliminary study on one-year rotation with canola followed by vetch, *T. basicola* spore population was reduced to a very low level.

Growing rotation crops like canola that also have a biofumigation effect followed by a cereal crop would additionally increase the time that the remaining spores of *T. basicola* would have to survive until the next cotton crop and provide good soil structure and microbial diversity in addition to economic yield. Growing biofumigation crops for two years followed by a winter cover crop in a severely infested field may reduce the *T. basicola* population to a large extent.

Rotation with cereals delays, but does not prevent, the build-up of black root rot. Two or more consecutive cereal crops can reduce the severity of black root rot in the following cotton crop.

**Solarization:** Solarization using polyethylene sheets can also reduce the spore population. Clear polyethylene can increase the soil temperature up to 10°C during daytime.

There is biodegradable polyethylene available in the market that would be easy to handle. When applied in conjunction with incorporation of biofumigation crops, it could accelerate anaerobic reactions and produce gases that may be toxic to the pathogen.

### (ii) Wilt, *Verticillium dahliae*

**(a) Symptoms:** Symptoms of the disease appear when the crop is in squares and progressively become more intense as the plants produce flowers and bolls. Normally it appears in the later part of October but its greatest severity is observed in December and January. The disease appears in circular patches which increase in size as the season progresses and over the years.

Affected plants show a characteristic chlorosis of the leaves. The veins turn brown and interveinal areas are chlorotic. This stage is followed by cupping of the lamina and the tissue between veins undergoes necrosis (Fig. 18.2). The leaves dry up at margins. Ultimately most of the leaves dropped. Where defoliation is severe, regrowth may occur. There is intense browning in the vascular tissue of the woody stem and root. When the petiole is cut across, brown spots marking the vascular elements are seen as characteristic of the disease.

**Fig. 18.2.** Verticillium wilt of cotton.

### (b) Management

**Solarization:** Soil solarization in the summer of 1986 brought *V. dahliae* population to negligible levels in fields A and B (Table 18.1). Furthermore, inoculum density (ID) values in solarized plots remained close to 0 propagules/g dry soil (p/g) shortly after planting time in 1987, whereas they had increased in the unsolarized control (Tables 18.1 and 18.2). Disease incidence in unsolarized controls was fairly high in both fields, but it was very low in solarized plots when assessed in July 28 (Table 18.2) and it reached 22 percent in field A on September 9. The beneficial effects of soil solarization were further indicated in field B, since seed cotton yield in solarized plots was increased by 131 percent as compared to unsolarized ones, where severe disease developed even with low ID values (Table 18.2).

In 1987, soil solarization eradicated *V. dahliae* from soils with a range of ID values to undetectable levels (Table 18.1). ID values in solarized soils remained at low levels both in the 0-20 cm and 20-40 cm soil layers shortly after planting time in 1988 (Table 18.2). Furthermore, plant stand in solarized plots increased by 21-108 percent as compared to unsolarized ones, where infections by *Rhizoctonia solani* caused damping-off. Disease incidence in solarized plots of fields C and D was significantly reduced with respect to unsolarized controls, for which high incidence was correlated with ID values (Table 18.2). Similarly, seed cotton yield increased significantly after solarization, and to a larger extent in the most severely affected field C (110 percent) than in the less affected field D (33 percent) (Table 18.2).

Table 18.1. Effect of soil solarization on the population of *Verticillium dahliae* in soil (1987)

| Field | Treatment[a] | Inoculum density (p/g) | |
|---|---|---|---|
| | | Before treatment | After treatment |
| A | Unsolarized | 23.3 | 16.1 |
| | Solarized | 27.3 | 0 |
| B | Unsolarized | 2.4 | 2.8 |
| | Solarized | 1.8 | 0.2 |
| C | Unsolarized | 20.8 | 68.3 |
| | Solarized | 20.0 | 0 |
| D | Unsolarized | 3.0 | 34.3 |
| | Solarized | 3.5 | 0 |
| E | Unsolarized | 7.3 | 42.0 |
| | Solarized | 7.5 | 0 |

a Plots were solarized from July 9 - August 31 in 1986, and from July 16-September 30, in 1987.

Table 18.2. Effect of soil solarization on Verticillium wilt and seed cotton yield (1988)

| Field | Treatment | Inoculum density at planting (p/g) | | Disease incidence (%) | Seed cotton yield (kg/ha) |
|---|---|---|---|---|---|
| | | Upper layer | Lower layer | | |
| A | Unsolarized | 51.3 | - | 90.5 | - |
| | Solarized | - | - | 5.0 | - |
| B | Unsolarized | 6.0 | - | 60.1 | 2093 |
| | Solarized | 0.1 | - | 4.5 | 4833 |
| C | Unsolarized | 28.8 | 5.8 | 87.7 | 1562 |
| | Solarized | 1.5 | 0.3 | 13.5 | 3279 |
| D | Unsolarized | 15.3 | 5.0 | 55.0 | 3425 |
| | Solarized | 0.3 | 0 | 1.8 | 4575 |

### 18.1.2. Nematodes

#### (i) Reniform Nematode, *Rotylenchulus reniformis*

The reniform nematode is known to cause economic losses in cotton. It has a wide host range and distributed all over the tropics including India (Varaprasad, 1986). It is a major pest of cotton in USA, Egypt and India.

**(a) Symptoms:** The nematode (Fig. 18.3) causes stunting, delayed maturity, reduction in boll size and yield and at times reduction in lint percentage in USA. Stunted patches in field having pale greenish leaves occur in *R. reniformis* – infested Greenland cotton plants in Tamil Nadu, India. A hybrid cotton cv. 'DHB-115' exhibit a pinkish discolouration of leaves, leaf shedding and affected boll bursting, when parasitized by reniform nematode. The leaves of reniform nematode-stressed cotton plants have less reflectance than leaves of non-stressed plants.

**Fig. 18.3.** Cotton root infected with *Rotylenchulus reniformis.*

The reniform nematode infecting most cotton growing areas in the southeast including Alabama, can reduce cotton yield by up to 75%.

### (b) Management

**Biofumigation:** Cover crops (two mustard cultivars [Pacific Gold (*Brassica juncea* L.) and Ida Gold (*Sinapis alba* L.)] and winter rye (*Secale cereale*, cultivar Maton) as biofumigants were evaluated in controlling reniform nematodes in cotton (*Gossypium hirsutum*) under conservation tillage system. The preliminary data showed that all cover crops used in the test significantly reduced reniform nematode populations (range <500 to 2,000/150cc soil) from the initial nematode level of >3000/150cc of soil sampled in early spring. Moreover, both mustard cultivars produced significant biomass before the first frost in 2003. The following spring, however, Pacific Gold and winter rye grew vigorously and produced large biomass but Ida Gold minimally recovered from winter frost. Additionally, plots planted to Ida Gold yielded lower nematode counts and sustained better cotton stand compared to Pacific Gold and winter rye. Based on these results, Ida Gold is a promising cover crop for controlling reniform nematodes in cotton.

A winter cover crop [wheat; hairy vetch; Indian mustard (*Brassica juncea*) cultivar Fumus or fallow]-cotton test was conducted in 2004 and 2005.

Table 18.3. Reniform populations for different cover-crop treatments.

| Treatment | Reniform nematode populations | | | |
|---|---|---|---|---|
| | 2004 | | 2005 | |
| | During cover crop | After planting | During cover crop | After planting |
| Telone | 12,045 ab | 12,424 ab | 12,273 a | 14,546 a |
| Vetch | 20,190 a | 27,803 a | 19,924 a | 27,348 a |
| Wheat | 13,750 ab | 19,053 a | 14,318 a | 19,735 a |
| Fallow | 8,561 b | 14,318 ab | 11,136 a | 15,909 a |
| Mustard | 17,235 ab | 5,644 b | 13,106 a | 4962 b |

In 2004 after planting cotton, mustard plots had reduced soil reniform nematode populations compared to either vetch or wheat plots (Table 18.3). The Telone and fallow plots had intermediate populations. In 2005 after cotton was planted, nematode populations following mustard were statistically lower than all other treatments.

If the ratio of the nematode population prior to cotton planting and after cotton planting was examined only the mustard plots had a ratio below 1.0, indicating the population was suppressed by this treatment. At the end of season when root segments were taken and observed for females (Table 18.4), both years showed significant differences among treatments. In both years, mustard plots had the lowest number of swollen females observed on a 5-cm section of cotton root. The soil population data and the lower number of reniform females in the root tissue indicate that Indian mustard had a season-long impact on the nematode compared to all other treatments.

Table 18.4. Reniform reproduction per 5-cm section of cotton root taken prior to harvest following different cover-crop treatments.

| Treatment | Reniform nematode reproduction | |
|---|---|---|
| | 2004 | 2005 |
| Telone | 11.0 a | 11.1 a |
| Vetch | 10.9 a | 11.1 a |
| Wheat | 12.2 a | 12.2 a |
| Fallow | 10.5 ab | 10.7 a |
| Mustard | 8.8 b | 8.6 b |

Seed cotton yields from the six plants mapped per plot in 2005 were greatest following a mustard cover crop (182 g), while the fallow plots had the lowest yield (82 g). Hand-picked yields for the Telone plots were intermediate (142 g).

## 18.2. TOBACCO, *NICOTIANA TABACUM*

### 18.2.1. Diseases

#### (i) Bacterial Wilt, *Ralstonia solanacearum*

#### (a) Management

**Biofumigation:** Preliminary experiments have been conducted to evaluate the effectiveness of Brassica residues [cabbage, broccoli, cauliflower, white Chinese cabbage (WCC), green Chinese cabbage (GCC) and weed (wild *Brassica nigra*)] as biofumigants to control bacterial wilt of tobacco in laboratory and glasshouse. Tobacco plants wilted within 3 DAT and died rapidly in untreated soil. No disease symptoms appeared in soil added with broccoli or cauliflower residues.

The disease symptoms in soil added with residue of WCC or weed appeared 14 DAT with severity reduced by 10%. The height and dry mass of plants were higher and heavier in treated soil compared to untreated soil. Total population of bacteria, actinomycetes, and fungi increased significantly in treated soil. Result from this preliminary experiment indicated that incorporation of residue of Brassicas, especially broccoli and cauliflower offer potential control for bacterial wilt of tobacco and source of nutrition for the plant. (Yulianti, 2008).

## 18.3. WHEAT, *TRITICUM* SPP.

### 18.3.1. Diseases

#### (i) Take-all, *Gaeumannomyces graminis*

#### (a) Management

**Crop Rotation:** The use of Brassicas such as canola (*Brassica napus*) as break crops to help control take-all (*G. graminis*) in cereal rotations in Australia, is an example of biofumigation.

**Intercropping:** The effect of rape/wheat strip intercropping with different strip width on disease resistance, yield and agronomic characters of winter wheat were studied in field experiments. Results indicated that suitable intercropping width of rapeseed/wheat could improve not only disease resistance but also yield of winter wheat. Compared with control, incidence of wheat disease was reduced by 3.5% and 8.7%, respectively after artificial inoculation with root-rot and take-all. The spike numbers per plant, spike length and yield increased by 6.5 to 11.5%, 4.5 to 7.5% and 6.1 to 13.3%, respectively, but spikelet rate decreased by 1.0 to 2.4%. Intercropping width should not more than 8m. Analysis on soil nutrients after harvesting showed that N, P and K content in intercropping strip were higher than control. The laboratory experiment showed that glucosinolate and their breakdown products which excreted from rapeseed have a significant inhibition on wheat root rot and take-all disease (Chunlei *et al.*, 2008).

**Biofumigation:** In pot studies, Smith *et al.* (1999) showed that incorporation of ground, freeze-dried canola root residues at realistic field rates (0.1 – 0.5% w/w) caused a 50% reduction in the infection of wheat seedlings by several root fungal pathogens. In another study, Kirkegaard *et al.* (1998) grew several different crops in pots of soil with added inoculum of the take-all fungus (*G. graminis* var. *tritici*), and incorporated the root residues after harvesting the tops at flowering. Wheat seedlings were sown into the pots eight weeks later and the level of take-all infection was assessed. The results showed lower levels of infection on wheat after the brassicas compared with linseed and higher root GSL levels increased the level of suppression. These results demonstrate that significant suppression of fungal inoculum can be achieved in soil by incorporated canola roots or root residues, and that higher levels of GSL increase suppression.

In the field, canola roots are generally left undisturbed after harvest and there is a 4-5 month period from canola harvest (December) until the following wheat crops are sown (May). The impact of a range of crops including brassicas with high and low root GSLs on the levels of take-all inoculum in soil during the growth of the crops, at maturity, and during the summer fallow was investigated. The experiments were conducted at sites where high initial levels of take-all inoculum had been established in the previous year. The results showed that at harvest, the *Brassica* crops were more suppressive to the take-all fungus than linseed and that the level of suppression was greater for the high root GSL variety. These differences were not apparent during crop growth suggesting that the suppression occurred as a result of root decay around maturity rather than exudation from live roots. This is consistent with the fact that most 2PE-GSL is present in the large tap and lateral roots and is likely to be released during the decay of these larger roots as the crop matures (Kirkegaard and Sarwar 1999).

### (ii) Crown Rot

#### (a) Management

**Crop Rotation:** In studies investigating the impacts of different preceding crops on crown rot in durum wheat, Kirkegaard *et al.* (2004) showed that *Brassica* break crops led to lower levels of crown rot and higher yield in durum wheat compared with wheat following chickpea. There was no evidence that the effect was related to biofumigation, but several plausible explanations were suggested including (1) more rapid breakdown of residual wheat stubble under dense canola canopies reducing crown rot inoculum; (2) higher soil N status following chickpea increasing crown rot severity and (3) altered soil/residue biology that was less conducive to crown rot inoculum survival. The latter possibility was supported by the higher levels of *Trichoderma* spp., known antagonists of crown rot, which were isolated from wheat following brassicas.

### 18.3.2. Nematodes

#### (i) Cereal Cyst Nematode, *Heterodera avenae*

**(a) Symptoms:** The disease is characterized by patchy growth of stunted and yellow plants. The patches tend to coalesce with monoculture within 3-4 years. Infested plants exhibit stiff, thin and narrow leaves, reduced tillering, delayed emergence of ears, reduced number of spikelets and grains. Such effects can be attributed to increased nutrient and moisture stresses, affecting developmental processes. Root system becomes bushy, bunchy and shallow due to their proliferation and has slight swelling near the root tip. Such plants can be easily pulled out of the ground. Presence of cysts on the roots is the only confirmatory indication of nematode infestation.

**(b) Management**

**Solarization:** Deep summer ploughing (2-3 times) of the infested field during May-June to expose cysts to harsh sunlight that in turn kills the cysts thereby reduces the population to 40% in each season.

Summer ploughing reduces nematode population to a great extent (up to 42.4%) in the population of *H. avenae* and consequent yield increase by up to 97.5% of wheat in Rajasthan.

**Crop Rotation:** Crop rotation with coriander, fenugreek, onion, mustard, gram and carrot brings down the cyst population to 47-55% level.

## 18.4. RICE, *ORYZA SATIVA*

### 18.4.1. Nematodes

#### (i) White Tip Nematode, *Aphelenchoides besseyi*

**(a) Symptoms:** The characteristic symptoms are exhibited when the rice is in tillering stage. The tips (3-5 cm) of upper leaves turn light yellow or white. The flag leaves are characteristically shortened and twisted. Chlorotic patches develop in the interveinal areas of sheaths and blades of the leaves are often invaded by fungi such as *Corticium sasaki* and *Acrocylindrium oryzae*. In severe cases of infestation, the panicle emergence from boot leaf is inhibited. The infected panicle become short and terminal portion of the spikelet produce shriveled grain. Abnormal development of floral parts has also been noticed.

### (b) Management

**Solarization:** Sun drying of seeds for 6 hours on 4 consecutive days was found effective against seed-borne infection of *A. besseyi* on paddy seeds.

## 18.5. CORN, *ZEA MAYS*

### 18.5.1. Diseases

### (i) Stalk Rot, *Fusarium moniliforme, Macrophomina phaseolina*

### (a) Management

**Solarization:** Seven weeks solarization of irrigated soil raised its temperature by 11.5°C over non-solarized soil at 10 cm depth and effectively controlled weeds (98.5%), stalk borer (8.9%) and stalk rot disease (69.1%) in corn. Solarization also reduced symptoms of fungal pathogens, *Fusarium moniliforme* and *Macrophomina phaseolina* significantly by 64.2% and 78.4%, respectively, and completely controlled *M. phaseolina* in corn cultivars, viz. Pool-10, Shaheen and Gauher. Growth of crop planted in solarized plots was better and it yielded almost one to three times more grains in the tested cultivars. Conclusively, the soil solarization had reduced stalk breakage and was found more economical since it provided essential nutrients to the plants in simpler forms and enabled them to withstand the disease without application of any pesticide.

## 18.6. SUGAR BEET, *BETA VULGARIS*

### 18.6.1. Diseases

### (i) Root Rot, *Rhizoctonia solani*

### (a) Management

**Crop Rotation:** During 2007, a field study was carried out with differing management regimes of a *Brassica juncea* (mustard) catch crop, grown within a sugar beet-winter wheat rotation, to analyse its action on sugar beet root rot caused by *R. solani*. The study involved three treatments; mustard pulled out at flowering (MP) and representing the first phase, mustard crushed at flowering and incorporated into soil (MC) representing the first and second phases, and bare soil (BS) as a control. Biofumigation mainly affected the parameters associated with primary infection resulting in a net reduction in the primary phase of the epidemic where control by MC was greater than that of MP. However, a significant, albeit much reduced effect was detected for parameters associated with secondary infection that resulted in a net increase in the intrinsic growth parameter of secondary infection of MP and MC treatments compared to control (Motisi *et al.*, 2008).

Control value of sugar beet root rot caused by *R. solani* in pot trials after cultivation of brown mustard (Exp.Y-008) was as high as 90, which is the same as wild oats (Sakuma) (Hashizume *et al.*, 2008).

### 18.6.2. Nematodes

#### *(i) Cyst Nematode, Heterodera schachtii*

The sugarbeet cyst nematode is one of the most serious pests of this crop which is widespread and occurs in most of the regions where sugar beets are grown causing significant yield loss.

**(a) Symptoms:** Sugar beets from infested fields were generally stunted and covered with hairy roots. Small, white, lemon-shaped female bodies were observed on these hairy roots.

#### (b) Management

**Trap Cropping:** Management of sugarbeet cyst nematode, *H. schachtii* with *Brassica* crops is a common strategy in North-western Europe and in some parts of the US (Hafez and Sundararaj, 2004; Muller, 1999). However, rather than using the crops primarily as biofumigants, they are grown as trap crops, attracting the nematodes to their roots and allowing root invasion by the nematodes, but preventing their multiplication.

The quality improvement of sugarbeet crop, decreasing cyst nematode infestations below the threshold value and increasing both tap root weight and sucrose, was effectively achieved through the study of rotation effects, including intercrops of *Brassicaceae* species, selected for high glucosinolate content. Nematicidal *Brassicaceae* can accumulate the majority of glucosinolates in the root system (catch effect), which is most suitable to control cyst nematodes.

Oil radish is used as a trap crop for the sugarbeet cyst nematode. Its roots exude chemicals stimulate hatching of nematode eggs. The larvae that emerge are unable to develop into reproductive females reducing the population densities for the following crop (Hafez, 1998).

**Biofumigation:** *Brassicaceae* cover crops attract the juvenile stages of endoparasitic nematodes working as a trap crop, since these, after root penetration, are poisoned by hydrolysis products and are not successful in completing their developmental cycle in 10-12 weeks, i.e. the intercropping time. Consequently, the nematode population in soil progressively decreases. At full flowering, the plants are chopped and immediately incorporated at around 12 cm depth by means of a stalk cutter and a miller, working at some meters distance from each other. A light irrigation sprinkled after incorporation in soil, aims at promoting the glucosinolate hydrolysis and the subsequent isothiocyanate release (Lazzeri *et al.*, 2004).

The nematicidal effect of catch crop is produced during the whole cultivation time, while its incorporation as green manure shows an overall amendment effect, increasing the organic matter amount and improving soil fertility, being the biofumigant effect during incorporation only secondary.

The studies on the effect of *Brassicaceae* on the developmental stages of the cyst nematode revealed the biocidal plant ability interrupting the *H. schachtii* life cycle to $J_3$ or $J_4$ female, without any formation of adult female and cysts. On the contrary, juvenile males developed to adults, changing the sex ratio of the nematode populations.

The main glucosinolates effective against *H. schachtii* in Northern Italy agronomic conditions derive either from radish (*Raphanus sativus* s. sp. *oleiformis*) or white mustard (*Sinapis alba*) varieties. Radish cvs. Nemex and Pegletta and white mustard cv. Maxi and other varieties with high nematicidal power were at first tested as catch crops and used as intercrops in quadriennial rotation schemes (Tacconi *et al.*, 1998; Tacconi and Venturi, 1990) (Tables 18.5 and 18.6).

Table 18.5. Host status of fodder radish and white mustard biocidal selections vs. sugar beet cyst nematode in a field test in Northern Italy (Tacconi *et al.*, 1998)

| Plant species | Female specimens/ 10 g roots | | | | Male specimens/ 10 g roots | | Host status* |
|---|---|---|---|---|---|---|---|
| | J2 | J3-J4 | Adult | Cyst | J3-J4 | Adult | |
| *Beta vulgaris* ssp. *saccharifera* cv. Sigma | 6.5 | 12.0 | 14.0 | 2.6 | 3.0 | 2.0 | 5 |
| *Raphanus sativus* ssp. *oleiformis* cv. Sereno | 1.9 | 1.8 | 0.7 | 0.1 | 2.8 | 4.0 | 3 |
| *R. sativus* ssp. *oleiformis* cv. Pegletta | 3.2 | 1.4 | 0.0 | 0.0 | 3.9 | 2.9 | 1 |

| | | | | | | | |
|---|---|---|---|---|---|---|---|
| *R. sativus* ssp. *oleiformis* cv. Levana | 5.4 | 5.5 | 0.8 | 1.1 | 6.4 | 4.1 | 3 |
| *R. sativus* ssp. *oleiformis* cv. Nemex | 2.7 | 0.8 | 0.0 | 0.0 | 2.5 | 1.9 | 1 |
| *Sinapis alba* cv. Emergo | 1.7 | 1.5 | 0.2 | 0.0 | 2.4 | 2.3 | 2 |
| *Sinapis alba* cv. Maxi | 1.9 | 0.9 | 0.0 | 0.0 | 0.1 | 0.6 | 1 |

* Based on BIOTEST; a catch crop shows nematicidal effects, at a host status included from 1 to 3.

Table 18.6. Reproduction factor of *H. schachtii* population, between rotation end and beginning, with and without nematicidal catch crops (Tacconi and Venturi, 1990)

| Rotations (1983-89)* | Reproduction factor (Pf/Pi) without catch crop | Reproduction factor (Pf/Pi) with catch crop |
|---|---|---|
| Biennial – (B-Ba) + (B-W) + (B-W) + B | 3.78 | 2.38 |
| Triennial - Biennial – (B-Ba-W) + (B-S-W) + B | 3.17 | 1.58 |
| Quadriennial – (O-M-W-B) | 0.11 | 0.07 |
| Quadriennial + Biennial (B-Ba-S-W) + (B-W) + B | 2.25 | 2.35 |

* B - sugarbeet; Ba – barley; M – maize; O – oats; S – soybean; W - wheat

In studies carried out on some selections of other plant species such as *Clemone spinosa* (family Capparaceae), *Eruca sativa* cv. Prisca and *Reseda luteola* (family Resedaceae) resulted effective against *H. schachtii* (Tacconi *et al.*, 1998) (Table 18.7).

Table 18.7. Host status of biocidal selections vs. sugarbeet cyst nematode (Tacconi *et al.*, 1998)

| Plant species | Female specimens/ 10 g roots | | | | Male specimens/ 10 g roots | | Host status* |
|---|---|---|---|---|---|---|---|
| | J2 | J3-J4 | Adult | Cyst | J3-J4 | Adult | |
| *Beta vulgaris* ssp. *saccharifera* cv. Dima | 2.6 | 10.4 | 19.9 | 0.8 | 11.5 | 8.3 | 5 |
| *Raphanus sativus* ssp. *oleiformis* cv. Pegletta | 5.6 | 3.3 | 0.0 | 0.0 | 3.1 | 0.3 | 1 |
| *Cleome spinosa* Italian ecotype | 3.7 | 3.5 | 0.2 | 0.0 | 1.3 | 1.1 | 1 |
| *Eruca sativa* cv. Prisca | 4.4 | 1.6 | 0.1 | 0.0 | 1.6 | 0.6 | 1 |
| *Reseda luteola* | 0.1 | 0.0 | 0.0 | 0.1 | 0.0 | 0.0 | 0 |

| *Sinapis alba* Srilanka ecotype | 1.4 | 1.5 | 3.2 | 0.2 | 2.8 | 1.3 | 5 |
|---|---|---|---|---|---|---|---|

* Based on BIOTEST (Behringer et al., 1984); a catch crop shows nematicidal effects, at a host status included from 1 to 3.

Following the major breakthrough in 1985 that oil radish variety RSO1841 significantly reduced sugar beet cyst nematode (SBCN) population, field studies were conducted with Pagletta, Nemex and R184 and more reduction in nematode population was achieved. Higher quality of sugarbeet crop was generally produced on soils planted with oilseed radish or white mustard residues than with the regular management practices. In the recent greenhouse studies it was found that two new oil radish varieties Defender and Comet significantly reduced the population of *H. schachtii* (95%). In the first experiment, mustard 'Concerta' produced 35% more above ground biomass than radish 'Colonel' and the viable cysts declined 29% and 19% in oil radish and mustard treatments, respectively. In the second experiment, radish 'Adagio' produced significantly more above ground biomass than mustard 'Metex'. In the third experiment, biomass (top, root and total) production of oil radish 'Dacapo' was significantly higher than mustard 'Metex'. A fourth experiment indicated that maximum beet yield (MT/acre) was with the Luna (37.0) planted plots followed by the Defender (36.0) plots. Recent studies in 2007 proved that among six varieties of green manure crops, maximum beet yield (MT/acre) was from Defender (36.8) Colonel (36.1) and Arugula (36.4) planted pots. At present Defender is the most economical variety highly suitable for the SBCN management (Hafez and Sundararaj, 2008).

In Wyoming, oilseed radish (*Raphanus sativus*) and yellow mustard (*Sinapis alba*) reduced the sugar beet cyst nematode populations by 19-75%, with greater suppression related to greater amount of cover crop biomass (Koch, 1995).

## 18.7. SOYBEAN, *GLYCINE MAX*

### 18.7.1. Diseases

#### (i) Stem and Root Rot, *Phytophthora sojae*

#### (a) Management

**Biofumigation:** Three types of ITC (Allyl, Ethyl, Phenyl) inhibited germination of oospores and encysted zoospore of *P. sojae,* causal agent of soybean *Phytophthora* stem and root rot, and the same effect was shown by treatment of brown mustard (Kondo) (Hashizume *et al.*, 2008).

### 18.7.2. Nematodes

#### (i) Root-knot Nematode, *Meloidogyne incognita*

##### (a) Management

**Biofumigation:** Davis *et al.* (1989) reported that phytolectins, particularly concavalin A, from jack bean (*Canavalia enciformis*) reduced *M. incognita* infestation in soybean.

#### (ii) Reniform Nematode, *Rotylenchulus reniformis*

##### (a) Management

**Biofumigation:** Alsayed *et al.* (1992) found that when leaf powders of *Canna indica,* castor, *Lantana camara, Cleodendrum incrme* and *Tagetes erecta* were added at 5, 10 or 15 g/kg soil, suppressed reproduction of *R. reniformis* and improved soybean shoot and root weights.

## 18.8. SUNFLOWER, *HELIANTHUS ANNUS*

### 18.8.1. Nematodes

#### (i) Reniform Nematode, *Rotylenchulus reniformis*

##### (a) Management

**Biofumigation:** El-Nagar *et al.* (1993) reported that addition of dry ground leaves of *Melia azedirach, Asparagus sprengeri, Eucalyptus* sp. and *Psidium guajava* at 2.5 and 5.0 g/pot significantly reduced the number of egg masses of *R. reniformis* on sunflower roots.

## 18.9. HYACINTH AND CROCUS

### Diseases

### 18.9.1. Root Rot, *Pythium* spp.

*Pythium* spp. cause root rot in bulb crops like hyacinth and crocus.

## (a) Management

**Biofumigation:** In a field experiment with hyacinth and crocus, treatment with *Brassica juncea* ISCI20 was compared to *Raphanus sativus* and untreated (fallow). Incorporation of both *B. juncea* and *R. sativus* resulted in increase of bulb yield compared to the fallow treatment. Moreover, yield losses caused by *Pythium* root rot were diminished. In infested field soil, no reduction of yield loss was observed compared to the non-infested control, whereas in untreated soil more than 20% yield loss occurred due to *Pythium* root rot.

# REFERENCES


Abdel-Rahim, M.F., Satour, M.M., El-Eraki, S.A., Elwakil, H.R., Grinstein, A., Chen, Y. and Katan, K. 1987. Soil solarization for controlling soil borne pathogens and weeds in furrow-irrigated Egyptian soils. *Proc. 7th Conf. of the Mediterranean Phytopathological Union,* pp. 76-77.

Abragan, F.N., Justo, V.P., Minguez, L.T., Abalde, J.A., Salvani, J.B. and Maghanoy, C.C. Jr. 2008. Evaluating biofumigation for soil-borne disease management in white potato. *Third Int. Biofumigation Symp.*, Canberra, Australia.

Adams, S. 1997. Seeing red colored mulch starves nematodes. *Agric. Res.* October, p. 18.

Aird, P. and Long, D. 2008. The successful integration of biofumigant crops into farming systems for soil improvement and soil-borne disease management in Tasmania, Australia. *Third Int. Biofumigation Symp.*, Canberra, Australia.

Aires, A., Rosa, E., Carvalho, R. and Rodrigues, C. 2006. Potential role of Brassicaceae plant as biofumigation agent in suppression of potato cyst nematodes (*Globodera rostochiensis* Woll). *Second Int. Biofumigation Symp.,* Idaho, USA.

Aiyappa, K.M., Srivastava, K.C. and Nanaya, K.A. 1970. Development of better agrotechniques for citrus fruits (a) water control, (b) inter-crops, (c) weed control. *Prog. Hort.* **13**: 16-17.

Alam, M.M. 1986. Possible use of weeds as soil amendment for the management of root-knot and stunt nematodes attacking eggplant. *Agric. Wastes* **16**: 97-102.

Alam, M.M., Khan, A.M. and Saxena, S.K. 1973. Some new host records of root-knot nematode, *Meloidogyne incognita* (Kofoid & White, 1919) Chitwood, 1949. *Indian J. Nematol.* **3**: 159-160.

Alam, M.M., Khan, A.M. and Saxena, S.K. 1977. Influence of different cropping sequences on soil population of plant parasitic nematodes. *Nematol. Medit.* **5**: 71-78.

Albuquerque, S.D.S. and Mohamed Ibrahim, P.A. 1956. *Calapogonium* for control of weeds in coconut gardens. *Agric. J.* **43**: 7-12.

Ali, S.S. 1982. Occurrence of root-knot nematodes in cardamom plantations of Tamil Nadu. *Proc. of Fifth Symp. on Plantation Crops*, Kasaragod, Kerala, India, pp. 615-620.

Ali, S.S. 1984. Preliminary observations on the effect of some systemic nematicides and neem oil cakes in a cardamom field infected with root-knot nematodes (Abstr.). *Proc. of the Sixth Symposium on Plantation Crops*, Kottayam, Kerala, India, pp. 215-223.

Ali, S.S. 1986. Root-knot nematode problem in cardamom and its management. *Proc. of the Second Group Discussion on the Nematological Problems of Plantation Crops,* Central Coffee Res. Inst., Balehonnur, pp. 10-12.

Ali, S.S. and Koshy, P.K. 1982. Occurrence and distribution of root-knot nematodes in cardamom plantations of Kerala. *Nematol. Medit.* **10**: 107-110.

Alsayed, A.A., Ahmad, S.S. and Abdul Hameed, S.H. 1992. Influence of decomposing dry leaf powders on *Rotylenchulus reniformis* and growth response of soybean. *Ann. Agri. Sci.,* Mashtohor **30**: 615-620.

Ameen, H.H. and Mohamed, M.M.M. 2006. Biofumigation as alternative method for management root-knot nematode *Meloidogyne incognita* affecting tomato under field conditions. *Second Int. Biofumigation Symp.,* Idaho, USA.

Amenduni, M., D'Amico, M., Colella, C. and Cirulli, M. 2004. Effect of *Brassicaceae* green manures and soil-solarization on corky root of tomato. *First Int. Biofumigation Symp.*, Florence, Italy.

Angus, J.F., Gardner, P.A., Kirkegaard, J.A. and Desmarchelier, J.M. 1994. Biofumigation: Isothiocyanates released from Brassica roots inhibit growth of the take-all fungus. *Plant and Soil* **16**: 107-112.

Anon, 1989. Biennial Report (1987-89) of AICRP on Plant Parasitic Nematodes with Integrated Approach for their Control, Indian Agri. Res. Inst., New Delhi, 80 pp.

Anon. 1990. Final Report of Research Scheme on Nematode Problems in Some Selected Temperate Fruit Crops and Their Management. Y. S. Parmar Univ. of Hort. & Forestry, Solan, 54 pp.

Anon. 1993. *QRT Report of AICRP on Nematodes (1989-93),* Dept. of Nematol., Haryana Agri. Univ., Hisar, India, 149 pp.

Anon. 2001. Oil radish green manure continues promise against nematodes. *The Grower,* June-July, p. 7.

Atwal, A.S. and Mangar, A. 1969. Repellent action of root exudates of *Sesamum orientale* against the root-knot nematode, *Meloidogyne incognita* (Heteroderidae: Nematoda). *Indian J. Ent.* **31**: 286.

Auger, J., Dugravot, S., Naudin, A., Abo-Ghalia, Pierre, D. and Thibout, E. 2002. Possible use of *Allium* allelochemicals in integrated control. IOBC WPRS Bulletin 25, 295.

Ayala, A., Roman, J. and Gonzalez, J. 1967. Pangola grass as a rotation crop for pineapple nematode control. *J. Agric. Sci., Univ. of Puerto Rico* **51**: 94-96.

Ayyar, P.N.K. 1926. A preliminary note on the root nematode, *Heterodera radicicola* Mitter and its economic importance in South India. *Sci. and Cult.* **22**: 391-393.

Baghel, P.P.S. 1995. Nematode problems in citrus. In *Nematode Pest Management - An Appraisal of Eco-friendly Approach,* (G. Swarup, D. R. Dasgupta and J.S. Gill, eds.), pp. 203-210. Nematol. Soc. of India, New Delhi.

Baghel, P.P.S. and Bhatti, D.S. 1983. Relative efficacy of nematicides for control of phytonematodes on grapevine varieties. *Third Nematol. Symp.,* Himachal Pradesh Agri. Univ., Solan, p. 39.

Baghel, P.P.S., Bhatti, D.S. and Chauhan, K.S. 1980. *Haryana J. Hort. Sci.* **9**: 136-137.

Baghel, P.P.S. and Gupta, D.C. 1986. Effect of intercropping on root-knot nematode (*Meloidogyne javanica*) infecting grapevine (var. Pearlette). *Indian J. Nematol.* **16**: 283-285.

Bajwa, M.S., Munshi, S.K. and Mann, M.S. 1978. Intercropping is an important feature of orchard management. *Punjab Hort. J.* **18**: 194-198.

Baker, K.F. 1987. Evolving concepts of biological control of plant pathogens. *Ann. Rev. Phytopathol.* **25**: 67-85.

Ball-Coelho, B., Bruin, A.J., Roy, R.C. and Riga, E. 2003. Forage pearl millet and marigold as rotation crops for biological control of root-lesion nematodes in potato. *Agronomy J.* **95**: 282-292.

Barber, C.A. 1901. *Department Land Records and Agriculture*. Madras Agri. Branch 2, Bull. No. 45, pp.227-234.

Barbercheck, M.E. and Broembsen, S.L. von. 1986. *Plant Dis.* **70**: 945-950.

Bélair G. and Coulombe, J. 2008. Green manure and biofumigation for root lesion nematode and Verticillium wilt management in strawberry production. *Third Int. Biofumigation Symp.*, Canberra, Australia.

Bell, C.E. and Elmore, C.L. 1983. Soil solarization as a weed control method in fall planted cantaloupes. *Proc. Western Soc. Weed Sci.*, Granada, Spain. **36**: 174-177.

Bello, A. 1998. Biofumigation and integrated crop management. In *Alternatives to Methyl Bromide for the Southern European Countries* (A. Bello, J.A. Gonzalez, M. Arias and R. Rodriguez-Kabana, eds.), pp. 99-126. Phytoma-Espania, DG XI EU, CSIC, Valencia, Spain.

Bello, A., Escuer, M., Sanz, R., López-Pérez, J.A. and Guirao, P. 1997. In: *Posibilidad de Alternativas Viables al Bromuro de Metilo en Pimiento de Invernadero* (A. López and J.A. Mora, eds.), pp. 67-108. Consejería de Medio Ambiente, Agricultura y Agua de Murcia, Spain.

Bello, A., López-Pérez, J.A., Garcia-Alvarez, A., Sanz, R. And Lacasa, A. 2004. Biofumigation and nematode control in the Mediterranean region.

In *Proc. of the Fourth Int. Cong. of Nematol.* (R.C. Cook and D.J. Hunt, eds.), pp. 133-149. Nematology Monographs and Perspectives, Vol. 2, Brill, Leiden and Boston.

Bello, A., López-Pérez, J.A., Sanz, R., Escuer, M. and Herrero, J. 2000. Biofumigation and organic amendments. *Regional Workshop on Methyl Bromide Alternatives for North Africa and Southern European Countries.* UNEP, Paris, France, pp. 113-141.

Beltrami, G., Zavanella, M. and Curto, G. 2006. Combattere il nematode bieticola in tre mosse. *L'Informatore Agrario* **62**(3): 44-48.

Beltrami, G., Zavanella, M. and Curto, G. 2007. Piante biocide e varieta tolleranti contro il nematode della bieticola. *L'Informatore Agrario* **2**: 57-61.

Bernardi, R., Mari, M., Leoni, O., Casalini, L., Cinti, S. and Palmieri, S. 2004. Biofumigation for controlling post-harvest fruit pathogens. *First Int. Biofumigation Symp.*, Florence, Italy.

Besri, M. 1982a. Solar heating (solarization) of tomato supports for control of *Didymella lycopersici* stem canker. (Abstr.). *Phytopathology* **72**: 939.

Besri, M. 1982b. Conservation et dissemination de *Didymella lycopersici* dans les cultures de tomate par les tuteurs. *Phytopath. Z.* **105**: 1-10.

Besri, M. 1983. Lutte contre le chancre *Didymella lycopersici* de la tomate par chauffage solaire (solarisation) des tuteurs. *Phytopath. Z.* **108**: 333-340.

Besri, M. and Drame, A. 1982. Control of Fusarium oxysporum f. sp. lycopersici and Verticillium dahliae by soil solarization. Proc. of the First Meeting of the Arab Soc. for Plant Prot., Amman, p. 113.

Bhagawati, B., Goswami, B.K. and Singh, C.S. 2000. Management of disease complex of tomato caused by *Meloidogyne incognita* and *Fusarium oxysporum* f. sp. *lycopersici* through bioagents. *Indian J. Nematol.* **30**: 16-22.

Bharadwaj, L.N. and Sharma, S.K. 2000. Y.S. Parmar University of Horticulture and Forestry, Solan, Himachal Pradesh. In Progress Report for the Period Sept. 1997 - Aug. 2000 of National Network Project on Phytophthora Diseases of Horticultural Crops (S.R. Sarma, ed.), pp. 26-40. Indian Inst. of Spices Research, Calicut, Kerala.

Bhat, K. 1989. M. Sc. (Agri.) thesis, Univ. of Agri. Sci., Dharwad, Karnataka.

Bhattacharya, R.K. and Rao, V.N.M. 1984. Effect of soil covers and soil moisture regimes on nematode population in soil and in roots of banana. *J. Res. Assam Agri. Univ.* **5**:206-209.

Bhattacharjee, S.K. and Swami Vivekananda. 1989. In *Commercial Flowers.* (T.K. Bose and L.P. Yadav, eds.). Nayaprakash, Calcutta, 510 pp.

Bhatti, D.S. and Jain, R.K. 1977. Estimation of loss in okra, tomato and brinjal yield due to *Meloidogyne incognita. Indian J. Nematol.* **7**: 37-41.

Bijay Kumar. 2009. *Indian Horticulture Database 2007-08.* National Hort. Board, Gurgaon, Haryana, India, 324 pp.

Block, E., Ahmad, S., Jain, M.K., Crecely, J.W., Apitz-castro, R. and Cruz, M.R. 1984. Ajoene: a potent antithrombic agent from garlic. *J. Am. Chem. Soc.* **106**: 82-95.

Bora, P. and Bora, L.C. 2009. Bioformulation of antagonists for management of bacterial wilt of ginger. *Int. Conf. on Hort.*, Bangalore.

Borek V., Elberson L.R., McCaffry J.P. and Morra M.J. 1997. Toxicity of rapeseed meal and methyl isothiocyanate to larvae of the black vine weevil (Coleoptera: Curculionidae). *J. Econ. Ent.* **90**: 109-112.

Borkakaty, D. 1993. Management of Root-knot Nematode, Meloidogyne incognita (Kofoid and White, 1919) Chitwood, 1949 on Brinjal by Integrated Approach. M. Sc. (Agri.) thesis, Assam Agri. Univ., Jorhat.

Boyd, D.W. and Alverson, D.R., 2000. Repellency effects of garlic extracts on two spotted spider mite, *Tetranychus urticae* Koch. *J. Entomol. Sci.* **35**: 85.

Boydston, R.A., Al-Khatib, K., Vaughn, S.F., Collins, H.P. and Alva, A.K. 2004. Weed suppression using cover crops and seed meals. *First Int. Biofumigation Symp.*, Florence, Italy.

Branca, F. and Furtado, A. 2004. *Brassica macrocarpa* for controlling root-knot nematode in tomato protected crops. *First Int. Biofumigation Symp.*, Florence, Italy.

Brown, D., Brown, J., Hamilton, M., Davis, J. and Seip, L. 2006. The effect of glucosinolate type and quantity on weed seed germination and growth in glasshouse studies. *Second Int. Biofumigation Symp.*, Idaho, USA.

Brown, J., Hamilton, M. and Brown, D.A. 2004. Using brassicaeae seed meal as an alternative to highly toxic soil fumigants in strawberry production. *First Int. Biofumigation Symp.*, Florence, Italy.

Brown, J., Hamilton, M., Davis, J., Brown, D. and Seip, L. 2006. Herbicidal and crop phytotoxicity of Brassicaceae seed meals on strawberry transplants and established crops. *Second Int. Biofumigation Symp.*, Idaho, USA.

Brown, P.D. and Matthew, J.M. 1997. Control of soil-borne plant pests using glucosinolate-containing plants. *Advances in Agronomy* **61**: 167-215.

Brown, P.D. and Morra, M.J. 1997. Control of soil-borne plant pests using glucosinolate -containing plants. *Adv. in Agronomy* **61**:192-204.

Butler, E.J. 1906. The wilt disease of pigeon-pea and pepper. *Agric. J. India* **1**: 25-36.

Byun, J.K., Do, J.H. and Chang, K.H. 1991. The effect of black polythene film and rice straw mulches in vegetative growth and mineral content in jujube. *J. Korean Soc. Hort. Sci.* **32**: 81-86.

Cardwell, D. and Ingham, R. 1996. Management practices to suppress Columbia root-knot nematode. *Pacific Northwest Sustainable Agri.*, Oct., p. 6.

Cartia, G. and Greco, N. 1987. Effetti della solarizzazione del suolo su una coltura di peperone in serra. *Colture Protette* **16**(5): 61-65.

Cartia, G., Greco, N. and Cipriano, T. 1989. Effect of soil solarization and fumigants on soil-borne pathogens of pepper in greenhouse. *Acta Hort.* **255**: 111-116.

Cartia, G., Greco, N. and Cirvilleri, G. 1988. Solarizzazione e bromuro di mettle nella difesa dai parassiti del pomodoro in ambiente protetto. *Proc. Giornate Fitopatologiche*, Lecce, Italy. **1**: 437-488.

Castillo, M.B. 1985. Some studies on the use of organic soil amendments for nematode control. *Philippine Agriculturist* **68**: 76-93.

Caveness, F.E. 1981. Root-knot nematodes on cassava. Proc. of the Third Research Planning Conference on Root-knot Nematodes, Meloidogyne spp., Region IV and V, Ibadan, Nigeria.

Chacko, E.K. and Reddy, A. 1981. Effect of planting distance and intercropping with cowpea on weed growth in banana. *Proc. Eighth APWSS Conf.,* Bangalore, pp. 137-143.

Charles, J.S.K. 1989. *Bioecology of the Cyst Nematode Infecting Banana*. Ph. D. thesis, Kerala Agri. Univ., Trichur, 122 pp.

Charles, J.S.K. and Venkitesan, T.S. 1993. Status Report on the Nematological Investigations on Banana in Kerala Agricultural University. Kerala Agri. Univ., Vellanikkara, 43 pp.

Charles, J.S.K., Venkitesan, T.S., Thomas, Y. and Varkey, P.A. 1985. Correlation of plant growth components to bunch weight in banana infested with burrowing nematode, *Radopholus similis* (Cobb). *Indian J. Nematol.* **15**: 186-190.

Charlton, B.A. 2006. Effects of oxamyl on suppression of the tobacco rattle virus vector Paratrichodorus allius and corky ringspot disease of potato in the Klamath Basin of south-central Oregon. M.S. Agronomy thesis, Oregon State University.

Charron, C.S. and Sams, C. E. 1998. Macerated Brassica leaves suppress Pythium ultimum and Rhizoctonia solani mycelial growth. Proc. of the Annual Int. Conf. on Methyl Bromide Alternatives and Emissions Reductions, Orlando, FL.

Charron, C.S. and Sams, C. E. 1999. Inhibition of *Pythium ultimum* and *Rhizoctonia solani* by shredded leaves of *Brassica* spp. *J. Am. Soc. Hort. Sci.* **124**: 462-467.

Cheema, G.S. and Dani, P.G. *Papaya Cultivation in the Bombay Presidency*. Bull. 162, Dept. of Agri., Bombay.

Chellemi, D.O. and Olson, S.M. 1994. Effects of soil solarization and fumigation on survival of soil-borne pathogens of tomato in Northern Florida. *Plant Dis.* **78**: 1167- 1172.

Chellemi, D.O. Olson, S.M. and Mitchell, D.J. 1994. Effects of soil solarization and fumigation on survival of soil-borne pathogens of tomato in northern Florida. *Plant Dis.* **78**: 1167-1172.

Chen, Y., Gamliel, A, Stapleton, J.J. and Aviad, T. 1991. Chemical, physical and microbial changes related to plant growth in disinfested soils. In:

*Soil Solarization* (J. Katan and J.E. DeVay, eds.), pp. 103-29. CRC Press, Boca Raton, FL.

Chona, B.L., Sethi, C.L., Swarup, G. and Gill, J.S. 1965. Citrus nematode (*Tylenchulus semipenetrans* Cobb, 1913) associated with die-back disease of citrus. *Indian J. Nematol.* **22**: 370-373.

Chundawat, B.S., Joshi, H.S. and Patel, N. 1984. Studies on intercropping in Basrai banana. *S. Indian Hort.* **32**: 23-25.

Chunlei, Z., Jun, L., Ni, M., Feng, L., Dan, L., Liping, Y. and Guangming, L. 2008. Effect of rape/wheat strip intercropping on disease resistance, yield and agronomic characters of winter wheat. *Third Int. Biofumigation Symp.*, Canberra, Australia.

Cohen, M. F., Yamasaki, H. and Mazzola, M. 2005. *Brassica napus* seed meal soil amendment modifies microbial community structure, nitric oxide production and incidence of Rhizoctonia root rot. *Soil Biol. Biochem.* **37**: 1215-1227.

Colombo, A., Cataldi, S. and Marano, G. 2008. Effectiveness of biofumigation technique to control the southern root-knot nematode (*Meloidogyne incognita*) in Sicily. *Third Int. Biofumigation Symp.*, Canberra, Australia.

Cook, R.J. 1993. Making greater use of introduced microorganisms for control of plant pathogens. *Ann. Rev. Phytopathol.* **31**: 53-80.

Corti, C., Crippa, L., Geneveni, P.L. and Centemero, M. 1998. Compost use in plant nurseries: Hydrological and Physicochemical characteristics. *Compost Sci. and Utilization.*Vol 6, **1**: 35-45.

Costello, R.A. and Gillespie, D.R. 1993. The pepper weevil, *Anthonomus eugenii* Cano as a greenhouse pest in Canada. *IOBC Bulletin: Working Group IPM in Glasshouses.* **16**(2): 31-34.

Crosslin, J.M., Thomas, P.E. and Brown, C.R. 1999. Distribution of tobacco rattle virus in tubers of resistant and susceptible potatoes and systemic movement of virus into daughter plants. *Ame. J. Potato Res.* **76**:191-197.

Crow, W.T. 2005. Diagnosis of *Trichodorus obtusus* and *Paratrichodorus minor* on turf grasses in the Southeastern United States. Online. *Plant Health Progress* doi:10.1094/PHP-2005-0121-01-DG.

Curto, G., Dallavalle, E. and Lazzeri, L. 2005. Life cycle duration of *Meloidogyne incognita* and host status of Brassicaceae and Capparaceae selected for glucosinolate content. *Nematology* **7**: 203-212.

Curto, G., Lazzeri, L., Dallavalle, E. and Santi, R. 2008. Management of *Meloidogyne incognita* (Kofoid *et* White) Chitw. in organic horticulture. *Third Int. Biofumigation Symp.*, Canberra, Australia.

Curto, G., Lazzeri, L., Dallavalle, E., Santi, R. and Malguti, L. 2006. Effectiveness of crop rotation with Brassicaceae species for the management of southern root-knot nematode *Meloidogyne incognita* (Abstr.). *Second Int. Biofumigation Symp.*, Idaho, USA, p. 51.

Curto, G., Lazzeri, L., Santi, R. and Dallavalle, E. 2006. Evaluation of alternative strategies for the control of root-knot nematode *Meloidogyne incognita* (Kofoid et White) Chitw. on tomato crop, in plastic greenhouse. *Second Int. Biofumigation Symp.*, Idaho, USA.

Dabbas, M.R., Singh, D.P., Yadav, J.R. and Katiyar, R.P. 2009. Management of *Rhizoctonia* root rot of cauliflower through IDM practice. *Int. Conf. on Hort.*, Bangalore.

Darekar, K.S. and Mahse, N.L. 1988. Assessment of yield losses due to root-knot nematode, *Meloidogyne incognita* Race 3 in tomato, brinjal and bitter gourd. *Int Nematol. Network Newsl.* **5**(4): 7-9.

Darekar, K.S. and Patil, N.G. 1985. *J. Maharashtra Agri. Univ.* **10**: 104.

Das, A.K. 1994. *Assessment of Yield Loss Due to* Meloidogyne incognita *on French Bean and its Management.* M. Sc. (Agri.) thesis, Assam Agri. Univ., Jorhat.

Das, N. and Sinha, A.K. 2005. Integrated management of root-knot nematode (*Meloidogyne incognita*) on okra [*Abelmoschus esculentus* (L.) Moench]. *Indian J. Nematol.* **35**: 175-182.

Davis, J.R., Huisman, O.C., Westerman, D.T., Hafez, S.L., Everson, D.O., Sorenson, L.H. and Schneider, A.T. 1996. Effects of green manure on Verticillium wilt of potato. *Phytopathology* **86**: 444-453.

Davis, J.R. and McDole, R.E. 1979. Influence of cropping sequences on soil-borne populations of *Verticillium dahliae* and *Rhizoctonia solani*. In *Soil-Borne Plant Pathogens* (B. Schipper and W. Gams, eds.), pp. 399-405. Academic Press, London, 685 pp.

Daugovish, O., Downer, A.J. and Mochizuki, M.J. 2006. Mustard-derived biofumigation for lettuce in coastal California. *Second Int. Biofumigation Symp.*, Idaho, USA.

Davis, E.L., Kaplen, D.T., Dickson, D.W. and Mitchels, D.J. 1989. Reproduction of lectin treated *Meloidogyne* spp. in two treated soybean cultivars. *Rev. Nematol.* **12**: 257-263.

Davis, J.R. 1981. *Verticillium* wilt of potato in southeastern Idaho. *Univ. Idaho Curr. Inf. Ser.* 564.

Debode, J., Clewes, E., De Backer, G. and Hofte, M. 2005. Lignin is involved in the reduction of *Verticillium dahliae* var. *longisporum* inoculum in soil by crop residue incorporation. *Soil Biol. Biochem.* **37**: 301-309.

Delaquis, P.J. and G. Mazza. 1995. Antimicrobial properties of isothiocyanates in food preservation. *Food Technol.* November, pp. 73-84.

De Mastro, G., D'Addabbo, T., Lazzeri, L., Verdini, L. and Brunetti, G. 2008. Biofumigation for the control of root-knot nematodes on flower crops. *Third Int. Biofumigation Symp.*, Canberra, Australia.

De Mastro, G., D'Addabbo, T., Lazzeri, L., Verdini, L. and Grassano, N. 2008. Effectiveness of biofumigating *Brassica* treatments for the control of root-knot nematode *Meloidogyne incognita* on melon and watermelon. *Third Int. Biofumigation Symp.*, Canberra, Australia.

De Mastro, G., D'Addabbo, T., Verdini, L. and Radicci, V. 2008a. Biofumigation in greenhouse for the control of root-knot nematodes. *Third Int. Biofumigation Symp.*, Canberra, Australia.

De Mastro G., D'Addabbo T., Verdini L. and Radicci V. 2008b. Control of carrot root-knot and cyst nematodes by biofumigating treatments. *Third Int. Biofumigation Symp.*, Canberra, Australia.

DeVay, J.E., Stapleton, J.J. and Elmore, C.L. (eds.). 1991. *Soil Solarization.* Plant Production and Protection Paper 109. Rome: FAO/UN, 396 pp.

Devi, Geetanjali. 1993. Pathogenicity, Crop Loss Assessment and Management of Meloidogyne incognita (Kofoid and White, 1919) Chitwood, 1949 on Carrot (Dacus carota L.). M. Sc. (Agri.) thesis, Assam Agri. Univ., Jorhat.

Devi, T.P. and Goswami, B.K. 1992. Effect of VA mycorrhiza on the disease incidence due to *Macrophomina phaseolina* and *Meloidogyne incognita* on cowpea. *Ann. Agric. Res.* **13**: 253-265.

Deyton, D., Karpinets, T., Lyons, M., Cummins, J. and Sams, C. 2006. Enhanced biofumigation and composting systems for strawberry. *Second Int. Biofumigation Symp.*, Idaho, USA.

Dhillon, B.S. and Bajwa, G.S. 1976. Weedicides and mulches in vineyards. *Indian Hort.* **21**: 23-24.

Dohroo, N.P. 1995. Integrated management of yellows of ginger. *Indian Phytopath.* **48**: 90-92.

Dugravot, S., Sanon, A., Thibout, E. and Huignard, J. 2002. Susceptibility of *Callosobruchus maculatus* (Coleoptera: Bruchidae) and its parasitoid *Dinarmus basalis* (Hymenoptera: Pteromalidae) to sulfur-containing compounds: Consequences on biological control. *Environ. Entomol.* **31**: 550.

Dutt, R. and Bhatti, D.S. 1986. Determination of effective doses and time of application of nematicides and castor leaves for controlling *Meloidogyne javanica* in tomato. *Indian J. Nematol.* **16**: 8-11.

Eapen, S.J. 1987. Plant parasitic nematode problem in small cardamom. *Proc. of the Third Group Discussion on the Nematological Problems of Plantation Crops*, Sugarcane Breeding Institute, Coimbatore, pp. 43-44.

Eapen, S.J. 1995. *Investigations on Plant Parasitic Nematodes Associated with Cardamom.* Final Report of the Project Nema I (813), Indian Inst. of Spices Research, Calicut, 39 pp.

Eapen, S.J. and Ramana, K.V. 1996. Biological control of plant nematodes of spices. In *Biological Control in Spices* (M. Anandraj and K.V. Peter, eds.), pp. 20-22. Indian Inst. of Spices Res., Calicut.

Eapen, S.J. and Venugopal, M.N. 1995. Field evaluation of *Paecilomyces lilacinus* and *Trichoderma* spp. in cardamom nurseries for the control of root-knot nematodes and rhizome rot disease. *National Symposium on Nematode Problems of India - An Appraisal of the Nematode*

*Management with Eco-friendly Approaches and Biocomponents*, Indian Agril. Res. Inst., New Delhi, p.43.

Egley, G.H. 1983. Weed seed and seedling reductions by soil solarization with transparent polyethylene sheets. *Weed Sci.* **31**: 404-409.

Egunjobi, O.A. 1985. Effect of cocoa pod husks soil amendment on cowpea infestation by *Meloidogyne* spp. *Pakistan J. Nematol.* **3**: 99-103.

Elad, Y., Katan, J. and Chet, I. 1980. Physical, biological, and chemical control integrated for soilborne diseases in potato. *Phytopathology* **70**: 418-422.

Elmore, C.L., Roncaroni, J.A. and Giraud, D.D. 1993. Perennial weeds respond to control by soil solarization. *Calif. Agri.* **47**(1): 19-22.

Elmore, C.L., Stapleton, J.J., Bell, C.E. and DeVay, J.E. 1997. *Soil Solarization: A Nonpesticidal Method for Controlling Diseases, Nematodes and Weeds*. UC DANR Pub. 21377, Oakland, CA, 14 pp.

El-Nagar, H.I., Hendy, H.H., Abdul Hameed, S.H., Farhat, A.A. and Osman, A.A. 1993. *The Role of Dry Ground Leaves of some Plants in Controlling the Reniform Nematode* Rotylenchulus reniformis *Infecting Sunflower*. Bull. 44, Faculty Agric., Univ. Cairo, pp. 205-216.

Fahey, J., Zalemann, A. and Talalay, P. 2001. The chemical diversity and distribution of glucosinolates and isothionates among plants. *Phytochem.* **56**: 50-51.

Ferary S. and Auger J. 1996. What is the true odor of cut *Allium*? Complementary of various hyphenated methods: Gas chromatography-mass spectrometry and high-performance liquid chromatography, mass spectrometry with particle beam and atmospheric pressure ionization interfaces in sulphenic acids rearrangement components discrimination. *J. Chromatogr.* **A 750**: 63.

Fery, R.L. and Dukes, P. 1996. *J. Amer. Soc. Hort. Sci.* **121**: 1024-1027.

Forge, T.A., Russel, E.I. and Diane, K. 1995. Winter cover crops for managing root lesion nematodes affecting small fruit crops in the Pacific Northwest. *Pacific Northwest Sustainable Agri.*, Oct., p. 6.

Frank, D.L. 2004. Evaluation of Living and Synthetic Mulches on Zucchini for Control of Homopteran Pests. M. S. thesis, Univ. of Florida, Gainesville, Florida, USA.

Frank, D.L. and Liburd, O.E. 2005. Effects of living and synthetic mulches on the population dynamics of whiteflies and aphids, their associated natural enemies and insect-transmitted plant diseases in zucchini. *Environmental Entomol.* **34**: 857-865.

Furlan, L., Bonetto, C., Patalano, G. and Lazzeri., L. 2004. Potential of biocidal meals to control wireworm populations. *First Int. Biofumigation Symp.*, Florence, Italy.

Galletti, S., Leoni, O., Cinti, S. and Cerato, C. 2008. An alternative pathway for glucosinolate degradation in soil. *Third Int. Biofumigation Symp.*, Canberra, Australia.

Gamliel, A. and Stapleton, J.J. 1993a. Characterization of antifungal volatile compounds evolved from solarized soil amended with cabbage residues. *Phytopathology* **83**: 899- 905.

Gamliel, A. and Stapleton J.J. 1993b. Effect of soil amendment with chicken compost or ammonium phosphate and solarization on pathogen control, rhizosphere microorganisms and lettuce growth. *Plant Dis.* **77**: 886-891.

Geary, B., Brown, B., Ransom, C. and Hafez, S. 2006. Biofumigants in commercial onion production to control weed, nematode, disease and pests. *Second Int. Biofumigation Symp., Idaho, USA.*

Gibbson, M. 1990. *The Book of Rose.* Mac Donald General Books, 246 pp.

Gnanapragasam, N.C., Dharmasena, W.A.M. and Liyanage, D.D. 1989. Development of age resistance in young tea to different species of *Meloidogyne. Int. Nematol. Network Newsl.* **6**: 4.

Gnanapragasam, N.C. and Manuelpillai, M.E.K. 1990. Influence of inoculum levels and temperature on the population build-up and pathogenicity of the root-lesion nematode, *Pratylenchus loosi* Loof on tea. *Tea Quarterly* **53**: 19-22.

Gopalakrishnan, C. and Ajith Kumar, K. 2006. Increasing the field efficacy of *Pseudomonas fluorescens* in the management of bacterial wilt in tomato, *Lycopersicon esculentum* Mill. *National Symp. on Improving Input Use Efficiency in Hort.*, Indian Inst. of Hort. Res., Bangalore, pp. 197-198.

Goswami, B.K., Uma Rao and Singh, S. 1998. Potentiality of some fungal bioagents against root-knot nematode, *Meloidogyne incognita* infecting tomato. In *Advances in IPM for Horticultural Crops* (P. Parvatha Reddy, N.K. Krishna Kumar and A. Verghese, eds.), pp. 304-307. AAPMHE, Bangalore.

Gouws, R. 2004. Biofumigation as alternative control measure for common scab on seed potatoes in South Africa. *First Int. Biofumigation Symp.*, Florence, Italy.

Gouws, R. and Wehner, F.C. 2006. Can biofumigation effectively control common scab of potato under field conditions? *Second Int. Biofumigation Symp.*, Idaho, USA.

Govindaiah, Dandin, S.B., Philip, T. and Datia, R.K. 1990. Effect of marigold (*Tagetes patula*) intercropping against *Meloidogyne incognita* infecting mulberry. *Indian J. Nematol.* **20**: 96-99.

Greco, N., Brandonisio, A. and Elia, F. 1985. Control of *Ditylenchus dipsaci, Heterodera carotae* and *Meloidogyne javanica* by solarization. *Nematol. Medit.* **13**: 191-197.

Grewal, P.S. 1989. Effect of leaf-matter incorporation on *Aphelenchoides composticola* (Nematode), mycofloral composition, mushroom compost quality and yield of *Agaricus bisporus. Ann. Appl. Biol.* **115**: 299-312.

Grinstein, A., Orions, D., Greenberger, A. and Katan, J. 1979. Solar heating of the soil for the control of *Verticillium dahliae* and *Pratylenchus thornei* in potatoes. In: *Soilborne Plant Pathogens.* (B. Schippers and W. Gams, eds.), pp. 431-433. Academic Press, New York, 686 pp.

Grossman, J. 1990. New crop rotations foil root-knot nematodes. *Common Sense Pest Control*, Winter, p. 6.

Grossman, J. 1999. ESA and APS Joint Meeting-Part 8. *IPM Practitioner*, Oct., p. 13.

Guerrero, M.M., Ros, C., Díez , M.A., López-Pérez, J.A., Martínez, M.Á., Lacasa, C.M., Bello, A. and Lacasa, A. 2008. Biosolarisation as an alternative to methyl bromide in protected pepper crops in Spain. *Third Int. Biofumigation Symp.*, Canberra, Australia.

Gupta, P.K., Agarwal, R.K. and Sharma, S.L. 1986. Efficacy of soil amendments in managing 'Stalk rot' (*Sclerotinia sclerotiorum*) of cauliflower. *Indian J. Mycol. Pl. Path.* **4**: 129-131.

Gupta, R. and Sharma, N.K. 1993. A study of the nematicidal activity of allicin – an active principle in garlic *Allium sativum* L. against root-knot nematode, *Meloidogyne incognita* (Kofoid and White, 1919) Chitwood. *Int. J. Pest Mangmt.* **39**: 390.

Gutal, B.B., Bhilare, R.M. and Takte, R.L. 1992. Mulching effect on yield of tomato crop. In *Int. Agri. Engg. Conf.*, Bangkok, Vol. **3**: 883-887.

Hackney, R.W. and Dickerson, O.J. 1975. Marigold, castor bean and chrysanthemum as controls of *Meloidogyne incognita* and *Pratylenchus alleni. J. Nematol.* **7**: 84-90.

Hafez, S.L. 1998. *Management of Sugarbeet Cyst Nematode.* Univ. of Idaho Co-op. Extn., CIS 1071, P.2.

Hafez, S.L. and Sundararaj, P. 2004. Biological and chemical management strategies in the sugarbeet cyst nematode management. *Proc. of the Winter Commodity Schools* - 2004, Univ. of Idaho, USA, pp. 243-248.

Hafez, S. and Sundararaj, P. 2006. Two decades of green manure crop research in Idaho for nematode management– an overview. *Second Int. Biofumigation Symp.*, Idaho, USA.

Halappa, G. and Srinivas, L. 1973. Effect of black polythene film mulch on soil conditions and yield of cauliflower (*Brassica oleracea* L. var. *botrytis*). *Mysore J. Agric. Sci.* **5**: 501-519.

Halloway, R.I.C. and White, G.C. 1968. Can apples be grassed down in the year of planting (UK East Malling). *Weed Abstr.* **17**: 204.

Hao, J., Subbarao, K.V. and Koike, S.T. 2003. Effects of broccoli rotation on lettuce drop caused by *Sclerotinia minor* on the population density of sclerotia in soil. *Plant Dis.* **87**: 66.

Harrison, B.D. and Robinson, D.J. 1986. Tobraviruses. In: *The Plant Viruses* Vol. 2: The Rod-shaped Plant Viruses (M.H.V. van Regenmortel and H. Fraenkel-Conrat, eds.), pp. 339-369. Plenum Press, N.Y.

Hartz, T.K., DeVay, J.E. and Elmore, C.L. 1993. Solarization is an effective soil disinfestations technique for strawberry production. *Hort. Sci.* **28**: 104-106.

Harvey, S.G., Hannahan, H.N. and Sams, C.E. 2002. Indian mustard and allyl isothiocyanate inhibit *Sclerotium rolfsii*. *J. Amer. Soc. Hort. Sci.* **127**:

Harvey, S.G. and Sams, C.E. 2000. Proc. Ann. Int. Res. Conf. Methyl Bromide Alternatives Emissions Reductions. 92-1

Harvey, S.G. and Sams, C.E. 2002. *Brassica* biofumigation increases tomato fruit production. *American J. Alternative Agri*. (In press).

Hasan, N. and Jain, R.K. 1998. Nematological research in Uttar Pradesh: An overview. In *Phytonematology in India* (P.C. Trivedi, ed.), pp. 150-169. C.B.S. Publishers and Distributors, New Delhi.

Haseeb, A. and Pandey, R. 1989. Observations on *Meloidogyne* spp. affecting Japanese mint - new disease records. *Nematropica* **19**: 93-97.

Haseeb, A. and Butool, F. 1990. Evaluation of different pesticides and oilcake on the control of root-knot nematode (*Meloidogyne incognita*) infesting davana (*Artemisia pallens*). *Curr. Nematol.* **2**: 201-204.

Hashizume, K., Mitsuhori, T., Zhang, X., Shimoji, I., Sano, Z., Sakuma, F., Kitamura, T., Kondo, N., Fujine, O., Sumino, S., Maeda, M. and Ubukata, M. 2008. Introduction of fumigation crop research against soil-borne diseases and nematodes in Japan. *Third Int. Biofumigation Symp.*, Canberra, Australia.

Hashmi, S. and Hashmi, G. 1990. Crop rotation and *Meloidogyne javanica* in vegetable production. *Int. Nematol. Network Newsl.* **7**(1): 19-21.

Haque, M.M. and Gaur, H.S. 1985. Effect of multiple cropping sequences on the dynamics of nematode population and crop performance. *Indian J. Nematol.* **15**: 262-263.

Hazarika, K. 1990. *Pathogenicity and Management of* Meloidogyne incognita *on Brinjal* (Solanum melongena). M. Sc. (Ag) thesis, Assam Agri. Univ., Jorhat.

Heald, C.M. and Robinson, A.F. 1987. Effect of soil solarization on *Rotylenchulus reniformis* in the lower Rio Grande Valley of Texas. *J. Nematol.* **19**: 93-103.

Heald, C.M. and Thomas, C.E. 1983. Nematode control by soil solarization. *J. Nematol*. **13**: 114-115.

Himelrick, D.G. and Dozier, W.A. 1991. Soil fumigation and soil solarization in strawberry production. *Adv. in Strawberry Production* **10**: 12-27.

Hoda, H., Ameen, M. and Mohamed, M.M. 2006. Biofumigation as alternative method for management of root-knot nematode *Meloidogyne incognita* affecting tomato under field conditions. *Second Int. Biofumigation Symp.,* Idaho, USA.

Hoitink, H.A.J. and Boehm, M.J. 1999. Biocontrol within the context of soil microbial communities: A substrate-dependent phenomenon. *Ann. Rev. Phytopathol*. **37**: 427-446.

Hoitink, H.A.J. and Fahy, P.C. 1986. Basis for the control of soilborne plant pathogens with composts. *Ann. Rev. Phytopathol*. **24**: 93-114.

Hoitink, H.A.J. and Grebus, M. 1994. Status of biological control of plant diseases with composts. *Compost Sci. and Utilization*.Vol 6, **1**:35-45.

Holm, L.G., Plucknett, D.L., Pancho, J.V. and Herberger, J.P. 1977. *The World's Worst Weeds. Distribution and Biology*. East-West Centre, Univ. of Hawaii, Honolulu.

Hooks, C.R.R., Valenzuela, H.R. and Defrank, J. 1998. Incidence of pest and arthropod natural enemies in zucchini grown in living mulches. *Agri. Ecosystems and Environment* **69**: 217-231.

Horiuchi, S. 1991. Solarization for greenhouse crops in Japan. *In Soil Solarization* (J. E. De Vay, J. J. Stapleton and C. L. Elmore, eds.). U. N. Food & Agric. Org. Plant Production and Protec. Paper 109.

Horowitz, M., Roger, Y. and Herlinger, G. 1983. Solarization for weed control. *Weed Sci.* **31**: 170-179.

Hussain, Z. and Bora, B.C. 2008. Integrated management of *Meloidogyne incognita* and *Ralstonia solanacearum* complex in brinjal. *Indian J. Nematol.* **38**: 159-164.

Ichniohe, M. 1975. *Japanese J. Nematol.* **5**: 36-40.

Ichniohe, M. 1980. Studies on the root-knot nematode of black pepper plantation in Amazon. *Annual Report of the Society of Plant Protection, North Japan* **31**: 1-8.

Inada, K. 1973. Photo-selective plastic film for mulch. *J. Agri. Res. Quar.* **7**: 252-256.

Ingham, E. 1996. The soil foodweb: Its importance in ecosystem health, 13 pp.

Ingham, R.E., David, N.L., Horneck, D.A., Dillon, M. and Delgado, J. 2006. Strategies for using green manure crops to suppress Columbia root-knot nematode. *Second Int. Biofumigation Symp.*, Idaho, USA.

Isshiki, K., Tokuoka, K., Mori, R. and Chiba, S. 1992. Preliminary examination of allyl isothiocyanate vapor for food preservation. *Biosci. Biotech. Biochem.* **56**: 1476-1477.

Jain, R.K. and Bhatti, D.S. 1987. Population development of root-knot nematode (*Meloidogyne javanica*) and tomato yield as influenced by summer ploughing. *Trop. Pest Mangmt.* **33**: 122-125.

Jain, R.K. and Gupta, D.C. 1991. Integrated management of root-knot nematode, *Meloidogyne javanica* infecting *Solanum melongena*. *Afro-Asian J. Nematol.* **1**: 156-160.

Jain, R.K. and Hasan, N. 1986. Efficacy of "Neem" cake on fodder production, photosynthetic pigments and nematodes associated with oats and its residual effect on cowpea. *Indian J. Nematol.* **16**: 98-100.

Jain, R.K., Paruthi, I.J. and Gupta, D.C. 1990. Influence of intercropping on the incidence of root-knot nematode (*Meloidogyne javanica*) in brinjal. *Indian J. Nematol.* **20**: 236-238.

Jain, R.K., Paruthi, I.J., Gupta, D.C. and Dhankar, B.S. 1986. Appraisal of losses due to root-knot nematode, *Meloidogyne javanica* on okra under field conditions. *Trop. Pest Mangmt.* **32**: 341-342.

Jatala, P. 1985. Biological control of nematodes. In *Advanced Treatise on* Meloidogyne. *Biology and Control,* Vol. 1 (J.N. Sasser and C.C. Carter, eds.), pp. 303-308. North Carolina State Univ. Graphics, Raleigh.

Jatala, P. 1988. Nematodes in tuber and root crops and their management – Improvement of sweet potato (*Ipomea batatas*) in East Africa with some references of other tuber and root crops. *Report of the Workshop on Sweet Potato Improvement in Africa.* ILRAD, Nairobi, Kenya, UNDP Project CIAT – CIP – IITA, pp. 91-100.

Jauhari, O.S. and Srivastava, G.P. 1972. Peaches will do well in Eastern UP. *Indian Hort.* **15**: 13-15.

Jenkins, S.F. and Averre, C.W. 1986. Problems and progress in integrated control of southern blight of vegetables. *Plant Dis.* **70**: 614-619.

Jenkins, W.R. and Coursen, B.W. 1957. The effect of root-knot nematodes, *Meloidogyne incognita acrita* and *M. hapla*, on *Fusarium* wilt of tomato. *Plant Dis. Reptr.* **41**: 182-186.

Jian, P., Dan-ying, W., Chun-mei, X. and Xiu-fu, Z. 2008. Effects on paddy field environment and rice growth using winter rape as green manure. *Third Int. Biofumigation Symp.*, Canberra, Australia.

Jonathan, E.I. and Cannayane, I. 2002. Field application of biocontrol agents for management of spiral nematode, *Helicotylenchus multicinctus* in banana (Abstr.). *Global Conf. on Banana and Plantain,* Bangalore, p. 193.

Jonathan, E.I., Nagalakshmi, S. and Padmanabhan, D. 1990. Estimation of yield losses in betel vine due to *Meloidogyne incognita. Int. Nematol. Network Newsl.* **7** (4): 26.

Jonathan, E.I. and Rajendran, G. 2000. Assessment of avoidable yield loss in banana due to root-knot nematode, *Meloidogyne incognita. Indian J. Nematol.* **30**: 162-164.

Joaquim, T., Smith, V.L. and Rowe, R.C. 1988. Seasonal variation and the effects of wheat rotation on populations of *Verticillium dahliae* Kleb. in Ohio potato field soils. *Ame. Potato J.* **65**: 439–447.

Kaith, D.S. 1981. Intercropping and its after effects on the yield of *kala jira. Indian Cocoa, Arecanut and Spices J.* **18**: 6-7.

Kanwar, R.S. and Bhatti, D.S. 1992. Controlling root-knot nematode, *Meloidogyne javanica* through crop rotation. *Curr. Nematol.* **3**: 65-70.

Katan J. 1987. Soil solarization. In: *Innovative Approaches to Plant Disease Control* (I. Chet, ed.). John Wiley & Sons, New York.

Katan, J., and DeVay, J. 1991. *Soil Solarization.* CRC Press Inc., Boca Raton, Florida.

Katan, J., Greenburger, A., Alon, H. and Grinstein, A. 1976. Solar heating by polyethylene mulching for control of diseases caused by soil-borne pathogens. *Phytopathology* **66**: 683-688.

Kaur, D.J. 1987. Studies on Nematodes Associated with Ginger *(Zingiber officinale* Rosc.) in Himachal Pradesh. Ph. D. thesis, Himachal Pradesh Univ., Shimla, India, 198 pp.

Kaur, P. 1999. *Integrated Management Studies on Pea Diseases*. M. Sc. thesis, Himachal Pradesh Agri. Univ., Palampur, 75 pp.

Kerr, A. and Vythilingam, M.K. 1966. Replanting eelworm infested areas. *Tea Quarterly* **37**: 67-72.

Khan, A.M., Saxena, S.K. and Siddiqi, Z.A. 1971. Efficacy of *Tagetes erecta* in reducing root infesting nematodes of tomato and okra. *Indian Phytopath.* **24**: 166-169.

Khan, A.M., Saxena, S.K., Siddiqi, Z.A. and Upadhyay, R.S. 1975. Control of nematodes by crop rotation. *Indian J. Nematol.* **5**: 214-221.

Khan, R.M. and Parvatha Reddy, P. 1992. Nematode pests of ornamental crops. In *Nematode Pests of Crops* (D. S. Bhatti and R. K. Walia, eds.), pp. 250-257. CBS Publishers & Distributors, Delhi, 381 pp.

Khan, R.M. and Parvatha Reddy, P. 1994. Nematode problems of ornamental crops and their management. In *Floriculture - Technology, Trades and Trends,* (J. Prakash and K. R.Bhandary, eds.), pp. 468-472. Oxford & IBH Publishing Co. Pvt. Ltd., New Delhi.

Kirkegaard, J.A. and Matthiessen, J. 2004. Developing and refining the biofumigation concept. *Agroindustria* **3**: 233-239.

Kirkegaard, J.A. and Sarwar, M. 1998. Biofumigation potential of brassicas. I. Variation in glucosinolate profiles of diverse field-grown brassicas. *Plant and Soil* **201**, 71 - 89.

Kirkegaard, J.A., Sarwar, M., Wong, P.T.W. and Mead, A. 1998. Biofumigation by brassicas reduces take-all infection. In "Agronomy – Growing a Greener Future" *Proc. of the 9th Australian Agronomy Conf.* (D.L. Michalk and J.E. Pratley, eds.) pp. 465-468.

Kirkegaard, J.A., Simpfendorfer, S., Holland, J., Bambach, R., Moore, K.J. and Rebetzke, G.J. 2004. Effect of previous crops on crown rot and yield of durum and bread wheat in northern NSW. *Aust. J. Agric. Res.* **55**: 321-334.

Koch, D.W. 1995. Brassica utilization in sugarbeet rotations for biological control of cyst nematode. SARE project report LW91-022. Western Region SARE, Logan, Utah.

Koshy, P.K. 1986. The burrowing nematode, *Radopholus similis* (Cobb, 1893) Thorne, 1949. In *Plant Parasitic Nematodes of India-Problems and Progress.* (G. Swarup and D.R. Dasgupta, eds.), pp. 223-248. Indian Agri. Res. Inst., New Delhi, India.

Koshy, P.K. and Geetha, S.M. 1992. Nematode pests of palms and cocoa. In *Nematode Pests of Crops* (D.S. Bhatti and R.K. Walia, eds.), pp. 214-227. CBS Publishers & Distributors, Delhi.

Koshy, P.K., Nair, R.C.P., Sosamma, V.K. and Sundararaju, P. 1976. On the incidence of root-knot nematode in cardamom nurseries. *Indian J. Nematol.* **6**: 174-175.

Koshy, P.K. and Sosamma, V.K. 1987. Pathogenicity of *Radopholus similis* on coconut (*Cocos nucifera* L.) seedlings under greenhouse and field conditions. *Indian J. Nematol.* **17**: 108-118.

Koshy, P.K., Sosamma, V.K. and Sundararaju, P. 1979. Reactions of thirty one areca germplasm collections to the burrowing nematode, *Radopholus similis*. *Plant Dis. Reptr.* **63**: 433-435.

Koshy, P.K., Sundararaju, P. and Sosamma, V.K. 1978. Occurrence and distribution of *Radopholus similis* (Cobb, 1893) Thorne, 1949 in South India. *Indian J. Nematol.* **8**: 49-58.

Koenig, R.L. 2002. APSnet feature (website). U.S. National Organic Program Standards: Implications for Researchers.

Koike, S. and Subbarao, K.V. 2000. Broccoli residues can control *verticillium* wilt of cauliflower. *Calif. Agric.* **54**(3): 30-33.

Konar, A., Singh, N.J. and Paul, R. 2010 Influence of intercropping on population dynamics of major insect pests and vectors of potato. *J. Entomol. Res.* 34(2):

Krikun, J. and Orion, D. 1979. *Verticillium* wilt of potato: Importance and control. *Phytoparasitica* **7**: 107-116.

Kumar, A.C., Viswanathan, P.R.K. and D'Souza, G.I. 1971. A study of plant parasitic nematodes of certain commercial crops in coffee tracts of South India. *Indian Coffee* **36**: 1-3.

Lacasa, A., Guirao, P., Guerrero, M.M., Ros, C., López-Pérez, J.A., Bello, A. and Bielza, P. 2000. *Proc. Int. Workshop on Alternatives to MB for the Southern European Countries*. Agric. Minist. of Greece-DGXII, 133-135.

Lal, B.S. and Ansari, M.N.A. 1960. Field studies on the root-knot nematodes (*Meloidogyne* spp.) (Nematoda: Heteroderidae). *Sci. and Cult.* **26**: 279-281.

Lamberti, F. 1981. Combating nematode vectors of viruses. *Plant Dis.* **65**: 113-117.

LaMondia, J.A. and Brodie, B.B. 1984. Control of *Globodera rostochiensis* by solar heat. *Plant Dis.* **68**: 474-476.

LaMondia, J.A., Gent, M.P.N., Ferrandino, F.J., Elmer, W.H. and Stoner, K.A. 1999. Effect of compost amendment or straw mulch on potato early dying disease. *Plant Dis.* **83**: 361-366.

Lancaster J.E. and Collin H.A. 1981. Presence of alliinase in isolated vacuoles and of alkyl cysteines sulfoxides in the cytoplasm of bulbs of onion (*Allium cepa*). *Plant Sci. Letters* **22**: 169.

Larkin, R.P. and Griffin, T.S. 2007. Control of soilborne potato diseases with *Brassica* green manures. *Crop Prot.* **26**: 1067–1077.

Larkin, R.P., Griffin, T.S. and Honeycutt, C.W. 2006. Crop rotation and cover crop effects on soilborne diseases of potato. *Phytopathology* **96**: S64.

Lazzeri, L., Leoni, O., Bernardi, R., Malaguti, L. and Cinti, S. 2004. Plants, techniques and products for optimising biofumigation in full field. *Agroindustria* **3**: 181-188.

Lazzeri, L., Leoni, O., Nicola, G.D., Malaguti, L. and Bernardi, R. 2008. A simple analytical method for dhurrin content evaluation in cyanogenic plants for their application as biofumigant. *Third Int. Biofumigation Symp.*, Canberra, Australia.

Leihner, D.C. 1980. Cultural method of weed control in cassava. *Cassava cultural Practices Workshop*. Int. Devp. Centre, Brazil, pp. 107-111.

Lifshitz, R., Tabachnik, M., Katan, J. and Chet, I. 1983. The effect of sublethal heating on sclerotia of *Sclerotium rolfsii*. *Can J. Microbiol.* **29**: 1607-1610.

Lockeretz, W. 2000. Organic farming research, today and tomorrow. *Proc. 13th Int. IFOAM Scientific Conf.* (T. Alfoldi, W. Lockeretz and U. Niggli, eds.). vdf Hochschulverlag AG an der ETH, Zurich.

Loos, C.A. 1961. Eradication of the burrowing nematode, *Radopholus similis*, from bananas. *Plant Dis. Reptr.* **45**: 457-461.

Lopez-Perez, J.A., Roubtsova, T., de Cara-Garcia, M. and Ploeg, A.T. 2008. Rotation and biofumigation with five winter-grown crops and subsequent root-knot nematode infestation and yield of summer-grown tomato. *Agri. Ecosystems and Envt.* **123** (In press).

Luna, J. 1993. Crop rotation and cover crops suppress nematodes in potatoes. *Pacific Northwest Sustainable Agri.*, March, pp. 4-5.

Lyons, M.T., Sams, C.E., Peacock, A.D. and White, D.C. 2001. Combining methyl isothiocyanate, composts, and *Brassica* amendments to control southern blight of tomato. *Proc. of the Annual Int. Conf. on Methyl Bromide Alternatives and Emissions Reductions*, San Diego, CA.

Mace, M.E., Bell A. A. and Beckman, C.H. (eds.). 1981. *Fungal Wilt Diseases of Plants*. Academic Press. New York. 640 pp.

Maheswari, S.K., Dahatonde, B.N., Yadav, S. and Gandrade, S.K. 1985. Intercropping of *Rauwolfia serpentina* for higher monetary returns. *Indian J. Agric. Sci.* **55**: 332-334.

Mahrer, Y. and Katan, J. 1981. Spatial soil temperature regime under transparent polyethylene mulch: Numerical and experimental studies. *Soil Sci.* **131**(2):82-87.

Mai, W.F., Brodie, B.B., Harrison, M.B. and Jatala, P. 1981. Nematodes. In *Compendium of Potato Diseases* (W.J. Hooker, ed.), pp. 93-101. American Phytopathol. Soc.

Majumdar, P.K. and Kahali, A.S. 1986. An improved method of cultivation of *Cymbopogon winterianus*. *Indian Perfumer* **30**: 373-375.

Mallek, S.B., Prather, T.S. and Stapleton, J.J. 2007. Interaction effects of *Allium* spp. residues, concentrations and soil temperature on seed germination of four weedy plant species. *Appl. Soil Ecol.* **37**: 9.

Manduric, S. 2008. Cover crops in the year before planting potatoes - a key factor in a spraing disease management strategy. *Third Int. Biofumigation Symp.*, Canberra, Australia.

Mangoensoekarjo, S.N. and Kadnan, N. 1978. The use of herbicides for cover crop establishment in immature rubber. *Weed Abstr.* **28**: 133.

Mani, A. 1986. Occurrence of *Meloidogyne javanica* on citrus in Andhra Pradesh (India). *Int. Nematol. Network Newsl.* **3**(2): 9-10.

Mani, A. 1988. Effect of interculture of marigold and mustard with acid lime on citrus nematode, *Tylenchulus semipenetrans. Int. Nematol. Network Newsl.* **5**(4): 14-15.

Mani, A., Naidu, P.H. and Madhavachari, S. 1987. Occurrence and control of *Meloidogyne incognita* on turmeric in Andhra Pradesh, India. *Int. Nematol. Network Newsl.* **4**: 13-18.

Marchetti, R., Lazzeri, L., Malaguti, L., Orsi, A. and Sghedoni, L. 2008. Potentially mineralizable nitrogen in soil amended with biocidal and non-biocidal plant materials. *Third Int. Biofumigation Symp.*, Canberra, Australia.

Masood Ali. 1986. *Trop. Pest Mangmt.* **34**: 384-387.

Mathews, V., Aravindakshan, M. and Parameshwaran, N.K. 1987. Intercropping in rainfed banana Musa (AAB) "Palay Kondan". *Agric. Res. J.* Kerala **25**: 201-205.

Mattner, S., Gounder, R., Porter, I. and Allen, D. 2008. Biofumigation against soilborne pathogens and weeds of strawberry. *Third Int. Biofumigation Symp.*, Canberra, Australia.

Maqbool, M.A., Hashmi, S. and Ghaffar, A. 1982. Effect of latex extract from *Euphorbia caducifolia* and *Calatropis procera* on root-knot nematode *Meloidogyne incognita* infesting tomato and eggplant. *Pakistan J. Nematol.* **5**: 43-47.

Mazzola, M. and Abi Ghanem, R. 2006. Application of brassicaceae seed meals for the management of apple replant disease. *Second Int. Biofumigation Symp.,* Idaho, USA.

MBTOC. 1998. Report of the Methyl Bromide Technical Options Committee. Assessment of Alternatives to Methyl Bromide. UNEP, Kenya, 354 pp.

McCaffrey, J.P., Morra, M.J. and Main, M.E. 2006. *Brassica juncea* seed meal for fungus gnats control associated with house plants with consideration of phytotoxicity to specific plant species. *Second Int. Biofumigation Symp.*, Idaho, USA.

McGuire, A.M. 2003. Mustard green manures replace fumigant and improve infiltration in potato cropping system. Online. Crop Management. doi:10.1094JCM-2003-0822-01-RS.

McLeod, R.W. and Steel, C.C. 1999. Effects of Brassica-leaf green manures and crops on activity and reproduction of *Meloidogyne javanica*. *Nematology* **1**: 613-624.

McSorley, R. and Parrado, J.L. 1986. Application of soil solarization to Rockdale soils in a subtropical environment. *Nematropica* **16**: 125-140.

Medhane, N.S., Jagdale, G.B., Pawar, A.B. and Darekar, K.S. 1985. Effect of *Tagetes erecta* on root-knot nematodes infecting betel vine. *Int. Nematol. Network Newsl.* **2**(3): 11-12.

Melakeberham, H., Xu, A., Kravehenko, A., Mennan, S. and Riga, E. 2006. Potential use arugula (*Eruca sativa* L.) as a trap crop for *Meloidogyne hapla*. *Nematology* **8**: 793-799.

Menon, K.K. 1949. The survey of pollu and root diseases of pepper. *Indian J. Agric. Sci.* **19**: 89-136.

Mithen, R.F., Lewis, B.G. and Fenwick, G.R. 1986. *In vitro* activity of glucosinolates and their products against *Leptosphaeria maculans*. *Trans. Br. Mycol. Soc.* **87**: 433-440.

Mohandas, C. 2001. Nematode diseases of tuber crops and their management. *Natl. Cong. on Centenary of Nematol. in India – Appraisal & Future Plans*, Indian Agri. Res. Inst., New Delhi, pp. 35-36.

Mohandas, C. and Ramana, K.V. 1987. Slow wilt disease of black pepper and its control. *Indian Cocoa, Areca nut and Spices J.* **11**: 10-11.

Mohandas, C. and Ramana, K.V. 1991. Pathogenicity of *Meloidogyne incognita* and *Radopholus similis* on black pepper (*Piper nigrum* L.). *J. Plantation Crops* **19**: 41-43.

Mohanty, D.C., Sharma, Y.N. and Panda, B.S. 1991. Influence of mulch materials and cover crop on yield of turmeric crop under rainfed condition. *Indian Cocoa, Arecanut and Spices J.* **14**: 97-99.

Mojtahedi, H., Santo, G.S., Hang, A.N. and Wilson, J.H. 1991. Suppression of root-knot nematode population with selected rapeseed cultivars as green manure. *J. Nematol.* **23**: 170.

Mojtahedi, H., Santo, G.S., Wilson, J.H. and Hang, A.N. 1993. Managing *Meloidogyne chitwoodi* on potato with rapeseed as green manure. *Plant Dis.* **77**: 42-46.

Mojumder, V. and Mishra, S.D. 1991. Nematicidal efficacy of some plant products and management of *Meloidogyne incognita* in pulse crops by soaking seeds in their aqueous extracts. *Curr. Nematol.* **2**: 27-32.

Motisi, N., Bailey, D.J., Filipe, J.A.N., Lucas, P., Doré, T. and Montfort, F. 2008. Analysing disease progress curves to understand the effects of biofumigation on rhizoctonia root-rot of sugar beet. *Third Int. Biofumigation Symp.*, Canberra, Australia.

Motisi, N., Montfort, F., Faloya, V., Doré, T. and Lucas, P. 2008. Growing crops of *Brassica juncea* and then incorporating their residues give complementary control of Rhizoctonia root rot of sugar beet. *Third Int. Biofumigation Symp.*, Canberra, Australia.

Muller, J. 1999. The economic importance of *Heterodera schactii* in Europe. *Helminthologia* **36**: 205-213.

Muthuswamy, S. and Thamburaj, S. 1964. A note on effect of mulching with black polythene film on vegetative vigour of grape, Anab-e-Shahi. *S. Indian Hort.* **12**: 95-98.

Mutnal, S.M. 1984. M.Sc. (Agri.) thesis, Univ. of Agri. Sci., Dharwad, Karnataka.

Nadakal, A.M. and Thomas, A.N. 1967. Observations on nematodes associated with dry rot of *Dioscorea alata* L. *Sci. and Cult.* **33**: 142-143.

Nagesh, M., Parvatha Reddy, P. and Rao, M.S. 1996. Integrated management of *Meloidogyne incognita* on tuberose using *Paecilomyces lilacinus* in combination with plant extracts. *Nematol. Medit.* **25**: 3-6.

Nageswari, S. and Mishra, S.D. 2005. Integrated nematode management schedule incorporating neem products, VAM and soil solarization against *Heterodera cajani* infecting pigeonpea. *Indian J. Nematol.* **35**: 68-71.

Naik, D. 2004. Biotechnological Approaches for the Management of Wilt Disease Complex in Capsicum (*Capsicum annum* L.) and Egg plant (Solanum melongena) with Special Emphasis on Biological Control. Ph. D. thesis, Kuvempu Univ., Shimoga.

Nair, P.C.S. 1982. Agronomy of ginger and turmeric. *Proc. National Seminar on Ginger and Turmeric*, Central Plantation Crops Res. Inst., Kasargod, pp. 63-68.

Nammour, D., Auger, J. and Huignard, J. 1989. Mise en évidence de l'effet insecticide de composes soufrés (disulfures et trisulfures) sur *Bruchidus atrolineatus* (Pic) (Coléoptère : Bruchidae). *Insect Sci. Applic.* **10**: 49.

Nanjappa, C.K., Panda, M. and Seshadri, A.R. 1978. Effects of *Meloidogyne incognita* and *Rotylenchulus reniformis* on the emergence and survival of seedlings. *Third Int. Cong. on Plant Pathol.*, Germany.

Nayar, K., Mathew, J., Kuriyan, J., Gnanamanickam, S.S. and Mahadevan, A.. 1988. Role and association of root-knot nematode (*Meloidogyne incognita*) in the bacterial wilt of brinjal incited by *Pseudomonas solanacearum*. In *Advances in Research on Pant Pathogenic Bacteria.* Based on the proceedings of the *National Symposium on Phytobacteriology*, University of Madras, Madras, India, pp. 45-48.

Nehl, D.B. 2008. Utilising green manures in irrigated agriculture: Brassicas and other options. *Third Int. Biofumigation Symp.*, Canberra, Australia.

Netscher, C. 1983. Control of *Meloidogyne incognita* in vegetable production by crop rotation in Ivory Coast. *Acta Hort.* **152**: 219-225.

Nigam, K.B., Bhagawat Prasad, G.S. and Kahal, L.S. 1986. Studies on intercropping of poppy with garlic in black cotton soil. *Indian J. Agron.* **31**: 219-222.

Nirula, K.K. 1963. Collateral host plants of root-knot nematodes. *Curr. Sci.* **32**: 221-222.

Nisha, M.S. and Sheela, M.S. 2006. Bio-management of *Meloidogyne incognita* on coleus, *Solenostemon rotundifolius* by integrating solarization, *Paecilomyces lilacinus, Bacillus macerans* and neem cake. *Indian J. Nematol.* **26**: 136-138.

Noble, R.R.P., Harvey, S.G. and Sams, C.E. 2002. Toxicity of Indian mustard and allylisothiocyanate to masked chafer beetle larvae. Online. Plant Health Progress doi:10.1094/PHP-2002-0610-01-RS.

Nyczepir, A.P. and Rodriguez-Kabana, R. 2004. Effectiveness of biofumigation for ring nematode control in a young peach orchard. *First Int. Biofumigation Symp.*, Florence, Italy.

Ogden, S. 1997. Marigolds bite back. *National Gardening*, March-April, p.21.

Os, G., Boer, M., Bijman, V. and Lazzeri, L. 2008. Biofumigation for soil-borne disease management in flower bulb culture. *Third Int. Biofumigation Symp.*, Canberra, Australia.

Ozores-Hampton, M., McSorley, R., Stansly, P.A., Roe, N.E. and Chellemi, D.O. 2004 Long term large scale soil solarization as a low-input production system for Florida vegetables. *Acta Hort.* **638**: 177-188.

Palanisamy, S. and Sivakumar, C.V. 1981. Assessment of avoidable yield loss in cowpea, black gram, maize and finger millet. *Second Nematol. Symp.*, Tamil Nadu Agricultural University, Coimbatore, p. 58.

Pandey, R. 2001. Influence of *Meloidogyne incognita, Pratylenchus thornei* and *Tylenchorhynchus vulgaris* on growth and oil yield of menthol mint. *Indian J. Nematol.* **31**: 111-114.

Pandey, R. 2005. Management of *Meloidogyne incognita* in *Artemisia pallens* with bioagents. *Phytoparasitica* **33**: 304-308.

Pandey, R. and Kalra, A. 2005. Root-knot disease of ashwagandha *Withania somnifera* and its eco-friendly cost effective management. *J. Mycol. Plant Pathol.* **33**: 240-245.

Pandey, R., Singh, H.B., Gupta, M.L., Singh, H.N., Sharma, S. and Kumar, S. 1998. Efficacy of VAM fungi, green manure, organic materials and pesticides in the management of root-knot nematode in menthol mint. *Proc. of National Symp. on Rational Approaches in Nematode Mangmt. for Sustainable Agri.* (S.C. Dhawan and K.K. Kaushal, eds.), pp. 47-50. Gujarat Agri. Univ., Anand.

Paruthi, I.J., Nandal, S.N. and Verma, K.K. 2010. Field evaluation of mustard and neem cake against root-knot nematode (*Meloidogyne* spp.) in bottle gourd. *National Conf. on Innovations in Nematological Research for Agricultural Sustainability – Challenges and A Roadmap Ahead.* Tamil Nadu Agri. Univ., Coimbatore, p. 84.

Paruthi, L.J., Jain, R.K. and Gupta, D.C. 1987. Effect of different periods of degradation of subabool leaves alone and in combination with nematicides in okra. *Indian J. Nematol.* **17**: 30-32.

Parvatha Reddy, P. 1985. Estimation of crop losses in peas due to *Meloidogyne incognita. Indian J. Nematol.* **15**: 226.

Parvatha Reddy, P. and Khan, R.M. 1991. Integrated management of root-knot nematodes infecting okra. *Curr. Nematol.* **2**: 115-116.

Parvatha Reddy, P., Rao, M.S. and Nagesh, M. 1996. Management of the citrus nematode on acid lime by integration of parasitic fungi and oil cakes. *Pest Mangmt. Hortil. Ecosystems* **2**: 15-18.

Parvatha Reddy, P. and Singh, D.B. 1981. Assessment of avoidable yield loss in okra, brinjal, French bean and cowpea due to root-knot nematodes. *Third Int. Symp. on Plant Pathol.*, New Delhi, pp.93-94.

Parvatha Reddy, P., Singh, D.B. and Ram Kishun. 1979. Effect of root-knot nematodes on the susceptibility of Pusa Purple Cluster brinjal to bacterial wilt. *Curr. Sci.* **48**: 915-916.

Patil, N.M., Dharne, P.K. and Deshmukh, S.S. 1999. Sucker dip treatment in monocrotophos with intercrops for the management of plant parasitic nematodes in banana. *Seminar on Technological Advancement in Banana Production, Handling and Processing Management*, Jalgaon, Maharashtra, pp. 126-128.

Pereira, C.M.F., Chaves, G.M., Zambolim, L., Matsuoka, K., Acuna, R.S. and DoVale, F.X.R. 1996. Control of *Sclerotium cepivorum* by the use of vermicompost, solarization, *Trichoderma harzianum* and *Bacillus subtilis*. *Summa Phytopathologica* **22**: 228-234.

Perwez, M.S., Rehman, M.F. and Haider, S.R. 1988. Effect of *Tagetes erecta* on *Meloidogyne javanica* infecting lettuce. *Int. Nematol. Network Newsl.* **5**(3): 18-19.

Ploeg, A.T. and Stapleton, J.J. 2001. Glasshouse studies on the effects of time, temperature and amendment of soil with broccoli plant residues on the infestation of melon plants by *Meloidogyne incognita* and *M. javanica*. *Nematology* **3**: 855-861.

Porras, M., Barrau, C., Romero, E., Zurera, C. and Romero, F. 2009. Effect of biofumigation with *brassica carinata* and soil solarization on *phytophthora* spp. and strawberry yield. *Acta Hort.* **842**: 969-972.

Porter, I.J. and Merriman, P.R. 1983. Effects of solarization of soil on nematode and fungal pathogens at two sites in Victoria. *Soil Biol. Biochem.* **15**: 39-44.

Porter, L.A. 1962. Polythene sheet as mulch for strawberries. *New Zealand J. Agric.* **105**: 247-255.

Potter, M.J., Davies, K. and Rathjen, A.J. 1998. Suppressive impact of glucosinolates in *Brassica* vegetative tissues on root lesion nematode *Pratylenchus neglectus*. *J. Chem. Ecol.* **24**: 67-80.

Powelson, M.L. and Rowe, R.C. 1993. Biology and management of early dying of potatoes. *Ann. Rev. Phytopath.* **31**: 111-126.

Prasad, S.K. and Chawla, M.L. 1965. Observations on the population fluctuations of citrus nematode, *Tylenchulus semipenetrans* Cobb, 1913. *Indian J. Entomol.* **27**: 450-455.

Prasad, S.K. and Dasgupta, D.R. 1964. Plant parasitic nematodes associated with roses in Delhi. *Indian J. Entomol.* **26**: 121-122.

Pullman, G., DeVay, J., Garber, R. and Weinhold, A. 1981. Soil solarization for the control of Verticillium wilt of cotton and the reduction of soilborne populations of *Verticillium dahliae, Pythium spp., Rhizoctonia solani,* and *Thielaviopsis basicola. Phytopathology* **71**: 954-959.

Punja, Z.K. 1985. The biology, ecology, and control of *Sclerotium rolfsii. Ann. Rev. Phytopathol.* **23**: 97-127.

Putnam, A.R. and DeFrank, J. 1983. Use of phytotoxic plant residues for selective weed control. *Crop Prot.* **2**: 81.

Raj, B.T. and Nirula, K.K. 1969. Trials on the eradication of root-knot nematodes from potato fields with DD (Abstr.). *All India Nematol. Symp.,* Indian Agri. Res. Inst., New Delhi, India.

Rajagopalan, P. and Naganathan, T.G. 1977. Studies on nematode parasites of grapevine. *Tamil Nadu Agri. Univ. Ann. Rept.* **6**: 129.

Rajendran, G., Naganathan, T.G. and Sivagami, V. 1979. Studies on banana nematodes. *Indian J. Nematol.* **9**: 54.

Rajendran, G., Vadivelu, S. and Muthukrishnan, T.S. 1976. Pathogenicity of *Meloidogyne incognita* on crossandra. *Indian J. Nematol.* **6**: 115-116.

Ram, S. and Gupta, D.C. 2001. Studies on the effect of intercropping of onion on the growth of tomato and population dynamics of root-knot nematode, *Meloidogyne javanica. National Conf. on Centenary of Nematol. in India – Appraisal & Future Plans*, Indian Agri. Res. Inst., New Delhi, p. 94.

Ramana, K.V., Mohandas, C. and Balakrishnan, R. 1987. Role of plant parasitic nematodes in the slow wilt disease complex of black pepper (*Piper nigrum* L.) in Kerala. *Indian J. Nematol.* **17**: 225-230.

Ramappa, H.K. 1988. *Studies on Nematode Parasites of French Bean and their Control*. M. Sc. (Agri.) thesis, Univ. of Agri. Sci., Bangalore.

Ramirez-Villapudua, J. and Munnecke, D.M. 1987. Control of cabbage yellows *(Fusarium oxysporum* f. sp. *conglutinans)* by solar heating of fields amended with dry cabbage residues. *Plant Dis.* **71**: 217-221.

Randhawa, G.S. and Mukhopadhyay, A. 1986. *Floriculture in India.* Allied Publishers Pvt. Ltd., Bangalore.

Rangahau, M.K. 2003. Potato early dying (Verticillium wilt) – Strategies for control. Crop and Food Research (Broad sheet) No. 126. http://www.crop.cri.nz/psp/books/bradsheets.htm.

Rangaswamy, S.D., Parvatha Reddy, P. and Nanje Gowda, D.N. 1999. Management of root-knot nematode, *Meloidogyne incognita* in tomato by intercropping with marigold and mustard. *Pest Mangnt. in Hortl. Ecosystems* **5**: 118-121.

Rao, M.S. and Parvatha Reddy, P. 1992. Studies on the comparative efficacy of certain plant leaves and carbofuran in the management of *Meloidogyne incognita* on tomato. *Curr. Nematol.* **3**: 5-6.

Rao, M.S. and Parvatha Reddy, P. 1993. Interactive effect of *Verticillium chlamydosporium* and castor cake on the control of root-knot nematode on egg plant. In *"Soil Organisms and Sustainability"* (D. Rajagopal, R.D. Kale and K. Bano, eds.), pp. 114-118. Indian Soc. Soil Biol. Ecol., Univ. of Agri. Sci., Bangalore.

Rao, M.S., Mohandas, S. and Parvatha Reddy, P. 1993. Management of reniform nematode on egg plant by integrating botanical (castor cake), mycorrhiza (*Glomus fasciculatum*) and bioagent (*Paecilomyces lilacinus*). *Golden Jubilee Symposium on Horticultural Research - Changing Scenario,* Bangalore, pp.259-260.

Rao, M.S., Parvatha Reddy, P. and Mohandas, S. 1996. Effect of integration of *Calotropis procera* leaf and *Glomus fasciculatum* on the management of *Meloidogyne incognita* infesting tomato. *Nematol. Medit.* **24**: 59-61.

Rao, M.S., Parvatha Reddy, P., Somasekhar, N. and Nagesh, M. 1997. Management of root-knot nematodes, *Meloidogyne incognita* in tomato nursery by integration of endomycorrhiza, *Glomus fasciculatum* with castor cake. *Pest Mangmt. Hortil. Ecosystems* **3**: 31-35.

Rao, M.S., Parvatha Reddy, P., Sukhada, M., Nagesh, M. and Pankaj. 1998. Management of root-knot nematodes on eggplant by integrating endomycorrhiza (*Glomus fasciculatum*) and castor (*Ricinus communis*) cake. *Nematol. Medit.* **26**: 217-219.

Rao, M.S., Tewari, R.P. and Pandey, M. 1992. Occurrence of mushroom nematodes in South India. *Mushroom Res.* **1**: 135-136.

Rao, M.S., Tewari, R.P., Reddy, P.P. and Pandey, M. 1991. Comparative efficacy of *Glyricidia maculata* and *Pedilanthus tithymaloides* in the management of nematode *Aphelenchoides sacchari* attacking *Agaricus bisporus*. *Proc. National Symp. Mushrooms*, Trivandrum, pp. 245-247.

Rathaiah, Y. 1986. Turmeric cultivation in Assam. *Indian Cocoa, Arecanut and Spices J.* **9**: 91-93.

Reddy, D.D.R. 1985. Analysis of crop losses in tomato due to *Meloidogyne incognita. Indian J. Nematol.* **15**: 55-59.

Rethinam, R., Sankaran, N. and Sankaran, S. 1977. Chemical weed control in turmeric. *Proc. Ann. Conf.,* ISWS, Hyderabad, p. 218.

Riga, E., Mojtahedi, H., Ingham, R. and McGuire, A. 2004. Green manure amendments and management of root-knot nematodes on potato in Pacific North West in USA. *Proc. of the Fourth Int. Cong of Nematol.* (R.C. Cook and D.J. Hunt, eds.), pp. 151-158. Brill, Leiden and Boston.

Riga, E., Pierce, F. and Collins, H. 2006. The use of arugula on its own and in combination with systemic nematicides against plant parasitic nematodes of potatoes (Abstr.). *Second Int. Biofumigation Symp.,* Idaho, USA, p. 42.

Ristaino, J.B., Perry, K.B. and Lumsden, R.D. 1991. Effect of solarization and *Gliocladium virens* on sclerotia of *Sclerotium rolfsii*, soil microflora, and the incidence of southern blight of tomato. *Phytopathology* **91**: 1117-1124.

Ristaino, J.B., Perry, K.B. and Lumsden, R.D. 1996. Soil solarization and *Gliocladium virens* reduce the incidence of southern blight (*Sclerotium rolfsii*) in bell pepper in the field. *Biocontrol Sci. Technol.* **6**: 583-593.

Robertson, B. and Grant, C. 2008. Specialised Australian cover crops - the Seedco experience. *Third Int. Biofumigation Symp.*, Canberra, Australia.

Rongai, D., Lazzeri, L., Cerato, C., Palmieri, S. and Patalano, G. 2006. A liquid formulation as biopesticide for the control of California red scale (*Aonidiella aurantii* Maskell). *Second Int. Biofumigation Symp.*, Idaho, USA.

Rosa, E.A.S. and Rodrigues, P.M.F. 1999. Towards a more sustainable agriculture system: The effect of glucosinolates on the control of soil-borne diseases. *J. Hort. Sci. Biotechnol.* **74**: 667-674.

Rowe, R.C. 1983. Early Dying – east and west. *Am. Veg. Grow.* **31**(3): 8-10.

Rowe, R.C. and Powelson, M.R. 2002. Potato Early Dying: Management challenges in a changing production environment. *Plant Dis.* **86**: 1184-1193.

Rubin, B. and Benjamin, A. 1984. Solar heating of the soil: involvement of environmental factors in the weed control process. *Weed Sci.* **32**:138-142.

Sahoo, S., Mohanty, K.C., Sahoo, N.K., Mohapatra, A.K.B. and Mishra, H.N. 2004. Effect of different cropping sequences on root-knot nematode (*Meloidogyne incognita*) and wilt incidence in tomato. *National Symp. on Paradigms in Nematol. Res. for Biodynamic Farming,* Univ. of Agri. Sci., Bangalore, p. 37.

Saikia, B. 1992. Pathogenicity, Crop Loss Assessment and Biological Control of Meloidogyne incognita (Kofoid and White, 1919) Chitwood, 1949 on Betel Vine (*Piper betle* L.). M. Sc. (Agri.) thesis, Assam Agri. Univ., Jorhat.

Sanders, D. 2005. *Growers Guidelines.* American Vegetable Grower, October 2005. University of California SAREP Online Cover Crop Database: Mustards.

Sano, Z. 2002. Nematode management strategies in East Asian countries. *Nematology* **4**: 129-130.

Sarma, Y.R., Anandraj, M. and Venugopal, M.N. 1996. Biological control of diseases of spices. In *Biological Control in Spices* (M. Anandraj and K.V. Peter, eds.), pp. 1-19. Indian Inst. of Spices Res., Calicut.

Sasser, J.N. 1989. *The Plant-parasitic Nematodes, The Farmers Hidden Enemy.* Dept. of Plant Path., North Carolina State Univ. & Consortium for Intnl. Crop Prot., 115 pp.

Sawant, I.S. and Sawant, S.D. 1989. Coffee fruit skin and cherry husk as substrates for mass multiplication of *Trichoderma harzianum* an antagonist of citrus *Phytophthora* (Abstr.). *Indian Phytopath.* **42**: 336.

Saxena, R. and Reddy, D.D.R. *Indian J. Nematol.* **17**: 91-94.

Scholte, K. 1990. Causes of differences in growth pattern, yield and quality of potatoes (*Solanum tuberosum* L.) in short rotations on sandy soil as affected by crop rotation, cultivar and application of granular nematicides. *Potato Res.* **33**: 181-190.

Seamons, C., Brown, J., McCaffrey, J. and Lloyd, J. 2004. Using *Brassicaceae* seed meal as a soil fumigant in horticultural situations. *First Int. Biofumigation Symp.*, Florence, Italy.

Seenivasan, N and Devarajan, K. 2008. Integrated approach for the management of root-knot nematode, *Meloidogyne incognita* in medicinal coleus. *Indian J. Nematol.* **38**: 154-158.

Seman-Varner, R., McSorley, R., and Gallaher, R.N. 2008. Soil nutrient and plant responses to solarization in an agroecosystem utilizing an organic nutrient source. *Renewable Agriculture and Food Systems* **23**: 149-154.

Senthamarai, M., Poornima, K. and Subramanian, S. 2006. Assessment of avoidable yield loss on *Coleus forskohlii* due to *Meloidogyne incognita. Indian J. Nematol.* **36**: 296-297.

Shalini, M. and Patil, V.S. 2004. Effect of integrated weed management practices on vegetative, reproductive and yield parameters in gerbera (*Gerbera jamesonii* H. Bolus). *J. Orn. Hort.* **7**: 144-147.

Shanthi, A. 2003. *Management of* Radopholus similis, Helicotylenchus multicinctus *and* Pratylenchus coffeae *in Banana.* Ph. D. thesis, Tamil Nadu Agri. Univ., Coimbatore.

Sharma, G.L. 1989. Estimated losses due to root-knot nematodes, *Meloidogyne incognita* and *M. javanica* in pea crop. *Int. Nematol. Network Newsl.* **6**(1): 28-29.

Sharma, N.K., Thapa, C.D. and Nath, A. 1981. Pathogenicity and identity of myceliophagus nematode infesting *Agaricus bisporus* (Lange) Sing. in Himachal Pradesh (India). *Indian J. Nematol.* **11**: 230-231.

Sharma, N.K., Thapa, C.D. and Damanjeet. 1984. Estimation of loss in mushroom yield due to myceliophagus nematode, *Aphelenchoides sacchari*. *Indian J. Nematol.* **14**: 121-124.

Sharma, S.B. 1985. A world list of nematode pathogens associated with chickpea, groundnut, pearl millet, pigeonpea and sorghum. ICRISAT Pulse Pathology Progress Report, 42 pp.

Sharma, S.B., Sharma, H.K. and Pankaj. 2002. Nematode problems in India. In *Crop Pest and Disease Management: Challenges for the Millennium* (D. Prasad and S.N. Puri, eds.), pp.267-275. Jyoti Publishers, New Delhi, India.

Sharma, S.K., Singh, I. and Sakhuja, P.K. 1980. Influence of different cropping sequences on the population of root-knot nematode, *Meloidogyne incognita* and the performance of the subsequent mungbean crop. *Indian J. Nematol.* **10**: 53-58.

Sheela, M.S., Jiji, T. and Nisha, M.S. 2002. Evaluation of different control strategies for the management of nematodes associated with vegetables (brinjal) (Abstr.). *Int. Conf. on Vegetables,* Bangalore, p. 268.

Sheela, M.S. and Nisha, M.S. 2004. Impact of biological control agents for the management of root-knot nematode in brinjal. *National Symp. on Paradigms in Nematolo. Res. for Biodynamic Farming,* Univ. of Agri. Sci., Bangalore, p. 63-64.

Shetty, K.G., Subbarao, K.V., Huisman, O.C. and Hubbard, J.C. 2000. Mechanism of broccoli-mediated *Verticillium* wilt reduction in cauliflower. *Phytopathology* **90**: 305-310.

Siddiqi, M.A. and Alam, M.M. 1988a. Control of root-knot and reniform nematodes by bare-root dip in leaf extracts of margosa and Persian lilac. *Zeitschrift fur Pflanzenkrankheiten und Pflanzenschutz* **95**: 138-142.

Siddiqi, M.A. and Alam, M.M. 1988b. Effect of seed treatment with azadirachtin on root-knot development and growth of some vegetables. *Ann. Appl. Biol.* **112**: 20-21.

Siddiqi, M.A. and Alam, M.M. 1988c. Effect of latex seed dressing on *Rotylenchulus reniformis* and plant growth of some vegetables. *Nematol. Medit.* **16**: 129-131.

Siddiqi, M.A., Haseeb, A. and Alam, M.M. 1987. Evaluation of nematicidal properties in some latex bearing plants. *Indian J. Nematol.* **7**: 99-102.

Siddiqi, M.A. and Saxena, S.K. 1987a. Effect of interculture of margosa and Persian lilac with tomato and eggplant on root-knot and reniform nematodes. *Int. Nematol. Network Newsl.* **4**(2): 5-8.

Siddiqi, M.A. and Saxena, S.K. 1987b. Studies on the control of the stunt nematode, *Tylenchorhynchus brassicae* by interculture of margosa and Persian lilac. *Int. Nematol. Network Newsl.* **4**(4): 27-29.

Siddiqui, Z.A., Khan, A.M. and Saxena, S.K. 1973. Studies on *Tylenchorhynchus brassicae* II - Effect of temperature and moisture on multiplication of nematode. *Indian J. Entomol.* **26**: 139-147.

Siddiqi, Z.A., Khan, A.M. and Saxena, S.K. 1976. Nematode population and yield of certain vegetables as influenced by oil-cake amendments. *Indian J. Nematol.* **6**: 179-182.

Singh, D. 1991. Studies on the Effect of Tagetes species on *Meloidogyne javanica* Infectivity in Brinjal (*Solanum melongena*). Ph. D. thesis, Haryana Agri. Univ., Hisar, 83 pp.

Singh, K., Grewal, K.S. and Jawanda, J.S. 1961. Improved cultural practices for citrus and mangoes. *Punjab Hort. J.* **1**: 135-146.

Singh, M.M., Awasthi, D.N. and Upadhyay, S.N. 1982. Effect of different mulches and irrigation interval on survival of apple grafts. *Prog. Hort.* **14**: 15-18.

Singh, R.K., Kumar, N.Y., Roy Burman, K.N. and Roy, A.C. 1982. Intercropping in areca gardens in North Bengal. *Indian Fmg.* **32**(9): 13-15.

Singh, R.S. and Sitaramaiah, K. 1966. Incidence of root-knot of okra and tomato in oil cake amended soil. *Plant Dis. Reptr.* **50**: 668-672.

Singh, R.S. and Sitaramaiah, K. 1967. Effect of decomposing green leaves, sawdust and urea on the incidence of root-knot of okra and tomato. *Indian Phytopath.* **20:** 349-355.

Singh, R.S. and Sitaramaiah, K. 1971. Control of root-knot through organic and inorganic amendments of soil: Effect of saw dust and inorganic nitrogen. *Indian J. Nematol.* **1**: 80-84.

Singh, R.V., Midha, S.K. and Kumar, V. 2003. Major nematode pests of economically important crops in India and their management. In *Advances in Nematology* (P.C. Trivedi, ed.), pp. 233-256. Scientific Publishers (India), Jodhpur.

Singh, S.B. and Mishra, R.S. 1976. Effect of various mulches on the growth and yield of cauliflower (*Brassica oleraces* L. var. *botrytis*). *Prog. Hort.* **7**: 65-67.

Sitaramaiah, K. and Naidu, H. 2003. Plant parasitic nematodes-management opportunities and challenges. In *Nematode Management in Plants* (P.C. Trivedi, ed.), pp. 57-80. Scientific Publishers (India), Jodhpur.

Siti, E., Cohn, E., Katan, J. and Mordechai, M. 1982. Control of *Ditylenchus dipsaci* in garlic by bulb and soil treatments. *Phytoparasitica* **10**: 93-100.

Sivakumar, M. and Marimuthu, T. 1986. Efficacy of different organic amendments against phytonematodes associated with betel vine. *Indian J. Nematol.* **16**: 278.

Sivakumar, M. and Marimuthu, T. 1987. Preliminary studies on the effect of soil solarization on phytonematodes of betel vine. *Indian J. Nematol.* **17**: 58-59.

Sivapalan, P. 1972. Nematode pests of tea. In *Economic Nematology* (J.M. Webster, ed.), pp. 285-310. Academic Press, New York and London.

Smith, B.J., Sarwar, M., Wong, P.T.W. and Kirkegaard, J.A. 1999. Suppression of cereal pathogens by canola root tissues in soil. *Proc. 10th Int. Rapeseed Cong.*, Canberra (In press).

Snyder, A.S., Johnson-Maynard, J.L. and Morra, M.J. 2006. Brassicaceae seed meals as a nitrogen source in carrot and strawberry production. *Second Int. Biofumigation Symp.*, Idaho, USA.

Srinivasan, A. and Muthurksihnan, T.S. 1975. A note on nematode fungal complex in crossandra (*Crossandra undulaefolia*) in Coimbatore. *Curr. Sci.* **44**: 743-744.

Srirama Rao, T. 1978. Turmeric cultivation in Andhra Pradesh. *Indian Cocoa, Arecanut and Spices J.* **2**: 31-32.

Srivastava, P.K., Parikh, M.M., Sawani, N.G. and Raman, S. 1995. Effect of drip irrigation and mulching on tomato yield. *Hort. Abstr.* **65**: 1318.

Srivastava, R.P. and Agarwal, N.G. 1965. A note on mulching of strawberry. *Indian J. Hort.* **22**: 211-213.

Standifer, L.C. 1984. Effects of solarization on soil weed seed populations. *Weed Sci.* **32**: 569-573.

Stapleton, J.J. 1997. Solarization: An implementable alternative for soil disinfestation. In: *Biological and Cultural Control of Plant Diseases* 12: 1-6. St. Paul, MN: APS Press.

Stapleton, J.J. 1998. Modes of action of solarization and biofumigation. In: *Soil Solarization and Integrated Management of Soilborne Pests* (J.J. Stapleton, J.E. DeVay, C.L. Elmore, eds.), pp. 78-88. Plant Production and Protection Paper 147. Rome: FAOIUN.

Stapleton, J.J. 2000. Soil solarization in various agricultural production systems. *Crop Prot.* **19**: 837-41.

Stapleton, J.J. 2006. Biocidal and allelopathic properties of gramineous crop residue amendments as influenced by soil temperature. *Proc. Calif. Conf. Biol. Control* **5**: 179-181.

Stapleton, J.J. and DeVay, J.E. 1984. Thermal components of soil solarization as related to changes in soil and root microflora and increased plant growth response. *Phytopathology* **74**: 255-259.

Stapleton, J.J. and Devay, J.E. 1986. Soil solarization: A non-chemical approach for management of plant pathogens and pests. *Crop Prot.* **5**: 190-198.

Stapleton, J.J. and DeVay, J.E. 1995. Soil solarization: A natural mechanism of integrated pest management. *In Novel Approaches to Integrated Pest Management* (R. Reuveni, ed.), pp 309-322. CRC Press, Boca Raton, Florida.

Stapleton, J.J, McKenry, M.V. and Ferguson, L. 1999. Methyl bromide alternatives: CDFA approves a solarization technique to ensure against nematode pest infestation of containerized nursery stock. *UC Plant Prot. Quarterly* **9**(2): 14.

Stapleton, J.J., Paplomatus, E., Wakeman, R. and DeVay, J. 1993. Establishment of apricot and almond trees using soil mulching with transparent (solarization) and black polyethylene film: Effects on *Verticillium* wilt and tree health. *Plant Path.* **42**: 333-338.

Stapleton, J.J., Quick, J. and DeVay, J.E. 1985. Soil solarization: Effect on soil properties, crop fertilization, and plant growth. *Soil Biol. Biochem.* **17**: 369-373.

Stark, J.C. 1995. Development of sustainable potato production systems for the Pacific Northwest. SARE Final Rept.

Stark, J.C., and Love, S.L. (eds.). 2003. *Potato Production Systems*. University of Idaho Extension, Moscow, ID, 426 pp.

Stirling, G.R. and Stirling, A.M. 2003. The potential of Brassica green manure crops for controlling root-knot nematode (*Meloidogyne javanica*) on horticultural crops in a subtropical environment. *Australian J. Exp. Agri.* **43**: 623-630.

Strobel, G. 2008. Discovery and development of *Muscodor* species for mycofumigation. *Third Int. Biofumigation Symp.*, Canberra, Australia.

Subbarao, K.V. and Hubbard, J.C. 1996. Interactive effects of broccoli residue and temperature on *Verticillium dahliae* microsclerotia in soil and on wilt in cauliflower. *Phytopathology* **86**: 1303-1310.

Subbarao, K.V., Hubbard, J.C. and Koike, S.T. 1999a. Evaluation of broccoli residue incorporation into field soil for *verticillium* wilt control in cauliflower. *Plant Dis.* **83**: 124-129.

Subbarao, K.V., Hubbard, J.C. and Koike, S.T. 1999b. Evaluation of broccoli residue incorporation into field soil for Verticillium wilt control in cauliflower. *Phytopathology* **88**: 1046-1055.

Subramanyam, S., Rajendran, G. and Vadivelu, S. 1990. Estimation of loss in tomato due to *Meloidogyne incognita* and *Rotylenchulus reniformis*. *Indian J. Nematol.* **20**: 239-240.

Sud, U.C., Varaprasad, K.S., Seshadri, A.R. and Kher, K. 1984. *Indian J. Nematol.* **14**: 148-151.

Sukumaran, S. and Sundararaju, P. 1986a. Pathogenicity of *Meloidogyne incognita* on ginger (*Zingiber officinale* Rosc.). *Indian J. Nematol.* **16**: 258.

Sukumaran, S. and Sundararaju, P. 1986b. Effect of root-knot nematode, *Meloidogyne incognita* on the growth of turmeric (Abstr.). *Seventh Symposium of Plantation Crops*, Coonoor, Nilgiris, Tamil Nadu, India, p. 33.

Sukumaran Pillay, V., Alexander, D. and Ibrahim, K.K. 1987. Effect of different type of mulches on yield in black pepper. *Indian Cocoa, Arecanut and Spices J.* **11**: 29-30.

Summers, C.G., Mitchell, J.P., Prather, T.S. and Stapleton, J.J. 2009. Sudex cover crops can kill and stunt subsequent tomato, lettuce and broccoli transplants through allelopathy. *Calif. Agri.* **63**: 40.

Sundarababu, R. and Vadivelu, S. 1988. *Indian J. Nematol.* **18**: 139.

Sundararaju, P., Cannayane, I. and Thangavelu, R. 2002. Biomanagement of *Pratylenchus coffeae* and *Meloidogyne incognita* on banana by using endophytic fungi. *Global Conference on Banana and Plantain*, Bangalore, p. 191

Tacconi, R., De Vincentis, F., Lazzeri, L. and Malaguti, L. 1998. Riproduzione su piante oleaginose di *Heterodera schactii. L'Informatore Agrario* **4**: 57-59.

Tacconi, R., Mambelli, S., Menichetti, P. and Pola, R. 1989. Osservazioni sul ciclo di *Heterodera schactii* su piante resistenti. *Nematol. Medit.* **17**: 21-25.

Tacconi, R. and Venturi, G. 1990. Mezzi agronomici di lotta contro il nematode *Heterodera schactii* in Italia. *L'Informatore Agrario* **1**: 443-448.

Tada, M., Hiroe, Y., Kiyohara, S. and Suzuki, S. 1988. Nematicidal and antimicrobial constituents from *Allium grayi* Regel and *Allium fistulosum* L. var. *caespitosum*. *Agric. Biol. Chem.* **52**: 2383-2385.

Talam, D.K. and 0tieno, W. 2002. Temperature relations in *Armillaria* spp. associated with root rot of tea in Kenya: Implications in disease management by soil solarization. *Tea* **23**(1): 22-28.

Tanda, A.S. and Atwal, A.S. 1988. Effect of sesame intercropping against the root-knot nematode (*Meloidogyne incognita*). *Nematologica* **34**: 484-492.

Taylor, A.L. and Sasser, J.N. 1978. *Biology, Identification and Control of Root-knot Nematodes* (Meloidogyne spp.). North Carolina State University Graphics, Raleigh, 111 pp.

Teotia, S.S. and Srivastava, R. 1971. Importance of mulching in fruit culture. *All India Symp. on Recent Advances in Hort.*

Thapa, C.D., Sharma, N.K and Nath, A. 1981. Mushroom nematodes and their source of infestation in mushroom beds in India. *Mushroom Sci.* **11**: 443-448.

Thompson, B.D. 1961. Response of strawberries to mulching with plastic. *Weed Abstr.* **10**: 676.

Tiyagi, S.A., Anver, S., Bano, M. and Siddiqi, M.A. 1986. Feasibility of growing zinnia as a 'mix-crop' along with tomato for control of root-knot and reniform nematodes. *Int. Nematol. Network Newsl.* **3**(3): 6-7.

Toba, H.H., Kishaba, A.N., Bohn, G.W. and Hield, H. 1977. Protecting muskmelon against aphid-borne viruses. *Phytopathology* **67**: 1418-1423.

Tollsten, L. and Bergström, G. 1988. Headspace volatiles of whole plants and macerated plant parts of *Brassica* and *Sinapis*. *Phytochem.* **27**: 4013-4018.

Tripathi, P.K. and Singh, C.S. 2006. Effect of some compatible biocontrol agents along with mustard cake and furadan on *Meloidogyne incognita* infecting tomato plant. *Indian J. Nematol.* **36**: 309-311.

Trivedi, P.C. and Tiagi, A. 1984. Effect of some plants on *Meloidogyne incognita* population in chilli soil. *Indian J. Botanical Sci.* **63**: 416-419.

Troncoso-Rojas, R., Espinoza, C., Sánchez-Estrada, A., Sergio-García, H., Pryor, B., Baez-Flores, M.E. and Tiznado-Hernández, M.E. 2008. Evaluation of isothiocyanates to control postharvest fruit and vegetable losses by fungal infections in Mexico. *Third Int. Biofumigation Symp.*, Canberra, Australia.

Tsror, L., Lebiush, S., Meshulam, M., Matan, E. and Lazzeri, L. 2006. Biofumigation for controlling soil-borne diseases of tomato, potato and olive. *Second Int. Biofumigation Symp.,* Idaho, USA, p. 46.

Tzortzakakis, E.A. and Goewn, S.R. 1994. Evaluation of *Pasteuria penetrans* alone and in combination with oxamyl, plant resistance and solarization for control of *Meloidogyne* spp. on vegetables grown in greenhouse in Crete. *Crop Prot.* **13**: 455-462.

Unamma, R.P.A., Ene, I.S.O., Odurukwe, S.O. and Enyinnia, T. 1986. Integrated weed management for cassava intercropped with maize. *Weed Res.* (UK) **36**: 9-17.

Upadhyay, K.D. and Dwivedi, K. 1987. Analysis of crop losses in pea and gram due to *Meloidogyne incognita*. *Int. Nematol. Network Newsl.* **4**(4): 6-7.

Urbasch, I. 1984. Production of C6-wound gases by plants and the effect on some phytopathogenic fungi. *Z. Naturforsch.* **39**c: 1003-1007.

USDA. 2000. *Economic Implications of the Methyl Bromide Phaseout*. U.S. Department of Agriculture, Agriculture Information Bulletin #756, pp. 77-105.

Usman, N.M., Balakrishnan, P. and Sarma, Y.R. 1996. Biocontrol of rhizome rot of ginger. *J. Plantation Crops* **24** (Suppl.): 184-191.

Vadivelu, R.S., Rajendran, G., Naganathan, T.G. and Jayaraj, S. 1987. Studies on the nematode pests of banana and their management - A review. *Third Group Discussion on the Nematological Problems of Plantation Crops*, Sugarcane Breeding Institute, Coimbatore, pp.32-33.

Van der Vetch, J. 1950. Plant parasitic nematodes. In *Diseases of Cultivated Plants in Indonesia* Vol. I (L.G.E. Karsh van and Jived. Vetch, eds.), pp. 16-45. W. van Wove, S'Gravenhagen.

Van Os, G.J., Bijman, V., de Boer, M., Breeuwsma, S., van der Bent, J. and Lazzeri, L. 2004. Biofumigation against soilborne fungal diseases in flower bulb. *First Int. Biofumigation Symp.*, Florence, Italy.

Van Os, G.J. and Lazzeri, L. 2006. Control of *Rhizoctonia solani* in lily by biofumigation with Brassica seed meal. *Second Int. Biofumigation Symp.*, Idaho, USA.

Venkata Ram, C.S. 1963. The Root-knot Eelworm of Mature Tea Meloidogyne brevicauda, its Importance in South India and an Evaluation of Certain Chemical Treatments for its Control. Tea Scientific Report for 1962-63, United Planters Association of South India, Appendix pp. 31-36.

Venkitesan, T.S. 1976. Studies on the Burrowing Nematode, Radopholus similis (Cobb, 1893) Thorne, 1949 on Pepper (*Piper nigrum* L.) and its Role in Slow Wilt Disease. Ph. D. thesis, Univ. of Agri. Sci., Bangalore.

Verma, A.C., Singh, H.K and Nehal Khan, M. 2005. Management of root-knot nematode, *Meloidogyne incognita* through antagonistic approaches in pointed gourd. *Indian J. Nematol.* **35**: 78-79.

Vernon, P. De. 1981. Control of *Eupatorium odoratum* in the Ivory Coast in village rubber plantations. *Weed Abstr.* **3**: 404.

Vyas, R.V., Patel, B.A., Patel, B.N. and Patel, J.G. 2009. Integrated management of root-knot nematode in brinjal under field conditions. *Indian J. Nematol.* **39**: 35-37.

Walker, J.C., Morell, S. and Foster, H.H. 1937. Toxicity of mustard oils and related sulfur compounds to certain fungi. *Am. J. Bot.* **24**: 536-541.

Wang, J.F., Luther, K., Ho, F.I., Lin, C.H. and Kirkegaard, J. 2008. Evaluation of Brassica accessions as potential biocidal green manure to control tomato bacterial wilt. *Third Int. Biofumigation Symp.*, Canberra, Australia.

Weingartner, D.P., Shumaker, J.R. and Smart, G.C. Jr. 1975. Incidence of corky ringspot disease on potato as affected by different fumigation rates, cultivars, and harvest dates. *Proc. Soil and Crop Sci. Soc. of Florida* **34**:194-196.

Weischer, B. 1967. *Report to the Government of India on Plant Parasitic Nematodes*. PL 24/52, Report No. TA 2332, FAO, Rome, 15 pp.

Whitehead, A.G. 1968. Taxonomy of *Meloidogyne* (Nematodea: Heteroderidae) with descriptions of four new species. *Trans. Zool. Soc. London* **31**: 263-401.

Wider, T.L. and Abawi, G.S. 2000. Mechanism of suppression of *Meloidogyne hapla* and its damage by green manure of Sudan grass. *Plant Dis.* **84**: 562-568.

Xiao, C.L., Subbarao, K.V., Schulbach, K.F. and Koike, S.T. 1998. Effects of crop rotation and irrigation on *Verticillium dahliae* microsclerotia in soil and wilt in cauliflower. *Phytopathology* **88**: 1046-1055.

Yulianti, T. 2008. Potency of Brassica residues as biofumigation for control of bacterial wilt of tobacco in Indonesia. *Third Int. Biofumigation Symp.*, Canberra, Australia.

Zaki, F.A. 1998. Biological control of *Meloidogyne javanica* in tomato by *Paecilomyces lilacinus* and castor leaves. *Indian J. Nematol.* **28**: 132-139.

Zaki, F.A. and Bhatti, D.S. 1986a. *Indian J. Nematol.* **16**: 30-35.

Zaki, F.A. and Bhatti, D.S. 1986b. *Indian J. Nematol.* **16**: 103-105.

Zaki, F.A. and Bhatti, D.S. 1989. Effect of castor (*Ricinus communis*) leaves in combination with different fertilizer doses on *Meloidogyne javanica* infecting tomato. *Indian J. Nematol.* **19**: 171-176.

# SUBJECT INDEX

**R**



**S**